Withdrawn
University of Waterloo

# Current Topics in Microbiology 237 and Immunology

Editors

R.W. Compans, Atlanta/Georgia
M. Cooper, Birmingham/Alabama
J.M. Hogle, Boston/Massachusetts · Y. Ito, Kyoto
H. Koprowski, Philadelphia/Pennsylvania · F. Melchers, Basel
M. Oldstone, La Jolla/California · S. Olsnes, Oslo
M. Potter, Bethesda/Maryland · H. Saedler, Cologne
P.K. Vogt, La Jolla/California · H. Wagner, Munich

Springer
Berlin
Heidelberg
New York
Barcelona
Budapest
Hong Kong
London
Milan
Paris
Singapore
Tokyo

# Vascular Growth Factors and Angiogenesis

Edited by Lena Claesson-Welsh

With 36 Figures and 3 Tables

Springer

Professor Dr. LENA CLAESSON-WELSH
University of Uppsala
Department of Medical Biochemistry & Microbiology
Biomedical Center
Box 575
S-75123 Uppsala
Sweden

*Cover Illustration:* Scanning electron microscopy at ×100 of a micro-vascular corrosion cast of a Wilms' tumor from a 13-month old boy. Numerous sprouting vessels are seen indicating a high angiogenic activity. The lack of regulation of angiogenesis in the tumor is apparent through the changes in vessel-calibre, blind endings and resin leakage. Photo: Dr. Erik Sköldenberg, Dept. of Anatomy, Biomedical Center, Uppsala, Sweden.

*Cover Design: design & production* GmbH, Heidelberg
ISSN 0070-217X
ISBN 3-540-64731-7 Springer-Verlag Berlin Heidelberg New York

This work is subject to copyright. All rights are reserved, whether the whole or part of the material is concerned, specifically the rights of translation, reprinting, reuse of illustrations, recitation, broadcasting, reproduction on microfilm or in any other way, and storage in data banks. Duplication of this publication or parts thereof is permitted only under the provisions of the German Copyright Law of September 9, 1965, in its current version, and permission for use must always be obtained from Springer-Verlag. Violations are liable for prosecution under the German Copyright Law.

© Springer-Verlag Berlin Heidelberg 1999
Library of Congress Catalog Card Number 15-12910
Printed in Germany

The use of general descriptive names, registered names, trademarks, etc. in this publication does not imply, even in the absence of a specific statement, that such names are exempt from the relevant protective laws and regulations and therefore free for general use.

Product liability: The publishers cannot guarantee the accuracy of any information about dosage and application contained in this book. In every individual case the user must check such information by consulting other relevant literature.

Typesetting: Scientific Publishing Services (P) Ltd, Madras

Production Editor: Angélique Gcouta

SPIN: 10575594     27/3020 – 5 4 3 2 1 0 – Printed on acid-free paper

# Preface

Currently, the cellular and molecular mechanisms governing the development and regulation of the vasculature are studied intensely and the field is rapidly progressing. Recently, novel growth factors and growth factor receptors specifically acting on endothelial cells have been discovered. Through these factors, communication networks are established between endothelial cells, the basement membrane and the pericytes; this interplay is critical for the regulated development and maintenance of the vasculature. The awareness that deregulated angiogenesis contributes to the progression of a number of diseases, such as cancer and inflammatory diseases, has clearly spurred the field to move forward. The focus of this book is on two important classes of endothelial cell specific growth factors, the vascular endothelial growth factor (VEGF) family, and the angiopoietins, and on their mechanisms of action. The reader will find up-to-date, focused reviews, which give the current picture, and indicate future directions.

I would like to honor Dr. Judah Folkman for his important contributions to the establishment of the field and thank him for his support.

Lena Claesson-Welsh

# List of Contents

N. Ferrara
Vascular Endothelial Growth Factor: Molecular
and Biological Aspects .......................... 1

M. G. Persico, V. Vincenti and T. DiPalma
Structure, Expression and Receptor-Binding Properties
of Placenta Growth Factor (PlGF) ................. 31

U. Eriksson and K. Alitalo
Structure, Expression and Receptor-Binding Properties
of Novel Vascular Endothelial Growth Factors......... 41

M. Shibuya, N. Ito and L. Claesson-Welsh
Structure and Function of Vascular Endothelial
Growth Factor Receptor-1 and -2 ................... 59

J. Taipale, T. Makinen, E. Arighi, E. Kukk,
M. Karkkainen and K. Alitalo
Vascular Endothelial Growth Factor Receptor-3........ 85

H. F. Dvorak, J. A. Nagy, D. Feng,
L. F. Brown and A. M. Dvorak
Vascular Permeability Factor/Vascular Endothelial
Growth Factor and the Significance of Microvascular
Hyperpermeability in Angiogenesis ................. 97

P. Carmeliet and D. Collen
Role of Vascular Endothelial Growth Factor
and Vascular Endothelial Growth Factor Receptors
in Vascular Development ......................... 133

J. Partanen and D. J. Dumont
Functions of Tie1 and Tie2 Receptor Tyrosine Kinases
in Vascular Development ......................... 159

S. Davis and G. D. Yancopoulos
The Angiopoietins: Yin and Yang in Angiogenesis ...... 173

Subject Index..................................... 187

# List of Contributors

(Their addresses can be found at the beginning of their respective chapters.)

| | | | |
|---|---|---|---|
| Alitalo, K. | 41, 85 | Ferrara, N. | 1 |
| Arighi, E. | 85 | Ito, N. | 59 |
| Brown, L.F. | 97 | Karkkainen, M. | 85 |
| Carmeliet, P. | 133 | Kukk, E. | 85 |
| Claesson-Welsh, L. | 59 | Makinen, T. | 85 |
| Collen, D. | 133 | Nagy, J.A. | 97 |
| Davis, S. | 173 | Partanen, J. | 159 |
| DiPalma, T. | 31 | Persico, M.G. | 31 |
| Dumont, D.J. | 159 | Shibuya, M. | 59 |
| Dvorak, A.M. | 97 | Taipale, J. | 85 |
| Dvorak, H.F. | 97 | Vincenti, V. | 31 |
| Eriksson, U. | 41 | Yancopoulos, G.D. | 173 |
| Feng, D. | 97 | | |

# Vascular Endothelial Growth Factor: Molecular and Biological Aspects

N. Ferrara

| | | |
|---|---|---|
| 1 | Introduction | 1 |
| 2 | Biological Activities of Vascular Endothelial Growth Factor | 2 |
| 3 | Organization of the VEGF Gene and Characteristics of the VEGF Proteins | 4 |
| 4 | Regulation of VEGF Gene Expression | 6 |
| 4.1 | Oxygen Tension | 6 |
| 4.2 | Cytokines | 6 |
| 4.3 | Differentiation and Transformation | 7 |
| 5 | The VEGF Receptors | 8 |
| 6 | The VEGFR-1 and VEGFR-2 Tyrosine Kinases | 8 |
| 6.1 | Binding Characteristics | 8 |
| 6.2 | Signal Transduction | 9 |
| 6.3 | Regulation | 11 |
| 7 | Role of VEGF and its Receptors in Physiological Angiogenesis | 11 |
| 7.1 | Distribution of VEGFR-1 and VEGFR-2 mRNA | 11 |
| 7.2 | The VEGFR-1, VEGFR-2 and VEGF Gene Knockouts in Mice | 12 |
| 8 | Role of VEGF in Corpus Luteum Angiogenesis | 13 |
| 9 | Role of VEGF in Pathological Angiogenesis | 14 |
| 9.1 | Tumor Angiogenesis | 14 |
| 9.2 | Angiogenesis Associated with Other Pathological Conditions | 16 |
| 10 | VEGF and Therapeutic Angiogenesis | 18 |
| 11 | Conclusions | 20 |
| References | | 21 |

## 1 Introduction

The development of a vascular supply is a fundamental requirement for organ development and differentiation during embryogenesis as well as for wound healing and reproductive functions in the adult (Folkman 1995). Angiogenesis is also implicated in the pathogenesis of a variety of disorders: proliferative retinopathies,

---

Department of Cardiovascular Research, Genentech, Inc., 460 Point San Bruno Boulevard, South San Francisco, CA 94080, USA

age-related macular degeneration, tumors, rheumatoid arthritis and psoriasis (FOLKMAN 1995; GARNER 1994).

The search for positive regulators of angiogenesis has yielded several candidates, including fibroblast growth factors a and b (aFGF, bFGF), transforming growth factors alpha and beta (TGF-α, TGF-β), hepatocyte growth factor (HGF), tumor necrosis factor alpha (TNF-α), angiogenin, interleukin-8 (IL-8), etc. (FOLKMAN and SHING 1992; RISAU 1997). However, in spite of extensive research, there is still significant debate as to their role as endogenous mediators of angiogenesis. The negative regulators identified so far include thrombospondin (GOOD et al. 1990; DIPIETRO 1997), the 16-kilodalton $N$-terminal fragment of prolactin (FERRARA et al. 1991), angiostatin (O'REILLY et al. 1994) and endostatin (O'REILLY et al. 1997).

This chapter discusses the molecular and biological properties of the vascular endothelial growth factor (VEGF) proteins. Over the last few years, several additional members of the VEGF gene family have been identified, including VEGF-B, VEGF-C, Placenta growth factor (PlGF) and VEGF-D. This chapter focuses primarily on VEGF, also referred to as "VEGF-A". For a description of the other members of the family, the reader is referred to the appropriate chapters in this book. Work done by several laboratories over the last few years has elucidated the pivotal role of VEGF and its receptors in the regulation of normal and abnormal angiogenesis (FERRARA and DAVIS-SMYTH 1997). The finding that the loss of even a single VEGF allele results in embryonic lethality points to an irreplaceable role played by this factor in the development and differentiation of the vascular system (FERRARA et al. 1996; CARMELIET et al. 1996). Furthermore, VEGF-induced angiogenesis has been shown to result in a therapeutic effect in animal models of coronary or limb ischemia and, most recently, in a human patient affected by critical leg ischemia (FERRARA and DAVIS-SMYTH 1997).

## 2 Biological Activities of Vascular Endothelial Growth Factor

Vascular endothelial growth factor (VEGF) is a mitogen for vascular endothelial cells derived from arteries, veins and lymphatics, but is devoid of consistent and appreciable mitogenic activity for other cell types (FERRARA and DAVIS-SMYTH 1997). VEGF promotes angiogenesis in tri-dimensional in vitro models, inducing confluent microvascular endothelial cells to invade collagen gels and form capillary-like structures (PEPPER et al. 1992). Also, VEGF induces sprouting from rat aortic rings embedded in a collagen gel (NICOSIA et al. 1994). VEGF also elicits a pronounced angiogenic response in a variety of in vivo models, including the chick chorioallantoic membrane (LEUNG et al. 1989), the primate iris (TOLENTINO et al. 1996) etc.

VEGF induces expression of the serine proteases urokinase-type and tissue-type plasminogen activators (PA), and also PA inhibitor 1 (PAI-1) in cultured bovine microvascular endothelial cells (PEPPER et al. 1991). Moreover, VEGF increases expression of the metalloproteinase interstitial collagenase in human umbilical-vein endothelial cells (HUVEC), but not in dermal fibroblasts (UNEMORI et al. 1992). Other studies have shown that VEGF promotes expression of the urokinase receptor (uPAR) in vascular endothelial cells (MANDRIOTA et al. 1995). Additionally, VEGF stimulates hexose transport in cultured vascular endothelial cells (PEKALA et al. 1990).

VEGF is known also as vascular permeability factor (VPF), based on its ability to induce vascular leakage in the guinea-pig skin (DVORAK et al. 1995). DVORAK and colleagues proposed that an increase in microvascular permeability is a crucial step in angiogenesis associated with tumors and wounds (DVORAK 1986). According to this hypothesis, a major function of VPF/VEGF in the angiogenic process is the induction of plasma-protein leakage. This effect would result in the formation of an extravascular fibrin gel, a substrate for endothelial and tumor cell growth (DVORAK et al. 1987). Recent studies have also suggested that VEGF may induce fenestrations in endothelial cells (ROBERTS and PALADE 1995, 1997). Topical administration of VEGF acutely resulted in the development of fenestrations in the endothelium of small venules and capillaries, even in regions where endothelial cells are not normally fenestrated, and was associated with increased vascular permeability (ROBERTS and PALADE 1995, 1997).

MELDER et al. (1996) have shown that VEGF promotes expression of VCAM-1 and ICAM-1 in endothelial cells. This induction results in the adhesion of activated natural killer (NK) cells to endothelial cells, mediated by specific interaction of endothelial VCAM-1 and ICAM-1 with CD18 and VLA-4 on the surface of NK cells.

VEGF has been reported to have certain regulatory effects on blood cells. CLAUSS et al. (1990) reported that VEGF may promote monocyte chemotaxis, while BROXMEYER et al. (1995) have shown that VEGF induces colony formation by mature subsets of granulocyte-macrophage progenitor cells. These findings may be explained by the common origin of endothelial and hematopoietic cells and the presence of VEGF receptors in progenitor cells as early as hemangioblasts in blood islands in the yolk sac. Furthermore, GABRILOVICH et al. (1996) have reported that VEGF may have an inhibitory effect on the maturation of host professional antigen-presenting cells, such as dendritic cells. VEGF was found to inhibit immature dendritic cells, without having a significant effect on the function of mature cells. These findings led to the suggestion that VEGF may also facilitate tumor growth by allowing the tumor to avoid the induction of an immune response (GABRILOVICH et al. 1996).

VEGF induces vasodilatation in vitro in a dose-dependent fashion (KU et al. 1993; YANG et al. 1996) and produces transient tachycardia, hypotension and a decrease in cardiac output when injected intravenously in conscious, instrumented rats (YANG et al. 1996. Such effects appear to be caused by a decrease in venous return, mediated primarily by endothelial cell-derived nitric oxide (NO), as assessed by the requirement for an intact endothelium and the prevention of the effects by $N$-methyl-arginine (YANG et al. 1996). Accordingly, VEGF has no direct effect on contractility or rate in the isolated rat heart in vitro (YANG et al. 1996). These hemodynamic effects, however, are not unique to VEGF: other angiogenic factors, such as aFGF and bFGF, also have the ability to induce NO-mediated vasodilatation and hypotension (CUEVAS et al. 1991, 1996).

## 3 Organization of the VEGF Gene and Characteristics of the VEGF Proteins

The human VEGF gene is organized in eight exons, separated by seven introns. The coding region spans approximately 14 kb (HOUCK et al. 1991; TISCHER et al. 1991). The human VEGF gene has been assigned to chromosome 6p21.3 (VINCENTI et al. 1996). It is now well established that alternative exon splicing of a single VEGF gene results in the generation of four different molecular species, having respectively 121, 165, 189 and 206 amino acids following signal sequence cleavage ($VEGF_{121}$, $VEGF_{165}$, $VEGF_{189}$, $VEGF_{206}$). $VEGF_{165}$ lacks the residues encoded by exon 6, while $VEGF_{121}$ lacks the residues encoded by exons 6 and 7. Compared with $VEGF_{165}$, $VEGF_{121}$ lacks 44 amino acids; $VEGF_{189}$ has an insertion of 24 amino acids, highly enriched in basic residues; and $VEGF_{206}$ has an additional insertion of 17 amino acids (HOUCK et al. 1991). Analysis of the VEGF gene promoter region reveals a single major transcription start which lies near a cluster of potential Sp1 factor binding sites.

$VEGF_{165}$ is the predominant molecular species produced by a variety of normal and transformed cells. Transcripts encoding $VEGF_{121}$ and $VEGF_{189}$ are detected in the majority of cells and tissues expressing the VEGF gene (HOUCK et al. 1991). In contrast, $VEGF_{206}$ is a very rare form, so far identified only in a human fetal liver complementary deoxyribonucleic acid (cDNA) library (HOUCK et al. 1991). The genomic organization of the murine VEGF gene has been also described (SHIMA et al. 1996). Similarly to the human gene, the coding region of the murine VEGF gene encompasses approximately 14kb and is comprised of eight exons interrupted by seven introns. Analysis of exons suggests the generation of three isoforms: $VEGF_{120}$, $VEGF_{164}$ and $VEGF_{188}$. Therefore, murine VEGFs are shorter than human VEGF by one amino acid. However, a fourth isoform comparable with $VEGF_{206}$ is not predicted, since an in-frame stop codon is present the region corresponding to the human $VEGF_{206}$ open reading frame. Analysis of the 3′ untranslated region of the rat VEGF messenger ribonucleic acid (mRNA) has revealed the presence of four potential polyadenylation sites (LEVY et al. 1996). A frequently used site is about 1.9kb further downstream from the previously reported transcription termination codon (CONN et al. 1990). The sequence within this 3′ untranslated region reveals a number of sequence motifs that are known to be involved in the regulation of mRNA stability (LEVY et al. 1996).

Native VEGF is a basic, heparin-binding, homodimeric glycoprotein of 45,000Da (FERRARA and HENZEL 1989). These properties correspond to those of $VEGF_{165}$, the major isoform (HOUCK et al. 1992). $VEGF_{121}$ is a weakly acidic polypeptide that fails to bind to heparin (HOUCK et al. 1992). $VEGF_{189}$ and $VEGF_{206}$ are more basic and bind to heparin with a greater affinity than $VEGF_{165}$ (HOUCK et al. 1992). Such differences in the isoelectric point and the affinity for heparin may affect the bioavailability of VEGF profoundly.

$VEGF_{121}$ is a freely diffusible protein; $VEGF_{165}$ is also secreted, although a significant fraction remains bound to the cell surface and the extracellular matrix (ECM). In contrast, $VEGF_{189}$ and $VEGF_{206}$ are almost completely sequestered in the ECM (PARK et al. 1993). However, these isoforms may be released in a soluble form by heparin or heparinase, suggesting that their binding site is represented by proteoglycans containing heparin-like moieties. The long forms may also be released by plasmin following cleavage at the carboxy (COOH) terminus. This action generates a bioactive proteolytic fragment with a molecular weight of $\sim$34,000Da (HOUCK et al. 1992).

Plasminogen activation and generation of plasmin have been shown to play an important role in the angiogenesis cascade. Thus, proteolysis of VEGF is likely also to occur in vivo. KEYT et al. (1996a) have shown that the bioactive product of plasmin action is comprised of the first 110 amino ($NH_2$)-terminal amino acids of VEGF. These findings suggest that the VEGF proteins may become available to endothelial cells by at least two different mechanisms: as freely diffusible proteins ($VEGF_{121}$, $VEGF_{165}$) or following protease activation and cleavage of the longer isoforms. However, loss of heparin binding, whether it is due to alternative splicing of RNA or plasmin cleavage, results in a substantial loss of mitogenic activity for vascular endothelial cells: compared with $VEGF_{165}$, $VEGF_{121}$ or $VEGF_{110}$ which demonstrate a 50- to 100-fold reduced potency when tested in endothelial cell growth assay (KEYT et al. 1996a).

It has been suggested that the stability of VEGF–heparan-sulfate-receptor complexes contributes to effective signal transduction and stimulation of endothelial cell proliferation (KEYT et al. 1996a). Thus, VEGF has the potential to express structural and functional heterogeneity to yield a graded and controlled biological response. Very recently, POLTORAK et al. (1997) provided evidence for the existence of an additional, alternatively spliced molecular species of VEGF. A VEGF isoform containing exons 1–6 and 8 of the VEGF gene was found to be expressed as a major VEGF mRNA form in several cell lines derived from carcinomas of the female reproductive system. This mRNA is predicted to encode a VEGF form of 145 amino acids ($VEGF_{145}$). Recombinant $VEGF_{145}$ induced the proliferation of vascular endothelial cells, albeit at much lower potency than $VEGF_{165}$. $VEGF_{145}$ binds to the kinase domain region (KDR) receptor, also denoted VEGF receptor-2 (VEGFR-2) on the surface of endothelial cells. It also binds to heparin with an affinity similar to that of $VEGF_{165}$.

Recently, MULLER et al. (1997) determined the crystal structure of VEGF at a resolution of 2.5A. Overall, the VEGF monomer resembles that of platelet-derived growth factor (PDGF), but its $N$-terminal segment is helical rather than extended. The dimerization mode of VEGF is similar to that of PDGF and very different from that of TGF-$\beta$. Functional analysis of the binding epitopes for two receptor-blocking antibodies reveal different binding determinants near each of the VEGFR-2 binding hot spots.

## 4 Regulation of VEGF Gene Expression

### 4.1 Oxygen Tension

Among the mechanisms that have been proposed to participate in the regulation of VEGF gene expression, oxygen tension plays a major role, both in vitro and in vivo. VEGF mRNA expression is rapidly and reversibly induced by exposure to low $pO_2$ in a variety of normal and transformed cultured-cell types (MINCHENKO et al. 1994; SHIMA et al. 1995). Also, ischemia caused by occlusion of the left anterior descending coronary artery results in a dramatic increase in VEGF RNA levels in the pig and rat myocardium, suggesting that VEGF may mediate the spontaneous revascularization that follows myocardial ischemia (BANAI et al. 1994b; HASHIMOTO et al. 1994). Furthermore, hypoxic upregulation of VEGF mRNA in neuroglial cells, secondary to the onset of neuronal activity, has been proposed to play an important physiological role in the development of the retinal vasculature (STONE et al. 1995).

Similarities exist between the mechanisms leading to hypoxic regulation of VEGF and erythropoietin (Epo) (GOLDBERG and SCHNEIDER 1994). Hypoxia-inducibility is conferred on both genes by homologous sequences. By deletion and mutation analysis, a 28-base sequence has been identified in the 5′ promoter of the rat and human VEGF gene, which mediates hypoxia-induced transcription (LEVY et al. 1995; LIU et al. 1995). Such a sequence reveals a high degree of homology and similar protein-binding characteristics as the hypoxia-inducible factor 1 (HIF-1) binding site, within the Epo gene (MADAN and CURTIN 1993). HIF-1 has been identified as a mediator of transcriptional responses to hypoxia and is a basic, heterodimeric, helix–loop–helix protein (WANG and SEMENZA 1995). When reporter constructs containing the VEGF sequences that mediate hypoxia-inducibility were co-transfected with expression vectors encoding HIF-1 subunits, reporter gene transcription was much greater than that observed in cells transfected with the reporter alone, both in hypoxic and normoxic conditions (FORSYTHE et al. 1996).

However, transcriptional activation is not the only mechanism leading to VEGF upregulation in response to hypoxia (IKEDA et al. 1995; LEVY et al. 1996). Increased mRNA stability has been identified as a significant post-transcriptional component. Sequences that mediate increased stability were identified in the 3′ untranslated region of the VEGF mRNA.

### 4.2 Cytokines

Various cytokines or growth factors may upregulate VEGF mRNA expression. Epidermal growth factor (EGF), TGF-β or keratinocyte growth factor (KGF) result in a marked induction of VEGF mRNA expression (FRANK et al. 1995) EGF also stimulates VEGF release by cultured glioblastoma cells (GOLDMAN et al. 1993). In addition, treatment of quiescent cultures of epithelial and fibroblastic cell lines with TGF-β resulted in induction of VEGF mRNA and release of VEGF protein in the medium (PERTOVAARA et al. 1994). Based on these findings, it has been proposed that VEGF may function as a paracrine mediator for indirectly acting angiogenic agents, such as TGF-β (PERTOVAARA et al. 1994). Furthermore, IL-1-β induces VEGF expression in aortic smooth muscle cells (LI et al. 1995). Both IL-1-α and prostaglandin $E_2(PGE_2)$ have been shown to induce expression of VEGF in cultured synovial fibroblasts, suggesting the participation of such inductive mechanisms in inflammatory angiogenesis (BEN-AV et al. 1995). IL-6 has also been shown to significantly induce VEGF expression in several cell lines (COHEN et al. 1996). IGF-1, a mitogen implicated in the growth of several malignancies, has also been shown to induce VEGF mRNA and protein in cultured colorectal carcinoma cells (WARREN et al. 1996). Such induction was mediated by a combined increase in the transcriptional rate of the VEGF gene and in the stability of the mRNA.

## 4.3 Differentiation and Transformation

Cell differentiation has been shown to play an important role in the regulation of VEGF gene expression (CLAFFEY et al. 1992). The VEGF mRNA is upregulated during the conversion of 3T3 pre-adipocytes into adipocytes or during the myogenic differentiation of C2C12 cells. Conversely, VEGF gene expression is repressed during the differentiation of the pheochromocytoma cell line PC12 into non-malignant, neuron-like cells.

Specific transforming events also result in induction of VEGF gene expression. A mutated form of the murine p53 tumor-suppressor gene has been shown to result in induction of VEGF mRNA expression in NIH 3T3 cells in transient transfection assays (KIESER et al. 1994). Likewise, oncogenic mutations or amplification of *ras* lead to VEGF upregulation (RAK et al. 1995; GRUGEL et al. 1995). Interestingly, expression of oncogenic *ras*, either constitutive or transient, potentiated the induction of VEGF by hypoxia (MAZURE et al. 1996). Moreover, the von Hippel-Lindau (VHL) tumor-suppressor gene has been recently implicated in the regulation of VEGF gene expression (SIEMEISTER et al. 1996; ILIOPOULOS et al. 1996; GNARRA et al. 1996).

The VHL tumor-suppressor gene is inactivated in patients with VHL disease and in most sporadic clear cell renal carcinomas. Although the function of the VHL protein remains to be fully elucidated, it is known that such a protein interacts with the elongin BC subunits in vivo, and regulates RNA polymerase-II elongation activity in vitro by inhibiting formation of the elongin ABC complex. Human renal cell carcinoma cells, either lacking endogenous wild-type VHL gene or expressing an inactive mutant, demonstrated altered regulation of VEGF gene expression, which was corrected by introduction of the wild-type VHL gene. Most of the endothelial cells' mitogenic activity released by tumor cells expressing the mutant

VHL gene was neutralized by anti-VEGF antibodies (SIEMEISTER et al. 1996). These findings suggest that VEGF is a key mediator of the abnormal vascular proliferations and solid tumors characteristic of VHL syndrome.

ILIOPULOS et al. (1996) have shown that one function of the VHL protein is to provide a negative regulation of a series of hypoxia-inducible genes, including the VEGF, PDGF-B chain and the glucose transporter GLUT1 genes. In the presence of a mutant VHL, mRNAs for such genes were produced under both normoxic and hypoxic conditions. Reintroduction of wild-type VHL resulted in inhibition of mRNA production under normoxic conditions and restored the characteristic hypoxia-inducibility of those genes (ILIOPOULOS et al. 1996). In addition, GNARRA et al. (1996) have shown that VHL regulates VEGF expression at a post-transcriptional level and that VHL inactivation in target cells causes a loss of VEGF suppression, leading to formation of a vascular stroma. Interestingly, despite fivefold differences in VEGF mRNA levels, VHL overexpression did not affect VEGF transcription initiation.

## 5 The VEGF Receptors

Two classes of high-affinity VEGF binding sites on the surface of bovine endothelial cells were described initially, with $K_d$ values of 10 pM and 100 pM (VAISMAN et al. 1990; PLOUET and MOUKADIRI 1990). Lower affinity binding sites on mononuclear phagocytes were subsequently described (SHEN et al. 1993). It has been suggested that such binding sites are involved in mediating the chemotactic effects of VEGF for monocytes (CLAUSS et al. 1990).

Ligand autoradiography studies on fetal and adult rat tissue sections demonstrated that high-affinity VEGF binding sites are localized to the vascular endothelium of large or small vessels in situ (JAKEMAN et al. 1992, 1993). VEGF binding was apparent not only on proliferating, but also on quiescent endothelial cells (JAKEMAN et al. 1992, 1993). Also, the earliest developmental identification of high-affinity VEGF binding was in hemangioblasts in the blood islands in the yolk sac (JAKEMAN et al. 1993).

## 6 The VEGFR-1 and VEGFR-2 Tyrosine Kinases

### 6.1 Binding Characteristics

Two VEGF receptor tyrosine kinases (RTKs) have been identified. The VEGFR-1 (fms-like-tyrosine kinase) (DE VRIES et al. 1992) and VEGFR-2 (KDR; TERMAN et al., 1992) receptors bind VEGF with high affinity. The murine homologue of

VEGFR-2 (also denoted fetal liver kinase-1; Flk-1), shares 85% sequence identity with human KDR (MATTHEWS et al. 1991). Both VEGFR-1 and VEGFR-2 have seven immunoglobulin (Ig)-like domains in the extracellular domain (ECD), a single transmembrane region and a consensus tyrosine kinase sequence, which is interrupted by a kinase-insert domain (SHIBUYA et al. 1990; TERMAN et al. 1991; MATTHEWS et al. 1991). VEGFR-1 has the highest affinity for rhVEGF$_{165}$, with a $K_d$ of approximately 10–20pM (DE VRIES et al. 1992). VEGFR-2 has a somewhat lower affinity for VEGF: the $K_d$ has been estimated to be approximately 75–125pM (TERMAN et al. 1992).

A cDNA coding an alternatively spliced soluble form of VEGFR-1 (sVEGFR-1), lacking the seventh Ig-like domain, the transmembrane sequence and the cytoplasmic domain, has been identified in HUVEC (KENDALL et al. 1996). This sVEGFR-1 receptor binds VEGF with high affinity ($K_d$ 10–20pM) and is able to inhibit VEGF-induced mitogenesis and may be a physiological negative regulator of VEGF's action (KENDALL et al. 1996).

An additional member of the family of RTKs with seven Ig-like domains in the ECD is VEGFR-3 (also denoted Flt-4; PAJUSOLA et al. 1992; GALLAND et al. 1992; FINNERTY et al. 1993) which, however, is not a receptor for VEGF, but rather binds a newly identified ligand called VEGF-C or VEGF-related peptide (VRP) (see Chapt. 3).

Recent studies have mapped the binding site for VEGF to the second immunoglobulin-like domain of VEGFR-1 and VEGFR-2. Deletion of the second domain of VEGFR-1 completely abolished the binding of VEGF. Introduction of the second domain of VEGFR-2 into an VEGFR-1 mutant lacking the homologous domain restored VEGF binding. However, the ligand specificity was characteristic of the VEGFR-2. To further test this hypothesis, chimeric receptors, where the first three or just the second Ig-like domains of VEGFR-1 replaced the corresponding domains in VEGFR-3, were created. Both swaps conferred upon VEGFR-3 the ability to bind VEGF with an affinity nearly identical to that of wild-type VEGFR-1. Furthermore, transfected cells expressing these chimeric VEGFR-3 receptors exhibited increased DNA synthesis in response to VEGF or PlGF (DAVIS-SMYTH et al. 1996).

An application of these structure–function studies is the generation of inhibitors of VEGF activity. The first three Ig-like domains of VEGFR-1 fused to a heavy chain Fc potently inhibits VEGF bioactivity across species. The Fc may confer sufficient half-life and stability when injected systemically (CHAMOW and ASHKENAZI 1996). Therefore, this agent may be a useful tool in determining the role of endogenous VEGF in several in vivo models.

## 6.2 Signal Transduction

VEGF has been shown to induce the phosphorylation of at least 11 proteins in bovine aortic endothelial cells (GUO et al. 1995). Phospholipase C (PLC)-γ, and two proteins that associate with PLC-γ were phosphorylated in response to VEGF.

Furthermore, immunoblot analysis to search for mediators of signal transduction that contain SH2 domains demonstrated that VEGF induces phosphorylation of phosphatidyl inositol 3-kinase, *ras* GTPase activating protein (GAP) and several other proteins. These findings suggest that VEGF promotes the formation of multimeric aggregates of VEGF receptors with proteins that contain SH2 domains. These studies, however, did not identify which VEGF receptor(s) are involved in these events. Recently, it was suggested that NO mediates, at least in part, the mitogenic effect of VEGF on cultured microvascular endothelium isolated from coronary venules (MORBIDELLI et al. 1996). The proliferative effect of VEGF was reduced by pre-treatment of the cells with NO-synthase inhibitors. Exposure of the cells to VEGF induced a significant increment in cyclic guanosine monophosphate (cGMP) levels. These findings suggest that VEGF stimulates proliferation of post-capillary endothelial cells through the production of NO and cGMP accumulation.

Several studies have indicated that VEGFR-1 and VEGFR-2 have different signal transduction properties (WALTENBERGER et al. 1994; SEETHARAM et al. 1995). Porcine aortic endothelial cells lacking endogenous VEGF receptors display chemotaxis and mitogenesis in response to VEGF when transfected with a plasmid coding for VEGFR-2 (WALTENBERGER et al. 1994). In contrast, transfected cells expressing VEGFR-1 lack such responses (WALTENBERGER et al. 1994; SEETHARAM et al. 1995). VEGFR-2 undergoes strong ligand-dependent tyrosine phosphorylation in intact cells, while VEGFR-1 reveals a weak or undetectable response (WALTENBERGER et al. 1994; SEETHARAM et al. 1995). In addition, VEGF stimulation results in weak tyrosine phosphorylation that does not generate any mitogenic signal in transfected NIH 3T3 cells expressing VEGFR-1 (SEETHARAM et al. 1995). These findings agree with other studies showing that PlGF, which binds with high affinity to VEGFR-1, but not to VEGFR-2, lacks direct mitogenic or permeability-enhancing properties or the ability to effectively stimulate tyrosine phosphorylation in endothelial cells (PARK et al. 1994).

It seems, then, that interaction with VEGFR-2 is a critical requirement to induce the full spectrum of VEGF biological responses. In further support of this conclusion, VEGF mutants that bind selectively to VEGFR-2 are fully active endothelial-cell mitogens (KEYT et al. 1996b). These findings led to cast doubt on the role of VEGFR-1 as a truly signaling receptor. However, more recent evidence indicates that VEGFR-1 indeed signals, although our understanding of these events is fragmentary. CUNNINGHAM et al. (1995) have demonstrated an interaction between VEGFR-1 and the p85 subunit of phosphatidyl inositol 3-kinase (CUNNINGHAM et al. 1995), suggesting that p85 couples VEGFR-1 to intracellular signal transduction systems and implicate elevated levels of phosphatidyl inositol (3,4,5) P3 levels in this process (CUNNINGHAM et al. 1995). Also, members of the *Src* family, such as *Fyn* and *Yes*, show an increased level of phosphorylation following VEGF stimulation in transfected cells expressing VEGFR-1, but not VEGFR-2 (WALTENBERGER et al. 1994). Furthermore, it has been shown that a specific biological response, the migration of monocytes in response to VEGF (or PlGF), is mediated by VEGFR-1 (BARLEON et al. 1996).

## 6.3 Regulation

The expression of VEGFR-1 and -2 genes is largely restricted to the vascular endothelium. The promoter region of VEGFR-1 has been cloned and characterized and a 1-kb fragment of the 5' flanking region, essential for endothelial-specific expression, was identified (MORISHITA et al. 1995). Likewise, a 4-kb 5' flanking sequence has been identified in the promoter of the human VEGFR-2 that confers endothelial cell-specific activation (PATTERSON et al. 1995).

Similarly to VEGF, hypoxia has been proposed to play an important role in the regulation of VEGF-receptor gene expression. Exposure of rats to acute or chronic hypoxia led to pronounced upregulation of both VEGFR-1 and VEGFR-2 genes in the lung vasculature (TUDER et al. 1995). Also, VEGFR-1 and -2 mRNAs were substantially upregulated throughout the heart, following myocardial infarction in the rat (LI et al. 1996). However, in vitro studies have yielded unexpected results. Hypoxia increases VEGF receptor number by 50% in cultured bovine retinal capillary endothelial cells, but the expression of VEGFR-2 is not induced although, paradoxically, shows an initial downregulation (TAKAGI et al. 1996). BROGI et al. (1996) have proposed that the hypoxic upregulation of VEGFR-2 observed in vivo is not direct, but requires the release of an unidentified paracrine mediator from ischemic tissues.

Recent studies have provided evidence of a differential transcriptional regulation of the VEGFR-1 and VEGFR-2 genes by hypoxia (GERBER et al. 1997). When HUVEC were exposed to hypoxic conditions, in vitro, increased levels of VEGFR-1 expression were observed. In contrast, VEGFR-2 mRNA levels were unchanged or slightly repressed. Promoter deletion analysis demonstrated that a 430-bp region of the VEGFR-1 promoter was required for transcriptional activation in response to hypoxia. This region includes a heptamer sequence matching the HIF-1 consensus binding site previously found in other hypoxia inducible genes. The element mediating the hypoxia response was further defined as a 40-bp sequence, including the putative HIF-1 binding site, but was not found in the VEGFR-2 promoter. These findings indicate that, unlike the VEGFR-2 gene, the VEGFR-1 receptor gene is directly upregulated by hypoxia via a hypoxia-inducible enhancer element located at position −976 to −937 of the VEGFR-1 promoter (GERBER et al. 1997). Also, recent studies have shown that both TNF-$\alpha$ (PATTERSON et al. 1996) and TGF-$\beta$ (MANDRIOTA et al. 1996) have the ability to inhibit the expression of the VEGFR-2 gene in cultured endothelial cells.

# 7 Role of VEGF and its Receptors in Physiological Angiogenesis

## 7.1 Distribution of VEGF, VEGFR-1 and VEGFR-2

The proliferation of blood vessels is crucial for a wide variety of physiological processes, such as embryonic development, normal growth and differentiation, wound healing and reproductive functions.

During embryonic development, VEGF expression is first detected within the first few days following implantation in the giant cells of the trophoblast (BREIER et al. 1992; JAKEMAN et al. 1993). At later developmental stages in the mouse or rat embryos, the VEGF mRNA is expressed in several organs, including heart, vertebral column, kidney and along the surface of the spinal cord and brain. In the developing mouse brain, the highest levels of mRNA expression are associated with the choroid plexus and the ventricular epithelium (BREIER et al. 1992). In the human fetus (16–22 weeks), VEGF mRNA expression is detectable in virtually all tissues and is most abundant in lung, kidney and spleen (SHIFREN et al. 1994).

In situ hybridization studies have shown that the VEGFR-2 mRNA is expressed in the yolk sac and intraembryonic mesoderm and later on in angioblasts, endocardium and small and large vessel endothelium (QUINN et al. 1993; MILLAUER et al. 1993). These findings strongly suggested a role for VEGFR-2 in the regulation of vasculogenesis and angiogenesis. Other studies have demonstrated that expression of VEGFR-2 mRNA is first detected in the proximal-lateral embryonic mesoderm, which gives rise to the heart (YAMAGUCHI et al. 1993). VEGFR-2 is then detectable in endocardial cells of the heart primordia and, subsequently, in the major embryonic and extraembryonic vessels (YAMAGUCHI et al. 1993). These studies have indicated that VEGFR-2 may be the earliest marker of endothelial-cell precursors. The VEGFR-1 mRNA is selectively expressed in vascular endothelial cells, in both fetal and adult mouse tissues (PETERS et al. 1993). Similarly to the high-affinity VEGF binding, the VEGFR-1 mRNA is expressed in both proliferating and quiescent endothelial cells, suggesting a role for VEGFR-1 in the maintenance of endothelial cells (PETERS et al. 1993).

VEGF expression is also detectable around microvessels in areas where endothelial cells are normally quiescent, such as kidney glomerulus, pituitary, heart, lung and brain (FERRARA et al. 1992; MONACCI et al. 1993). These findings raised the possibility that VEGF may be required not only to induce active vascular proliferation but, at least in some circumstances, also for the maintenance of the differentiated state of blood vessels (FERRARA et al. 1992). In agreement with this hypothesis, ALON et al. (1995) have shown that VEGF acts as a survival factor, at least for the developing retinal vessels. They propose that hyperoxia-induced vascular regression in the retina of neonatal animals is a consequence of inhibition of VEGF production by glial cells. Accordingly, intraocular administration of VEGF to newborn rats at the onset of hyperoxia was able to prevent cell apoptosis and regression of the retinal vasculature (ALON et al. 1995).

## 7.2 The VEGFR-1, VEGFR-2 and VEGF Gene Knockouts in Mice

Recent studies have demonstrated that both VEGFR-1 and VEGFR-2 are essential for normal development of embryonic vasculature. However, their respective roles in endothelial-cell proliferation and differentiation appear to be distinct (FONG et al. 1995; SHALABY et al. 1995). Mouse embryos homozygous for a targeted mutation in the VEGFR-1 locus died in utero between day 8.5 and day 9.5 (FONG et al.

1995). Endothelial cells developed in both embryonic and extraembryonic sites, but failed to organize in normal vascular channels. Mice in which the VEGFR-2 gene had been inactivated lacked vasculogenesis and also failed to develop blood islands. Hematopoietic precursors were severely disrupted and organized blood vessels failed to develop throughout the embryo or the yolk sac, resulting in death in utero between day 8.5 and day 9.5 (SHALABY et al. 1995).

However, these findings do not necessarily imply VEGF as being equally essential, since other ligands might potentially activate the VEGFR-1 and -2 and, thus, substitute VEGF's action. Very recent studies (CARMELIET et al. 1996; FERRARA et al. 1996) have generated direct evidence for the role played by VEGF in embryonic vasculogenesis and angiogenesis. Inactivation of the VEGF gene in mice resulted in embryonic lethality in heterozygous embryos, between day 11 and day 12. The VEGF+/− embryos were growth retarded and also exhibited a number of developmental anomalies. The forebrain region appeared significantly underdeveloped. In the heart region, the outflow region was grossly malformed; the dorsal aortae were rudimentary, and the thickness of the ventricular wall was markedly decreased. The yolk sac revealed a markedly reduced number of nucleated red blood cells within the blood islands. Also, the vitelline veins failed to fuse with the vascular plexus of the yolk sac. Significant defects in the vasculature of other tissues and organs, including placenta and nervous system, were observed. In situ hybridization confirmed expression of VEGF mRNA in heterozygous embryos. Thus, the VEGF+/− phenotype appears to be due to gene dosage and not to maternal imprinting.

While several heterozygous phenotypes have been described (BRANDON et al. 1995), this may be the first example of embryonic lethality following the loss of a single allele of a gene that is not maternally imprinted. Therefore, VEGF and its receptors are essential for blood island formation and angiogenesis, such that even reduced concentrations of VEGF are inadequate to support a normal pattern of development. However, inactivation of the PlGF gene does not result in embryonic lethality, even in the homozygous state (CARMELIET and COLLEN 1997). PlGF−/− mice are viable and fertile, although they may have some impairment of wound healing.

## 8 Role of VEGF in Corpus Luteum Angiogenesis

The development and endocrine function of the ovarian corpus luteum (CL) are dependent on the growth of new capillary vessels. Although several molecules have been implicated as mediators of CL angiogenesis, at present, there is no direct evidence for the involvement of any. The VEGF mRNA is temporally and spatially related to the proliferation of blood vessels in the rat, mouse and primate ovary and in the rat uterus, suggesting that VEGF is a mediator of the cyclical growth of blood vessels that occurs in the female reproductive tract (PHILLIPS et al. 1990; RAVINDRANATH et al. 1992; SHWEIKI et al. 1993; CULLINAN-BOVE and KOOS 1993).

Very recently, the hypothesis that VEGF is a mediator of CL angiogenesis has been examined in a rat model of hormonally induced ovulation (FERRARA et al. 1998). Treatment with Flt (1–3)-IgG resulted in virtually complete suppression of CL angiogenesis. This effect was associated with inhibition of CL development and progesterone release. Failure of maturation of the endometrium was also observed. Areas of ischemic necrosis were demonstrated in the CL of treated animals; however, no effect on the pre-existing ovarian vasculature was observed. These findings demonstrate that, in spite of the redundancy of potential mediators, VEGF is essential for CL angiogenesis. Furthermore, they have implications in the control of fertility and the treatment of ovarian disorders characterized by hypervascularity and hyperplasia, such as polycystic ovary syndrome.

## 9 Role of VEGF in Pathological Angiogenesis

### 9.1 Tumor Angiogenesis

In 1945, ALGIRE and CHALKLEY, on the basis of microscopic observations of the vascular development of tumor xenografts in transparent chambers in mice, proposed that the growth of solid tumors is dependent on the development of a new vascular supply derived from the host (ALGIRE and CHALKLEY 1945). In 1971, FOLKMAN proposed inhibition of angiogenesis as a novel strategy to treat cancer (FOLKMAN 1971). Since then, extensive research has been devoted to the identification of tumor angiogenesis factor(s).

Many tumor cell lines secrete VEGF in vitro (FERRARA et al. 1992). In situ hybridization studies have demonstrated that the VEGF mRNA is markedly upregulated in the vast majority of human tumors examined so far. These include: lung (VOLM et al. 1997a,b), breast (BROWN et al. 1995a; YOSHIJI et al. 1996), gastrointestinal tract (BROWN et al. 1993b; SUZUKI et al. 1996), kidney (BROWN et al. 1993a), bladder (BROWN et al. 1993a), ovary (OLSON et al. 1994), endometrium (GUIDI et al. 1996) and uterine cervix (GUIDI et al. 1995) carcinomas, angiosarcoma (HASHIMOTO et al. 1995), germ cell tumors (VIGLIETTO et al. 1996) and several intracranial tumors, including glioblastoma multiforme (SHWEIKI et al. 1992; PLATE et al. 1992; PHILLIPS et al. 1993) and sporadic, as well as VHL syndrome-associated capillary hemangioblastoma (BERKMAN et al. 1993; WIZIGMANN VOOS et al. 1995). In glioblastoma multiforme and other tumors with significant necrosis, the expression of VEGF mRNA is highest in hypoxic tumor cells adjacent to necrotic areas (SHWEIKI et al. 1992; PLATE et al. 1992; PHILLIPS et al. 1993). A correlation exists between the degree of vascularization of the malignancy and VEGF mRNA expression (BERKMAN et al. 1993; WIZIGMANN VOOS et al. 1995; GUIDI et al. 1995). In virtually all specimens examined, the VEGF mRNA was expressed in tumor cells, but not in endothelial cells. In contrast, the mRNAs for VEGFR-1 and -2 were upregulated in the endothelial cells associated with the tumor (BROWN et al.

1993b; PLATE et al. 1993). These findings are consistent with the hypothesis that VEGF is primarily a paracrine mediator (FERRARA et al. 1993).

Immunohistochemical studies have localized the VEGF protein not only to the tumor cells, but also to the vasculature (PLATE et al. 1992; BROWN et al. 1993b). This localization indicates that tumor-secreted VEGF accumulates in the target cells (QU et al. 1995). Interestingly, recent studies have suggested that the angiogenesis mediated by the human immunodeficiency virus (HIV-1) Tat protein (ALBINI et al. 1996a) requires activation of VEGFR-2 (ALBINI et al. 1996b). Tat induces growth of Kaposi's sarcoma (KS) spindle cells and has been implicated in the vascularity of the KS lesions (ALBINI et al. 1996b).

Elevations in VEGF levels have been detected in the serum of some cancer patients (KONDO et al. 1994). Also, a correlation has been noted between VEGF expression and microvessel density in primary breast cancer sections (TOI et al. 1996). A post-operative survey indicated that the relapse-free survival rate of patients with VEGF-positive tumors was significantly worse than that of VEGF-negative, suggesting that expression of VEGF is associated with stimulation of angiogenesis and with early relapse in primary breast cancer (GASPARINI et al. 1997). A similar correlation has been described in gastric-carcinoma patients (MAEDA et al. 1996). VEGF-positivity in tumor sections was correlated with vessel involvement, lymph node metastasis and liver metastasis. Furthermore, patients with VEGF-positive tumors had a worse prognosis than those with VEGF-negative tumors (MAEDA et al. 1996).

The availability of specific monoclonal antibodies capable of inhibiting VEGF-induced angiogenesis in vivo and in vitro (KIM et al. 1992) made it possible to generate direct evidence for a role of VEGF in tumorigenesis. In a study published by KIM et al. (1993), such antibodies were found to exert a potent inhibitory effect on the growth of three human tumor cell lines injected subcutaneously in nude mice, the SK-LMS-1 leiomyosarcoma, the G55 glioblastoma multiforme and the A673 rhabdomyosarcoma. The growth inhibition ranged between 70% and more than 95%. Subsequently, other tumor cell lines were found to be inhibited in vivo by this treatment (WARREN et al. 1995; MELNYK et al. 1996; ASANO et al. 1995; BORGSTROM et al. 1998a).

In agreement with the hypothesis that inhibition of neovascularization is the mechanism of tumor suppression, the density of blood vessels was significantly lower in sections of tumors from antibody-treated animals than in controls. Furthermore, neither the antibodies nor VEGF had any effect on the in vitro growth of the tumor cells (KIM et al. 1993). Intravital videomicroscopy techniques have allowed a more direct verification of the hypothesis that anti-VEGF antibodies indeed block tumor angiogenesis (BORGSTROM et al. 1996). Non-invasive imaging of the vasculature revealed a nearly complete suppression of tumor angiogenesis in anti-VEGF treated animals compared with controls, at all time points examined (BORGSTROM et al. 1996).

VEGF is a mediator of the in vivo growth of human colon carcinoma HM7 cells in a nude mouse model of liver metastasis (WARREN et al. 1995). Treatment with anti-VEGF monoclonal antibodies resulted in a dramatic decrease in the number and size of metastases. Similarly, administration of anti-VEGF neutralizing antibodies inhibited primary tumor growth and metastasis of A431 human epidermoid carcinoma cells in severe combined immune deficient (SCID) mice (MELNYK et al. 1996) or HT-1080 fibrosarcoma cells implanted in BALB/c nude mice (ASANO et al. 1995).

Recently, BORGSTROM et al. (1998b) have shown that a combination treatment that includes anti-VEGF monoclonal antibody and doxorubicin results in a significant enhancement of the efficacy of either agent alone and led, in some cases, to complete regression of tumors derived from MCF-7 breast carcinoma cells in nude mice.

Intravital fluorescence microscopy and video imaging analysis have also been applied to address the important issue regarding the effects of VEGF on permeability and other properties of tumor vessels (YUAN et al. 1996). Treatment with anti-VEGF monoclonal antibodies was initiated after tumor xenografts had already been established and vascularized, and resulted in time-dependent reductions in vascular permeability (YUAN et al. 1996). These effects were accompanied by striking changes in the morphology of vessels, with dramatic reduction in diameter and tortuosity. This reduction in diameter is expected to block the passage of blood elements and eventually stop the flow in the tumor vascular network. A regression of blood vessels was observed after repeated administrations of anti-VEGF antibody. These findings suggest that tumor vessels require constant stimulation with VEGF in order to maintain not only their proliferative properties, but also some key morphological features (YUAN et al. 1996).

An independent verification of the hypothesis that the VEGF action is required for tumor angiogenesis has been provided by the finding that retrovirus-mediated expression of a dominant negative VEGFR-2 mutant, which inhibits signal transduction through wild-type VEGFR-2, suppresses the growth of glioblastoma multiforme and other tumor cell lines in vivo (MILLAUER et al. 1994).

## 9.2 Angiogenesis Associated with Other Pathological Conditions

Diabetes mellitus, occlusion of central retinal vein or prematurity with subsequent exposure to oxygen can all be associated with intraocular neovascularization (GARNER 1994). The new blood vessels may lead to vitreous hemorrhage, retinal detachment, neovascular glaucoma and eventual blindness (GARNER 1994). Diabetic retinopathy is the leading cause of blindness in the working population (OLK and LEE 1993). All of these conditions are known to be associated with retinal ischemia (PATZ 1980). In 1948, MICHAELSON proposed that a key event in the pathogenesis of these conditions was the release by the ischemic retina of diffusible angiogenic factor(s) ("factor X") responsible for retinal and iris neovascularization into the vitreous (MICHAELSON 1948). VEGF, by virtue of its diffusible nature and hypoxia-inducibility, was an attractive candidate as a mediator of intraocular neovascularization. Accordingly, elevations of VEGF levels in the aqueous and vitreous of eyes with proliferative retinopathy have been described (AIELLO et al. 1994; ADAMIS et al. 1994; MALECAZE et al. 1994).

In a large series, a strong correlation was found between levels of immunoreactive VEGF in the aqueous and vitreous humors and active proliferative retinopathy. VEGF levels were undetectable or very low (<0.5ng/ml) in the eyes of patients affected by non-neovascular disorders or diabetes without proliferative retinopathy (AIELLO et al. 1994). In contrast, the VEGF levels were in the range 3–10ng/ml in the presence of active proliferative retinopathy associated with diabetes, occlusion of central retinal vein or prematurity. In agreement with these findings, in situ hybridization studies have demonstrated upregulation of VEGF mRNA in the retina of patients with proliferative retinopathies secondary to diabetes, central retinal vein occlusion, retinal detachment or intraocular tumors (PE'ER et al. 1996).

More direct evidence for a role of VEGF as a mediator of intraocular neovascularization has been generated in a primate model of iris neovascularization and in a murine model of retinopathy of prematurity (MILLER et al. 1994; PIERCE et al. 1995). In the former, intraocular administration of anti-VEGF antibodies dramatically inhibits the neovascularization that follows occlusion of central retinal veins (ADAMIS et al. 1996). Likewise, soluble VEGFR-1 or VEGFR-2 fused to an IgG suppresses retinal angiogenesis in the mouse model (AIELLO et al. 1995).

Neovascularization is also a major cause of visual loss in age-related macular degeneration (AMD), the overall leading cause of blindness (GARNER 1994). Most AMD patients demonstrate atrophy of the retinal pigment epithelia and characteristic formations called "drusen". A significant percentage of AMD patients (~20%) manifest the neovascular (exudative) form of the disease. In this condition, the new vessels stem from the extraretinal choriocapillary (GARNER 1994). Leakage and bleeding from these vessels may lead to damage to the macula and, ultimately, to loss of central vision. Because of the proximity of the lesions to the macula, laser photocoagulation or surgical therapy are of very limited value. Very recent studies have documented the immunohistochemical localization of VEGF in surgically resected choroidal neovascular membranes from AMD patients (LOPEZ et al. 1996; KVANTA et al. 1996). These findings suggests a role for VEGF in the progression of AMD-related choroidal neovascularization, raising the possibility that a pharmacological treatment with monoclonal antibodies or other VEGF inhibitors may constitute a therapy for this condition.

Two independent studies have suggested that VEGF is involved in the pathogenesis of rheumatoid arthritis (RA), an inflammatory disease in which angiogenesis plays a significant role (KOCH et al. 1994; FAVA et al. 1994). The RA synovium is characterized by the formation of pannus, an extensively vascularized tissue, which invades and destroys the articular cartilage (FASSBENDER and SIMLING-ANNENFELD 1983). Levels of immunoreactive VEGF were found to be high in the synovial fluid of RA patients, while they were very low or undetectable in the synovial fluid of patients affected by other forms of arthritis or by degenerative joint disease (KOCH et al. 1994; FAVA et al. 1994). Furthermore, anti-VEGF antibodies significantly reduced the endothelial cell chemotactic activity of the RA synovial fluid (KOCH et al. 1994).

It has been shown that VEGF expression is increased in psoriatic skin (DETMAR et al. 1994). Increased vascularity and permeability are characteristic of psoriasis. Also, VEGF mRNA expression has been examined in three bullous disorders with subepidermal blister formation, bullous pemphigoid, erythema multiforme and dermatitis herpetiformis (BROWN et al. 1995b).

Angiogenesis is also important in the pathogenesis of endometriosis, a condition characterized by ectopic endometrium implants in the peritoneal cavity. Recently, elevation of VEGF in the peritoneal fluid of patients with endometriosis has been reported (MCLAREN et al. 1996; SHIFREN et al. 1996). Immunohistochemistry indicated that activated peritoneal fluid macrophages, as well as tissue macrophages within the ectopic endometrium, are the main source of VEGF in this condition. (MCLAREN et al. 1996; SHIFREN et al. 1996). VEGF upregulation has also been implicated in the hypervascularity of the ovarian stroma that characterizes Stein-Leventhal syndrome (KAMAT et al. 1995). Moreover, SATO et al. (1995) proposed that VEGF may be responsible for the characteristic hypervascularity of Graves' disease. Thyroid-stimulating hormone (TSH), insulin phorbol ester, dibutiryl cAMP and Graves' IgG were found to stimulate VEGF mRNA expression in cultured human thyroid follicles (SATO et al. 1995).

## 10 VEGF and Therapeutic Angiogenesis

The availability of agents able to promote the growth of new collateral vessels would be, potentially, of major therapeutic value for disorders characterized by inadequate tissue perfusion, and might constitute an alternative to surgical reconstruction procedures. For example, chronic limb ischemia, most frequently caused by obstructive atherosclerosis affecting the superficial femoral artery, is associated with a high rate of morbidity and mortality, and treatment is currently limited to surgical revascularization or endovascular interventional therapy (GRAOR and GRAY 1991); no pharmacological therapy has been shown to be effective for this condition.

It has been shown that intraarterial or intramuscular administration of recombinant human (rh)VEGF$_{165}$ may significantly augment perfusion and development of collateral vessels in a rabbit model, in which chronic hindlimb ischemia was created by surgical removal of the femoral artery (TAKESHITA et al. 1994). These studies provided angiographic evidence of neovascularization in the ischemic limbs. Arterial gene transfer with cDNA encoding VEGF also led to revascularization in the same rabbit model to an extent comparable with that achieved with the recombinant protein (TAKESHITA et al. 1996a,b). In addition, the hypothesis that the angiogenesis initiated by the administration of VEGF improved muscle function in ischemic limbs was tested by WALDER et al. (1996). A single intraarterial injection of rhVEGF$_{165}$ augmented muscle function in this rabbit model of peripheral limb ischemia. This exercise-induced hyperemia was significantly improved in ischemic limbs treated with rhVEGF$_{165}$ (WALDER et al. 1996). Such improvement in perfusion was, however, not seen in other non-ischemic tissues, including the contralateral limb. Similarly, BAUTERS et al. (1994) have shown that both maximal flow velocity and maximal blood flow are significantly increased in ischemic limbs following VEGF administration.

Other studies have shown that VEGF administration also leads to a recovery of normal endothelial reactivity in dysfunctional endothelium. Following obstruction of a large artery and development of collateral vessels, the increase in blood flow that normally follows acetylcholine infusion is severely blunted; serotonin paradoxically leads to a decrease in blood flow (BAUTERS et al. 1995). Thirty days after a single intraarterial bolus of VEGF$_{165}$, restoration of the normal increase in blood flow was demonstrated in the ischemic rabbit hindlimb, following acetylcholine or serotonin infusion (BAUTERS et al. 1995).

BANAI et al. (1994a) have shown that VEGF administration results in increased coronary blood flow in a dog model of coronary insufficiency. Following occlusion of the left circumflex coronary artery, daily intraluminal injections of rhVEGF distal to the occlusion resulted in a significant enhancement of collateral blood flow over a 4-week period. In addition, HARADA et al. (1996) demonstrated that extraluminal administration of as little as 2μg of rhVEGF by an osmotic pump results in a significant increase in coronary blood flow in a pig model of chronic myocardial ischemia created by ameroid occlusion of the left proximal circumflex artery. Also, magnetic resonance imaging provided a non-invasive assessment of the benefits secondary to VEGF administration in the porcine model (PEARLMAN et al. 1995). Image series converted to a space–time map demonstrated a reduction in the size of the ischemic zone and a decreased delay in contrast arrival after VEGF treatment. These findings demonstrated improvement in cardiac global and regional function and reduced infarct size, resulting from enhanced collateral blood supply (PEARLMAN et al. 1995; WARE and SIMONS 1997).

A further potential therapeutic application of VEGF is the prevention of restenosis following percutaneous transluminal angioplasty (PTA). Between 15% and 75% of patients undergoing PTA for occlusive coronary or peripheral arterial disease develop restenosis within 6 months (GRAOR and GRAY 1991). It has been proposed that damage to the endothelium is a crucial event, triggering fibrocellular intimal proliferation (ESSED et al. 1983). Therefore, the induction of rapid re-endothelialization may be an effective strategy to prevent the cascade of events leading to neointima formation and ultimately to restenosis in patients. Recent evidence shows that VEGF accelerates re-endothelialization and also attenuates intimal hyperplasia in balloon-injured rat carotid artery or rabbit aorta (ASAHARA et al. 1995; CALLOW et al. 1994).

Recently, the hypothesis that VEGF may result in therapeutically significant angiogenesis in humans has been tested by ISNER et al. (1996b) in a gene-therapy trial in patients with severe limb ischemia. A case report of an interim analysis of this trial has been published (ISNER et al. 1996a). Arterial gene transfer of 2000μg naked plasmid DNA encoding VEGF$_{165}$, applied to the hydrogel polymer coating of an angioplasty balloon, resulted in angiographic and histologic evidence of angiogenesis in the knee mid-tibial and ankle levels 4 weeks after transfer. Such effects persisted at a 12-week view (ISNER et al. 1996a).

## 11 Conclusions

The recent findings that heterozygous mutations inactivating the VEGF gene result in profound deficits in vasculogenesis and blood island formation, leading to early intrauterine death, emphasize the pivotal role played by this molecule in the development of the vascular system. Future studies, using inducible gene knockout technology (KUHN et al. 1995) should help determine the timing, when the embryo is most vulnerable to VEGF deficiency.

The elucidation of the signal transduction properties of the VEGF receptors holds the promise to dissect the pathways leading to such fundamental biological events as endothelial cell differentiation, morphogenesis and angiogenesis. Furthermore, a more complete understanding of the signaling events involving other endothelial cell-specific tyrosine kinases as well as cell-adhesion molecules and their interrelation with the VEGF/VEGF receptor system should provide a more integrated view of the biology of the endothelial cell, both in normal and abnormal circumstances. In this context, recent studies have shown that VEGF-mediated angiogenesis requires a specific vascular integrin pathway, mediated by $\alpha v\beta 5$ (FRIEDLANDER et al. 1995). Furthermore, a ligand selective for the endothelial cell-specific tyrosine kinase Tie-2 has been recently identified and named angiopoietin (Ang)-1 (DAVIS et al. 1996). Gene knockout studies have shown that Ang-1 is required for the correct assembly of the vessel wall (SURI et al. 1996). Ang-1 seems to play a crucial role in mediating reciprocal interactions between the endothelium and surrounding matrix and mesenchyme, and plays a later role in angiogenesis than VEGF. Also, unlike VEGF, Ang-1 does not directly stimulate endothelial cell growth. Interestingly, very recent studies provide evidence for the existence of Ang-2, a natural antagonist for the Tie-2 receptor. (MAISONPIERRE et al. 1997). Transgenic expression of Ang-2 disrupted blood vessel formation. The interrelation between the VEGF and Ang systems is likely to be an area of intense investigation in vascular biology.

An attractive possibility is that recombinant VEGF or gene therapy with the VEGF gene may be used to promote endothelial cell growth and collateral vessel formation. This would represent a novel therapeutic modality for conditions that frequently are refractory to conservative measures and unresponsive to pharmacological therapy. rhVEGF$_{165}$ is already in clinical trials for the treatment of myocardial ischemia associated with coronary artery disease.

The high expression of VEGF mRNA in human tumors, the presence of the VEGF protein in ocular fluids of individuals with proliferative retinopathies and in the synovial fluid of RA patients, as well as the localization of VEGF in AMD lesions, strongly supports the hypothesis that VEGF is a key mediator of angiogenesis associated with various disorders. Therefore, anti-VEGF antibodies or other inhibitors of VEGF, used alone or in combination with other agents, may be of therapeutic value for a variety of malignancies and other disorders. Very recently, a humanized version of a high-affinity anti-VEGF monoclonal antibody, which retains the same affinity and efficacy as the original murine antibody, has been generated (PRESTA et al. 1997) and is being tested in humans as a treatment for solid tumors, alone or in combination with chemotherapy.

In conclusion, in spite of the plurality of factors potentially involved in angiogenesis, one specific factor, VEGF, appears to play an irreplaceable role in a variety of physiological and pathological circumstances.

# References

Adamis AP, Miller JW, Bernal MT, D'Amico DJ, Folkman J, Yeo TK, Yeo KT (1994) Increased vascular endothelial growth factor levels in the vitreous of eyes with proliferative diabetic retinopathy. Am J Ophthalmol 118:445–450

Adamis AP, Shima DT, Tolentino MJ, Gragoudas ES, Ferrara N, Folkman J, D'Amore PA, Miller JW (1996) Inhibition of vascular endothelial growth factor prevents retinal ischemia-associated iris neovascularization in a nonhuman primate. Arch Ophthalmol 114:66–71

Aiello LP, Avery RL, Arrigg PG, Keyt BA, Jampel HD, Shah ST, Pasquale LR, Thieme H, Iwamoto MA, Park JE, Nguyen H, Aiello LM, Ferrara N, King GL (1994) Vascular endothelial growth factor in ocular fluid of patients with diabetic retinopathy and other retinal disorders. N Engl J Med 331:1480–1487

Aiello LP, Pierce EA, Foley ED, Takagi H, Chen H, Riddle L, Ferrara N, King GL, Smith LE (1995) Suppression of retinal neovascularization in vivo by inhibition of vascular endothelial growth factor (VEGF) using soluble VEGF-receptor chimeric proteins. Proc Natl Acad Sci USA 92:10457–10461

Albini A, Benelli R, Presta M, Rusnati M, Ziche M, Rubartelli A, Paglialunga G, Bussolino F, Noonan D (1996a) HIV-tat protein is a heparin-binding angiogenic growth factor. Oncogene 12:289–297

Albini A, Soldi R, Giunciuglio D, Giraudo E, Benelli R, Primo L, Noonan D, Salio M, Camussi G, Rockl W, Bussolino F (1996b) The angiogenesis induced by HIV-1 tat protein is mediated by the Flk-1/KDR receptor on vascular endothelial cells. Nat Med 2:1371–1375

Algire GH, Chalkley HW (1945) Vascular reactions of normal and malignant tissues in vivo. I. Vascular recations of mice to wounds and to normal and neoplastic transplants. J Natl Cancer Inst 6:73–85

Alon T, Hemo I, Itin A, Pe'er J, Stone J, Keshet E (1995) Vascular endothelial growth factor acts as a survival factor for newly formed retinal vessels and has implications for retinopathy of prematurity. Nat Med 1:1024–1028

Asahara T, Bauters C, Pastore C, Kearney M, Rossow S, Bunting S, Ferrara N, Symes JF, Isner JM (1995) Local delivery of vascular endothelial growth factor accelerates reendothelialization and attenuates intimal hyperplasia in balloon-injured rat carotid artery (see comments). Circulation 91:2793–2801

Asano M, Yukita A, Matsumoto T, Kondo S, Suzuki H (1995) Inhibition of tumor growth and metastasis by an immunoneutralizing monoclonal antibody to human vascular endothelial growth factor/vascular permeability factor121. Cancer Res 55:5296–5301

Banai S, Jaktlish MT, Shou M, Lazarous D, Scheinowitz M, Biro S, Epstein S, Unger E (1994a) Angiogenic-induced enhancement of collateral blood flow to ischemic myocardium by vascular endothelial growth factor. Circulation 89:2183–2189

Banai S, Shweiki D, Pinson A, Chandra M, Lazarovich G, Keshet E (1994b) Upregulation of vascular endothelial growth factor expression induced by myocardial ischemia: implications for coronary angiogenesis. Cardiovasc Res 28:1176–1179

Barleon B, Sozzani S, Zhou D, Weich HA, Mantovani A, Marme D (1996) Migration of human monocytes in response to vascular endothelial growth factor (VEGF) is mediated via the VEGF receptor flt-1. Blood 87:3336–3343

Bauters C, Asahara T, Zheng LP, Takeshita S, Bunting S, Ferrara N, Symes JF, Isner JM (1994) Physiological assessment of augmented vascularity induced by VEGF in ischemic rabbit hindlimb. Am J Physiol 267:H1263–H1271

Bauters C, Asahara T, Zheng LP, Takeshita S, Bunting S, Ferrara N, Symes JF, Isner JM (1995) Recovery of disturbed endothelium-dependent flow in the collateral-perfused rabbit ischemic hindlimb after administration of vascular endothelial growth factor. Circulation 91:2802–2809

Ben-Av P, Crofford LJ, Wilder RL, Hla T (1995) Induction of vascular endothelial growth factor expression in synovial fibroblasts. FEBS Lett 372:83–87

Berkman RA, Merrill MJ, Reinhold WC, Monacci WT, Saxena A, Clark WC, Robertson JT, Ali IU, Oldfield EH (1993) Expression of the vascular permeability factor/vascular endothelial growth factor gene in central nervous system neoplasms. J Clin Invest 91:153–159

Borgstrom P, Hillan KJ, Sriramarao P, Ferrara N (1996) Complete inhibition of angiogenesis and growth of microtumors by anti-vascular endothelial growth factor neutralizing antibody: novel concepts of angiostatic therapy from intravital videomicroscopy. Cancer Res 56:4032–4039

Borgstrom P, Hillan K, Sriramarao P, Ferrara N (1998a) Anti-Vascular endothelial growth factor antibody inhibits angiogenesis and growth of human prostate cancer in vivo. Prostate (in press)

Borgstrom P, Hillan KJ, Sriraramao P, Ferrara N (1998b) Combination treatment with anti-vascular endothelial growth factor antibody and doxorubicin suppresses growth of breast carcinoma cells (submitted)

Brandon EP, Idzerda RL, McKnight GS (1995) Targeting the mouse genome: a compendium of knockouts (part III). Curr Biol 5:873–881

Breier G, Albrecht U, Sterrer S, Risau W (1992) Expression of vascular endothelial growth factor during embryonic angiogenesis and endothelial cell differentiation. Development 114:521–532

Brogi E, Schatteman G, Wu T, Kim EA, Varticovski L, Keyt B, Isner JM (1996) Hypoxia-induced paracrine regulation of vascular endothelial growth factor receptor expression. J Clin Invest 97:469–476

Brown LF, Berse B, Jackman RW, Tognazzi K, Manseau EJ, Dvorak HF, Senger DR (1993a) Increased expression of vascular permeability factor (vascular endothelial growth factor) and its receptors in kidney and bladder carcinomas. Am J Pathol 143:1255–1262

Brown LF, Berse B, Jackman RW, Tognazzi K, Manseau EJ, Senger DR, Dvorak HF (1993b) Expression of vascular permeability factor (vascular endothelial growth factor) and its receptors in adenocarcinomas of the gastrointestinal tract. Cancer Res 53:4727–4735

Brown LF, Berse B, Jackman RW, Tognazzi K, Guidi AJ, Dvorak HF, Senger DR, Connolly JL, Schnitt SJ (1995a) Expression of vascular permeability factor (vascular endothelial growth factor) and its receptors in breast cancer. Hum Pathol 26:86–91

Brown LF, Harrist TJ, Yeo KT, Stahle-Backdahl M, Jackman RW, Berse B, Tognazzi K, Dvorak HF, Detmar M (1995b) Increased expression of vascular permeability factor (vascular endothelial growth factor) in bullous pemphigoid, dermatitis herpetiformis, and erythema multiforme. J Invest Dermatol 104:744–749

Broxmeyer HE, Cooper S, Li ZH, Lu L, Song HY, Kwon BS, Warren RE, Donner DB (1995) Myeloid progenitor cell regulatory effects of vascular endothelial cell growth factor. Int J Hematol 62:203–215

Callow AD, Choi ET, Trachtenberg JD, Stevens SL, Connolly DT, Rodi C, Ryan US (1994) Vascular permeability factor accelerates endothelial regrowth following balloon angioplasty. Growth Factors 10:223–228

Carmeliet P, Collen D (1997) Genetic analysis of blood vessel formation: role of endothelial versus smooth muscle cells. Trends Cardiovasc Med 8:271–281

Carmeliet P, Ferreira V, Breier G, Pollefeyt S, Kieckens L, Gertsenstein M, Fahrig M, Vandenhoeck A, Harpal K, Eberhardt C, Declercq C, Pawling J, Moons L, Collen D, Risau W, Nagy A (1996) Abnormal blood vessel development and lethality in embryos lacking a single VEGF allele. Nature 380:435–439

Chamow SM, Ashkenazi A (1996) Immunoadhesins: principles and applications. Trends Biotechnol 14:52–60

Claffey KP, Wilkison WO, Spiegelman BM (1992) Vascular endothelial growth factor. Regulation by cell differentiation and activated second messenger pathways. J Biol Chem 267:16317–16322

Clauss M, Gerlach M, Gerlach H, Brett J, Wang F, Familletti PC, Pan YC, Olander JV, Connolly DT, Stern D (1990) Vascular permeability factor: a tumor-derived polypeptide that induces endothelial cell and monocyte procoagulant activity, and promotes monocyte migration. J Exp Med 172:1535–1545

Cohen T, Nahari D, Cerem LW, Neufeld G, Levi BZ (1996) Interleukin 6 induces the expression of vascular endothelial growth factor. J Biol Chem 271:736–741

Conn G, Bayne ML, Soderman DD, Kwok PW, Sullivan KA, Palisi TM, Hope DA, Thomas KA (1990) Amino acid and cDNA sequences of a vascular endothelial cell mitogen that is homologous to platelet-derived growth factor. Proc Natl Acad Sci USA 87:2628–2632

Cuevas P, Carceller F, Ortega S, Zazo M, Nieto I, Gimenez-Gallego G (1991) Hypotensive activity of fibroblast growth factor. Science 254:1208–1210

Cuevas P, Garcia-Calvo M, Carceller F, Reimers D, Zazo M, Cuevas B, Munoz-Willery I, Martinez-Coso V, Lamas S, Gimenez-Gallego G (1996) Correction of hypertension by normalization of endothelial levels of fibroblast growth factor and nitric oxide synthase in spontaneously hypertensive rats. Proc Natl Acad Sci USA 93:11996–12001

Cullinan-Bove K, Koos RD (1993) Vascular endothelial growth factor/vascular permeability factor expression in the rat uterus: rapid stimulation by estrogen correlates with estrogen-induced increases in uterine capillary permeability and growth. Endocrinology 133:829–837

Cunningham SA, Waxham MN, Arrate PM, Brock TA (1995) Interaction of the Flt-1 tyrosine kinase receptor with the p85 subunit of phosphatidylinositol 3-kinase. Mapping of a novel site involved in binding. J Biol Chem 270:20254–20257

Davis S, Aldrich TH, Jones PF, Acheson A, Compton DL, Jain V, Ryan TE, Bruno J, Radziejewski C, Maisonpierre PC, Yancopoulos GD (1996) Isolation of angiopoietin-1, a ligand for the TIE2 receptor, by secretion-trap expression cloning. Cell 87:1161–1169

Davis-Smyth T, Chen H, Park J, Presta LG, Ferrara N (1996) The second immunoglobulin-like domain of the VEGF tyrosine kinase receptor Flt-1 determines ligand binding and may initiate a signal transduction cascade. EMBO J 15:4919–4927

de Vries C, Escobedo JA, Ueno H, Houck K, Ferrara N, Williams LT (1992) The fms-like tyrosine kinase, a receptor for vascular endothelial growth factor. Science 255:989–991

Detmar M, Brown LF, Claffey KP, Yeo KT, Kocher O, Jackman RW, Berse B, Dvorak HF (1994) Overexpression of vascular permeability factor/vascular endothelial growth factor and its receptors in psoriasis. J Exp Med 180:1141–1146

DiPietro LA (1997) Thrombospondin as a regulator of angiogenesis. In: Rosen E, Goldberh I (eds) Regulation of angiogenesis. Springer, Berlin Heidelberg New York, pp 295–314

Dvorak HF (1986) Tumors: wounds that do not heal. Similarities between tumor stroma generation and wound healing. N Engl J Med 315:1650–1659

Dvorak HF, Brown LF, Detmar M, Dvorak AM (1995) Vascular permeability factor/vascular endothelial growth factor, microvascular hyperpermeability, and angiogenesis. Am J Pathol 146:1029–1039

Dvorak HF, Harvey VS, Estrella P, Brown LF, McDonagh J, Dvorak AM (1987) Fibrin containing gels induce angiogenesis. Implications for tumor stroma generation and wound healing. Lab Invest 57:673–686

Essed CD, Brand MVD, Becker AE (1983) Transluminal coronary angioplasty and early restenosis. Br Heart J 49:393–402

Fassbender HJ, Simling-Annenfeld M (1983) The potential aggressiveness of synovial tissue in rheumatoid arthritis. J Pathol 10:845–851

Fava RA, Olsen NJ, Spencer-Green G, Yeo KT, Yeo TK, Berse B, Jackman RW, Senger DR, Dvorak HF, Brown LF (1994) Vascular permeability factor/endothelial growth factor (VPF/VEGF): accumulation and expression in human synovial fluids and rheumatoid synovial tissue. J Exp Med 180:341–346

Ferrara N, Davis-Smyth T (1997) The biology of vascular endothelial growth factor. Endocr Rev 18:4–25

Ferrara N, Henzel WJ (1989) Pituitary follicular cells secrete a novel heparin-binding growth factor specific for vascular endothelial cells. Biochem Biophys Res Commun 161:851–858

Ferrara N, Clapp C, Weiner R (1991) The 16 K fragment of prolactin specifically inhibits basal or fibroblast growth factor stimulated growth of capillary endothelial cells. Endocrinology 129:896–900

Ferrara N, Houck K, Jakeman L, Leung DW (1992) Molecular and biological properties of the vascular endothelial growth family of proteins. Endocr Rev 13:18–32

Ferrara N, Winer J, Burton T, Rowland A, Siegel M, Phillips HS, Terrell T, Keller GA, Levinson AD (1993) Expression of vascular endothelial growth factor does not promote transformation but confers a growth advantage in vivo to Chinese hamster ovary cells. J Clin Invest 91:160–170

Ferrara N, Carver Moore K, Chen H, Dowd M, Lu L, O'Shea KS, Powell Braxton L, Hillan KJ, Moore MW (1996) Heterozygous embryonic lethality induced by targeted inactivation of the VEGF gene. Nature 380:439–442

Ferrara N, Chen H, Davis-Smyth T, Gerber H-P, Nguyen T-N, Peers D, Chisholm V, Hillan KJ, Schwall RH (1998) Vascular endothelial growth factor is essential for corpus luteum angiogenesis (submitted)

Finnerty H, Kelleher K, Morris GE, Bean K, Merberg DM, Kriz R, Morris JC, Sookdeo H, Turner KJ, Wood CR (1993) Molecular cloning of murine FLT and FLT4. Oncogene 8:2293–2298

Folkman J (1971) Tumor angiogenesis: therapeutic implications. N Engl J Med 285:1182–1186

Folkman J (1995) Angiogenesis in cancer, vascular, rheumatoid and other disease. Nat Med 1:27–31

Folkman J, Shing Y (1992) Angiogenesis. J Biol Chem 267:10931–10934

Fong GH, Rossant J, Gertsenstein M, Breitman ML (1995) Role of the Flt-1 receptor tyrosine kinase in regulating the assembly of vascular endothelium. Nature 376:66–70

Forsythe JA, Jiang BH, Iyer NV, Agani F, Leung SW, Koos RD, Semenza GL (1996) Activation of vascular endothelial growth factor gene transcription by hypoxia-inducible factor 1. Mol Cell Biol 16:4604–4613

Frank S, Hubner G, Breier G, Longaker MT, Greenhalg DG, Werner S (1995) Regulation of VEGF expression in cultured keratinocytes. Implications for normal and impaired wound healing. J Biol Chem 270:12607–12613

Friedlander M, Brooks PC, Shaffer RW, Kincaid CM, Varner JA, Cheresh DA (1995) Definition of two angiogenic pathways by distinct alpha v integrins. Science 270:1500–1502

Gabrilovich DI, Chen HL, Girgis KR, Cunningham HT, Meny GM, Nadaf S, Kavanaugh D, Carbone DP (1996) Production of vascular endothelial growth factor by human tumors inhibits the functional maturation of dendritic cells. Nat Med 2:1096–1103

Galland F, Karamysheva A, Mattei MG, Rosnet O, Marchetto S, Birnbaum D (1992) Chromosomal localization of FLT4, a novel receptor-type tyrosine kinase gene. Genomics 13:475–478

Garner A (1994) Vascular diseases. Pathobiology of ocular disease. Dekker, New York

Gasparini G, Toi M, Gion M, Verderio P, Dittadi R, Hanatani M, Matsubara I, Vinante O, Bonoldi E, Boracchi P, Gatti C, Suzuki H, Tominaga T (1997) Prognostic significance of vascular endothelial growth factor protein in node-negative breast carcinoma. J Natl Cancer Inst 89:139–147

Gerber HP, Condorelli F, Park J, Ferrara N (1997) Differential transcriptional regulation of the two VEGF receptor genes. Flt-1, but not Flk-1/KDR, is up-regulated by hypoxia. J Biol Chem 272:23659-23667

Gnarra JR, Zhou S, Merrill MJ, Wagner JR, Krumm A, Papavassiliou E, Oldfield EH, Klausner RD, Linehan WM (1996) Post-transcriptional regulation of vascular endothelial growth factor mRNA by the product of the VHL tumor suppressor gene. Proc Natl Acad Sci USA 93:10589–10594

Goldberg MA, Schneider TJ (1994) Similarities between the oxygen-sensing mechanisms regulating the expression of vascular endothelial growth factor and erythropoietin. J Biol Chem 269:4355–4361

Goldman C, Kim J, Wonf W-L, King V, Brock T, Gillespie Y (1993) Epidermal growth factor stimulates vascular endothelial growth factor production by malignant glioma cells. A model of glioblastoma multiforme pathophysiology. Mol Biol Cell 4:121–133

Good D, Polverini P, Rastinejad F, Beau M, Lemons R, Frazier W, Bouck N (1990) A tumor suppressor-dependent inhibitor of angiogenesis is immunologically and functionally indistinguishable from a fragment of thrombospaondin. Proc Natl Acad Sci USA 87:6624–6628

Graor RA, Gray BH (1991) Interventional treatment of peripheral vascular disease. Peripheral vascular diseases. Mosby, St Louis

Grugel S, Finkenzeller G, Weindel K, Barleon B, Marme D (1995) Both v-Ha-Ras and v-Raf stimulate expression of the vascular endothelial growth factor in NIH 3T3 cells. J Biol Chem 270:25915–25919

Guidi AJ, Abu-Jawdeh G, Berse B, Jackman RW, Tognazzi K, Dvorak HF, Brown LF (1995) Vascular permeability factor (vascular endothelial growth factor) expression and angiogenesis in cervical neoplasia. J Natl Cancer Inst 87:1237–1245

Guidi AJ, Abu Jawdeh G, Tognazzi K, Dvorak HF, Brown LF (1996) Expression of vascular permeability factor (vascular endothelial growth factor) and its receptors in endometrial carcinoma. Cancer 78:454–460

Guo D, Jia Q, Song HY, Warren RS, Donner DB (1995) Vascular endothelial cell growth factor promotes tyrosine phosphorylation of mediators of signal transduction that contain SH2 domains. Association with endothelial cell proliferation. J Biol Chem 270:6729–6733

Harada K, Friedman M, Lopez JJ, Wang SY, Li J, Prasad PV, Pearlman JD, Edelman ER, Sellke FW, Simons M (1996) Vascular endothelial growth factor administration in chronic myocardial ischemia. Am J Physiol 270:H1791–H1802

Hashimoto E, Ogita T, Nakaoka T, Matsuoka R, Takao A, Kira Y (1994) Rapid induction of vascular endothelial growth factor expression by transient ischemia in rat heart. Am J Physiol 267:H1948–H1954

Hashimoto M, Ohsawa M, Ohnishi A, Naka N, Hirota S, Kitamura Y, Aozasa K (1995) Expression of vascular endothelial growth factor and its receptor mRNA in angiosarcoma. Lab Invest 73:859–863

Houck KA, Ferrara N, Winer J, Cachianes G, Li B, Leung DW (1991) The vascular endothelial growth factor family: identification of a fourth molecular species and characterization of alternative splicing of RNA. Mol Endocrinol 5:1806–1814

Houck KA, Leung DW, Rowland AM, Winer J, Ferrara N (1992) Dual regulation of vascular endothelial growth factor bioavailability by genetic and proteolytic mechanisms. J Biol Chem 267:26031–26037

Ikeda E, Achen MG, Breier G, Risau W (1995) Hypoxia-induced transcriptional activation and increased mRNA stability of vascular endothelial growth factor in C6 glioma cells. J Biol Chem 270:19761–19766

Iliopoulos O, Levy AP, Jiang C, Kaelin WG Jr, Goldberg MA (1996) Negative regulation of hypoxia-inducible genes by the von Hippel-Lindau protein. Proc Natl Acad Sci USA 93:10595–10599

Isner JM, Pieczek A, Schainfeld R, Blair R, Haley L, Asahara T, Rosenfield K, Razvi S, Walsh K, Symes JF (1996a) Clinical evidence of angiogenesis after arterial gene transfer of phVEGF165 in patient with ischaemic limb (see comments). Lancet 348:370–374

Isner JM, Walsh K, Symes J, Pieczek A, Takeshita S, Lowry J, Rosenfield K, Weir L, Brogi E, Jurayj D (1996b) Arterial gene transfer for therapeutic angiogenesis in patients with peripheral artery disease. Hum Gene Ther 7:959–988

Jakeman LB, Winer J, Bennett GL, Altar CA, Ferrara N (1992) Binding sites for vascular endothelial growth factor are localized on endothelial cells in adult rat tissues. J Clin Invest 89:244–253

Jakeman LB, Armanini M, Philips HS, Ferrara N (1993) Developmental expression of binding sites and mRNA for vascular endothelial growth factor suggests a role for this protein in vasculogenesis and angiogenesis. Endocrinology 133:848–859

Kamat BR, Brown LF, Manseau EJ, Senger DR, Dvorak HF (1995) Expression of vascular permeability factor/vascular endothelial growth factor by human granulosa and theca lutein cells. Role in corpus luteum development. Am J Pathol 146:157–165

Kendall RL, Wang G, Thomas KA (1996) Identification of a natural soluble form of the vascular endothelial growth factor receptor, FLT-1, and its heterodimerization with KDR. Biochem Biophys Res Commun 226:324–328

Keyt BA, Berleau LT, Nguyen HV, Chen H, Heinsohn H, Vandlen R, Ferrara N (1996a) The carboxyl-terminal domain (111–165) of vascular endothelial growth factor is critical for its mitogenic potency. J Biol Chem 271:7788–7795

Keyt BA, Nguyen HV, Berleau LT, Duarte CM, Park J, Chen H, Ferrara N (1996b) Identification of vascular endothelial growth factor determinants for binding KDR and FLT-1 receptors. Generation of receptor-selective VEGF variants by site-directed mutagenesis. J Biol Chem 271:5638–5646

Kieser A, Weich H, Brandner G, Marme, D, Kolch W (1994) Mutant p53 potentiates protein kinase C induction of vascular endothelial growth factor expression. Oncogene 9:963–969

Kim KJ, Li B, Houck K, Winer J, Ferrara N (1992) The vascular endothelial growth factor proteins: identification of biologically relevant regions by neutralizing monoclonal antibodies. Growth Factors 7:53–64

Kim KJ, Li B, Winer J, Armanini M, Gillett N, Phillips HS, Ferrara N (1993) Inhibition of vascular endothelial growth factor-induced angiogenesis suppresses tumor growth in vivo. Nature 362:841–844

Koch AE, Harlow L, Haines GK, Amento EP, Unemori EN, Wong W-L, Pope RM, Ferrara N (1994) Vascular endothelial growth factor: a cytokine modulating endothelial function in rheumatoid arthritis. J Immunol 152:4149–4156

Kondo S, Asano M, Matsuo K, Ohmori I, Suzuki H (1994) Vascular endothelial growth factor/vascular permeability factor is detectable in the sera of tumor-bearing mice and cancer patients. Biochim Biophys Acta 1221:211–214

Ku DD, Zaleski JK, Liu S, Brock TA (1993) Vascular endothelial growth factor induces EDRF-dependent relaxation in coronary arteries. Am J Physiol 265:H586–H592

Kuhn R, Schwenk F, Aguet M, Rajewsky K (1995) Inducible gene targeting in mice. Science 269:1427–1429

Kvanta A, Algvere PV, Berglin L, Seregard S (1996) Subfoveal fibrovascular membranes in age-related macular degeneration express vascular endothelial growth factor. Invest Ophthalmol Vis Sci 37:1929–1934

Leung DW, Cachianes G, Kuang WJ, Goeddel DV, Ferrara N (1989) Vascular endothelial growth factor is a secreted angiogenic mitogen. Science 246:1306–1309

Levy AP, Levy NS, Wegner S, Goldberg MA (1995) Transcriptional regulation of the rat vascular endothelial growth factor gene by hypoxia. J Biol Chem 270:13333–13340

Levy AP, Levy NS, Goldberg MA (1996) Post-transcriptional regulation of vascular endothelial growth factor by hypoxia. J Biol Chem 271:2746–2753

Li J, Perrella MA, Tsai JC, Yet SF, Hsieh CM, Yoshizumi M, Patterson C, Endego WO, Zhou F, Lee M (1995) Induction of vascular endothelial growth factor gene expression by interleukin-1 beta in rat aortic smooth muscle cells. J Biol Chem 270:308–312

Li J, Brown LF, Hibberd MG, Grossman JD, Morgan JP, Simons M (1996) VEGF, flk-1, and flt-1 expression in a rat myocardial infarction model of angiogenesis. Am J Physiol 270:H1803–H1811

Liu Y, Cox SR, Morita T, Kourembanas S (1995) Hypoxia regulates vascular endothelial growth factor gene expression in endothelial cells. Identification of a 5' enhancer. Circ Res 77:638–643

Lopez PF, Sippy BD, Lambert HM, Thach AB, Hinton DR (1996) Transdifferentiated retinal pigment epithelial cells are immunoreactive for vascular endothelial growth factor in surgically excised age-related macular degeneration-related choroidal neovascular membranes. Invest Ophthalmol Vis Sci 37:855–868

Madan A, Curtin PT (1993) A 24-base-pair sequence 3' to the human erythropoietin gene contains a hypoxia-responsive transcriptional enhancer. Proc Natl Acad Sci USA 90:3928–3932

Maeda K, Chung YS, Ogawa Y, Takatsuka S, Kang SM, Ogawa M, Sawada T, Sowa M (1996) Prognostic value of vascular endothelial growth factor expression in gastric carcinoma. Cancer 77:858–863

Maisonpierre PC, Suri C, Jones PF, Bartunkova S, Wiegend SJ, Radziejewski C, Compton D, McClain J, Aldrich TH, Papadopulos N, Daly TJ, Davis S, Sato TN, Yancopoulos GD (1997) Angiopoietin-2, a natural antagonist for Tie-2 that disrupts in vivo angiogenesis. Science 277:55–60

Malecaze F, Clemens S, Simorer-Pinotel V, Mathis A, Chollet P, Favard P, Bayard F, Plouet J (1994) Detection of vascular endothelial growth factor mRNA and vascular endothelial growth factor-like activity in proliferative diabetic retinopathy. Arch Ophthalmol 112:1476–1482

Mandriota SJ, Seghezzi G, Vassalli JD, Ferrara N, Wasi S, Mazzieri R, Mignatti P, Pepper MS (1995) Vascular endothelial growth factor increases urokinase receptor expression in vascular endothelial cells. J Biol Chem 270:9709–9716

Mandriota SJ, Menoud PA, Pepper MS (1996) Transforming growth factor beta 1 down-regulates vascular endothelial growth factor receptor 2/flk-1 expression in vascular endothelial cells. J Biol Chem 271:11500–11505

Matthews W, Jordan CT, Gavin M, Jenkins NA, Copeland NG, Lemischka IR (1991) A receptor tyrosine kinase cDNA isolated from a population of enriched primitive hematopoietic cells and exhibiting close genetic linkage to c-kit. Proc Natl Acad Sci USA 88:9026–9030

Mazure NM, Chen EY, Yeh P, Laderoute KR, Giaccia AJ (1996) Oncogenic transformation and hypoxia synergistically act to modulate vascular endothelial growth factor expression. Cancer Res 56:3436–3440

McLaren J, Prentice A, Charnock-Jones DS, Smith SK (1996) Vascular endothelial growth factor (VEGF) concentrations are elevated in peritoneal fluid of women with endometriosis. Hum Reprod 11:220–223

Melder RJ, Koenig GC, Witwer BP, Safabakhsh N, Munn LL, Jain RK (1996) During angiogenesis, vascular endothelial growth factor and basic fibroblast growth factor regulate natural killer cell adhesion to tumor endothelium (see comments). Nat Med 2:992–997

Melnyk O, Shuman MA, Kim KJ (1996) Vascular endothelial growth factor promotes tumor dissemination by a mechanism distinct from its effect on primary tumor growth. Cancer Res 56:921–924

Michaelson IC (1948) The mode of development of the vascular system of the retina with some observations on its significance for certain retinal disorders. Trans Ophthalmol Soc UK 68:137–180

Millauer B, Wizigmann Voos S, Schnurch H, Martinez R, Moller NP, Risau W, Ullrich A (1993) High affinity VEGF binding and developmental expression suggest Flk-1 as a major regulator of vasculogenesis and angiogenesis. Cell 72:835–846

Millauer B, Shawver LK, Plate KH, Risau W, Ullrich A (1994) Glioblastoma growth inhibited in vivo by a dominant-negative Flk-1 mutant. Nature 367:576–579

Miller JW, Adamis AP, Shima DT, D'Amore PA, Moulton RS, O'Reilly MS, Folkman J, Dvorak HF, Brown LF, Berse B et al (1994) Vascular endothelial growth factor/vascular permeability factor is temporally and spatially correlated with ocular angiogenesis in a primate model. Am J Pathol 145:574–584

Minchenko A, Bauer T, Salceda S, Caro J (1994) Hypoxic stimulation of vascular endothelial growth factor expression in vivo and in vitro. Lab Invest 71:374–379

Monacci WT, Merrill MJ, Oldfield EH (1993) Expression of vascular permeability factor/vascular endothelial growth factor in normal rat tissues. Am J Physiol 264:C995–C1002

Morbidelli L, Chang CH, Douglas JG, Granger HJ, Ledda F, Ziche M (1996) Nitric oxide mediates mitogenic effect of VEGF on coronary venular endothelium. Am J Physiol 270:H411–H415

Morishita K, Johnson DE, Williams LT (1995) A novel promoter for vascular endothelial growth factor receptor (flt-1) that confers endothelial-specific gene expression. J Biol Chem 270:27948–27953

Muller YA, Li B, Christinger HW, Wells JA, Cunningham BC, de Vos AM (1997) Vascular endothelial growth factor: crystal structure and functional mapping of the kinase domain receptor binding site. Proc Natl Acad Sci USA 94:7192–7197

Nicosia R, Nicosia SV, Smith M (1994) Vascular endothelial growth factor, platelet-derived growth factor and indulin-like growth factor stimulate angiogenesis in vitro. Am J Pathol 145:1023–1029

O'Reilly MS, Holmgren L, Shing Y, Chen C, Rosenthal RA, Moses M, Lane WS, Cao Y, Sage EH, Folkman J (1994) Angiostatin: a novel angiogenesis inhibitor that mediates the suppression of metastases by a Lewis lung carcinoma (see comments). Cell 79:315–328

O'Reilly MS, Boehm T, Shing Y, Fukai N, Vasios G, Lane WS, Flynn E, Birkhead JR, Olsen BR, Folkman J (1997) Endostatin: an endogenous inhibitor of angiogenesis and tumor growth. Cell 88:277–285

Olk RJ, Lee CM (1993) Diabetic retinopathy: practical management. Lippincott, Philadelphia

Olson TA, Mohanraj D, Carson LF, Ramakrishnan S (1994) Vascular permeability factor gene expression in normal and neoplastic human ovaries. Cancer Res 54:276–280

Pajusola K, Aprelikova O, Korhonen J, Kaipainen A, Pertovaara L, Alitalo R, Alitalo K (1992) FLT4 receptor tyrosine kinase contains seven immunoglobulin-like loops and is expressed in multiple human tissues and cell lines. Cancer Res 52:5738–5743

Park JE, Keller H-A, Ferrara N (1993) The vascular endothelial growth factor isoforms (VEGF): differential deposition into the subepithelial extracellular matrix and bioactivity of extracellular matrix-bound VEGF. Mol Biol Cell 4:1317–1326

Park JE, Chen HH, Winer J, Houck KA, Ferrara N (1994) Placenta growth factor. Potentiation of vascular endothelial growth factor bioactivity, in vitro and in vivo, and high affinity binding to Flt-1 but not to Flk-1/KDR. J Biol Chem 269:25646–25654

Patterson C, Perrella MA, Hsieh CM, Yoshizumi M, Lee ME, Haber E (1995) Cloning and functional analysis of the promoter for KDR/flk-1, a receptor for vascular endothelial growth factor. J Biol Chem 270:23111–23118

Patterson C, Perrella MA, Endege WO, Yoshizumi M, Lee ME, Haber E (1996) Downregulation of vascular endothelial growth factor receptors by tumor necrosis factor-alpha in cultured human vascular endothelial cells. J Clin Invest 98:490–496

Patz A (1980) Studies on retinal neovascularization. Invest Ophthalmol Vis Sci 19:1133–1138

Pearlman JD, Hibberd MG, Chuang ML, Harada K, Lopez JJ, Gladstone SR, Friedman M, Sellke FW, Simons M (1995) Magnetic resonance mapping demonstrates benefits of VEGF-induced myocardial angiogenesis. Nat Med 1:1085–1089

Pe'er J, Folberg R, Itin A, Gnessin H, Hemo I, Keshet E (1996) Upregulated expression of vascular endothelial growth factor in proliferative diabetic retinopathy. Br J Ophthalmol 80:241–245

Pekala P, Marlow M, Heuvelman D, Connolly D (1990) Regulation of hexose transport in aortic endothelial cells by vascular permeability factor and tumor necrosis factor-alpha, but not by insulin. J Biol Chem 265:18051–18054

Pepper MS, Ferrara N, Orci L, Montesano R (1991) Vascular endothelial growth factor (VEGF) induces plasminogen activators and plasminogen activator inhibitor-1 in microvascular endothelial cells. Biochem Biophys Res Commun 181:902–906

Pepper MS, Ferrara N, Orci L, Montesano R (1992) Potent synergism between vascular endothelial growth factor and basic fibroblast growth factor in the induction of angiogenesis in vitro. Biochem Biophys Res Commun 189:824–831

Pertovaara L, Kaipainen A, Mustonen T, Orpana A, Ferrara N, Saksela O, Alitalo K (1994) Vascular endothelial growth factor is induced in response to transforming growth factor-beta in fibroblastic and epithelial cells. J Biol Chem 269:6271–6274

Peters KG, De Vries C, Williams LT (1993) Vascular endothelial growth factor receptor expression during embryogenesis and tissue repair suggests a role in endothelial differentiation and blood vessel growth. Proc Natl Acad Sci USA 90:8915–8919

Phillips HS, Hains J, Leung DW, Ferrara N (1990) Vascular endothelial growth factor is expressed in rat corpus luteum. Endocrinology 127:965–967

Phillips HS, Armanini M, Stavrou D, Ferrara N, Westphal M (1993) Intense focal expression of vascular endothelial growth factor mRNA in human intracranial neoplasms: Association with regions of necrosis. Int J Oncol 2:913–919

Pierce EA, Avery RL, Foley ED, Aiello LP, Smith LE (1995) Vascular endothelial growth factor/vascular permeability factor expression in a mouse model of retinal neovascularization. Proc Natl Acad Sci USA 92:905–909

Plate KH, Breier G, Weich HA, Risau W (1992) Vascular endothelial growth factor is a potential tumour angiogenesis factor in human gliomas in vivo. Nature 359:845–848

Plate KH, Breier G, Millauer B, Ullrich A, Risau W (1993) Up-regulation of vascular endothelial growth factor and its cognate receptors in a rat glioma model of tumor angiogenesis. Cancer Res 53:5822–5827

Plouet J, Moukadiri HJ (1990) Characerization of the receptors for vasculotropin on bovine adrenal cortex-derived capillary endothelial cells. J Biol Chem 265:22071–22075

Poltorak Z, Cohen T, Sivan R, Kandelis Y, Spira G, Vlodavsky I, Keshet E, Neufeld G (1997) VEGF145, a secreted vascular endothelial growth factor isoform that binds to extracellular matrix. J Biol Chem 272:7151–7158

Presta LG, Chen H, O'Connor SJ, Chisholm V, Meng YG, Krummen L, Winkler M, Ferrara N (1997) Humanization of an anti-VEGF monoclonal antibody for the therapy of solid tumors and other disorders. Cancer Res 57:4593–4599

Qu H, Nagy JA, Senger DR, Dvorak HF, Dvorak AM (1995) Ultrastructural localization of vascular permeability factor/vascular endothelial growth factor (VPF/VEGF) to the albuminal plasma membrane and vesiculovacuolar organelles of tumor microvascular endothelium. J Histochem Cytochem 43:381–389

Quinn TP, Peters KG, De Vries C, Ferrara N, Williams LT (1993) Fetal liver kinase 1 is a receptor for vascular endothelial growth factor and is selectively expressed in vascular endothelium. Proc Natl Acad Sci USA 90:7533–7537

Rak J, Mitsuhashi Y, Bayko L, Filmus J, Shirasawa S, Sasazuki T, Kerbel RS (1995) Mutant ras oncogenes upregulate VEGF/VPF expression: implications for induction and inhibition of tumor angiogenesis. Cancer Res 55:4575–4580

Ravindranath N, Little-Ihrig L, Phillips HS, Ferrara N, Zeleznik AJ (1992) Vascular endothelial growth factor messenger ribonucleic acid expression in the primate ovary. Endocrinology 131:254–260

Risau W (1997) Mechanisms of angiogenesis. Nature 386:671–674

Roberts WG, Palade GE (1995) Increased microvascular permeability and endothelial fenestration induced by vascular endothelial growth factor. J Cell Sci 108:2369–2379

Roberts WG, Palade GE (1997) Neovasculature induced by vascular endothelial growth factor is fenestrated. Cancer Res 57:765–772

Sato K, Yamazaki K, Shizume K, Kanaji Y, Obara T, Ohsumi K, Demura H, Yamaguchi S, Shibuya M (1995) Stimulation by thyroid-stimulating hormone and Grave's immunoglobulin G of vascular endothelial growth factor mRNA expression in human thyroid follicles in vitro and flt mRNA expression in the rat thyroid in vivo. J Clin Invest 96:1295–1302

Seetharam L, Gotoh N, Maru Y, Neufeld G, Yamaguchi S, Shibuya, M (1995) A unique signal transduction from FLT tyrosine kinase, a receptor for vascular endothelial growth factor VEGF. Oncogene 10:135–147

Shalaby F, Rossant J, Yamaguchi TP, Gertsenstein M, Wu XF, Breitman ML, Schuh AC (1995) Failure of blood-island formation and vasculogenesis in Flk-1-deficient mice. Nature 376:62–66

Shen H, Clauss M, Ryan J, Schmidt AM, Tijburg P, Borden L, Connolly D, Stern D, Kao J (1993) Characterization of vascular permeability factor/vascular endothelial growth factor receptors on mononuclear phagocytes. Blood 81:2767–2773

Shibuya M, Yamaguchi S, Yamane A, Ikeda T, Tojo A, Matsushime H, Sato M (1990) Nucleotide sequence and expression of a novel human receptor-type tyrosine kinase (flt) closely related to the fms family. Oncogene 8:519–527

Shifren JL, Doldi N, Ferrara N, Mesiano S, Jaffe RB (1994) In the human fetus, vascular endothelial growth factor is expressed in epithelial cells and myocytes, but not vascular endothelium: implications for mode of action. J Clin Endocrinol Metab 79:316–322

Shifren JL, Tseng JF, Zaloudek CJ, Ryan IP, Meng YG, Ferrara N, Jaffe RB, Taylor RN (1996) Ovarian steroid regulation of vascular endothelial growth factor in the human endometrium: implications for angiogenesis during the menstrual cycle and in the pathogenesis of endometriosis. J Clin Endocrinol Metab 81:3112–3118

Shima DT, Adamis AP, Ferrara N, Yeo KT, Yeo TK, Allende R, Folkman J, D'Amore PA (1995) Hypoxic induction of endothelial cell growth factors in retinal cells: identification and characterization of vascular endothelial growth factor (VEGF) as the mitogen. Mol Med 1:182–193

Shima DT, Kuroki M, Deutsch U, Ng YS, Adamis AP, D'Amore PA (1996) The mouse gene for vascular endothelial growth factor. Genomic structure, definition of the transcriptional unit, and

characterization of transcriptional and post-transcriptional regulatory sequences. J Biol Chem 271:3877–3883
Shweiki D, Itin A, Soffer D, Keshet E (1992) Vascular endothelial growth factor induced by hypoxia may mediate hypoxia-initiated angiogenesis. Nature 359:843–845
Shweiki D, Itin A, Neufeld G, Gitay-Goren H, Keshet E (1993) Patterns of expression of vascular endothelial growth factor (VEGF) and VEGF receptors in mice suggest a role in hormonally-mediated angiogenesis. J Clin Invest 91:2235–2243
Siemeister G, Weindel K, Mohrs K, Barleon B, Martiny Baron G, Marme D (1996) Reversion of deregulated expression of vascular endothelial growth factor in human renal carcinoma cells by von Hippel-Lindau tumor suppressor protein. Cancer Res 56:2299–2301
Stone J, Itin A, Alon T, Pe'er J, Gnessin H, Chan Ling T, Keshet E (1995) Development of retinal vasculature is mediated by hypoxia-induced vascular endothelial growth factor (VEGF) expression by neuroglia. J Neurosci 15:4738–4747
Suri C, Jones PF, Patan S, Bartunkova S, Maisonpierre PC, Davis S, Sato TN, Yancopoulos GD (1996) Requisite role of angiopoietin-1, a ligand for the TIE2 receptor, during embryonic angiogenesis. Cell 87:1171–1180
Suzuki K, Hayashi N, Miyamoto Y, Yamamoto M, Ohkawa K, Ito Y, Sasaki Y, Yamaguchi Y, Nakase H, Noda K, Enomoto N, Arai K, Yamada Y, Yoshihara H, Tujimura T, Kawano K, Yoshikawa K, Kamada T (1996) Expression of vascular permeability factor/vascular endothelial growth factor in human hepatocellular carcinoma. Cancer Res 56:3004–3009
Takagi H, King GL, Ferrara N, Aiello LP (1996) Hypoxia regulates vascular endothelial growth factor receptor KDR/Flk gene expression through adenosine A2 receptors in retinal capillary endothelial cells. Invest Ophthalmol Vis Sci 37:1311–1321
Takeshita S, Zhung L, Brogi E, Kearney M, Pu L-Q, Bunting S, Ferrara N, Symes JF, Isner JM (1994) Therapeutic angiogenesis: a single intra-arterial bolus of vascular endothelial growth factor augments collateral vessel formation in a rabbit ischemic hind-limb model. J Clin Invest 93:662–670
Takeshita S, Tsurumi Y, Couffinahl T, Asahara T, Bauters C, Symes J, Ferrara N, Isner JM (1996a) Gene transfer of naked DNA encoding for three isoforms of vascular endothelial growth factor stimulates collateral development in vivo. Lab Invest 75:487–501
Takeshita S, Weir L, Chen D, Zheng LP, Riessen R, Bauters C, Symes JF, Ferrara N, Isner JM (1996b) Therapeutic angiogenesis following arterial gene transfer of vascular endothelial growth factor in a rabbit model of hindlimb ischemia. Biochem Biophys Res Commun 227:628–635
Terman BI, Carrion ME, Kovacs E, Rasmussen BA, Eddy RL, Shows TB (1991) Identification of a new endothelial cell growth factor receptor tyrosine kinase. Oncogene 6:1677–1683
Terman BI, Dougher Vermazen M, Carrion ME, Dimitrov D, Armellino DC, Gospodarowicz D, Bohlen P (1992) Identification of the KDR tyrosine kinase as a receptor for vascular endothelial cell growth factor. Biochem Biophys Res Commun 187:1579–1586
Tischer E, Mitchell R, Hartman T, Silva M, Gospodarowicz D, Fiddes JC, Abraham JA (1991) The human gene for vascular endothelial growth factor. Multiple protein forms are encoded through alternative exon splicing. J Biol Chem 266:11947–11954
Toi M, Kondo S, Suzuki H, Yamamoto Y, Inada K, Imazawa T, Taniguchi T, Tominaga T (1996) Quantitative analysis of vascular endothelial growth factor in primary breast cancer. Cancer 77:1101–1106
Tolentino MJ, Miller JW, Gragoudas ES, Chatzistefanou K, Ferrara N, Adamis AP (1996) Vascular endothelial growth factor is sufficient to produce iris neovascularization and neovascular glaucoma in a nonhuman primate. Arch Ophthalmol 114:964–970
Tuder RM, Flook BE, Voelkel NF (1995) Increased gene expression for VEGF and the VEGF receptors KDR/Flk and Flt in lungs exposed to acute or to chronic hypoxia. Modulation of gene expression by nitric oxide. J Clin Invest 95:1798–1807
Unemori EN, Ferrara N, Bauer EA, Amento EP (1992) Vascular endothelial growth factor induces interstitial collagenase expression in human endothelial cells. J Cell Physiol 153:557–562
Vaisman N, Gospodarowicz D, Neufeld G (1990) Characterization of the receptors for vascular endothelial growth factor. J Biol Chem 265:19461–19466
Viglietto G, Romano A, Maglione D, Rambaldi M, Paoletti I, Lago CT, Califano D, Monaco C, Mineo A, Santelli G, Manzo G, Botti G, Chiappetta G, Persico MG (1996) Neovascularization in human germ cell tumors correlates with a marked increase in the expression of the vascular endothelial growth factor but not the placenta-derived growth factor. Oncogene 13:577–587
Vincenti V, Cassano C, Rocchi M, Persico G (1996) Assignment of the vascular endothelial growth factor gene to human chromosome 6p21.3. Circulation 93:1493–1495

Volm M, Koomagi R, Mattern J (1997a) Prognostic value of vascular endothelial growth factor and its receptor Flt-1 in squamous cell lung cancer. Int J Cancer 74:64–68

Volm M, Koomagi R, Mattern J, Stammler G (1997b) Angiogenic growth factors and their receptors in non-small cell lung carcinomas and their relationships to drug response in vitro. Anticancer Res 17:99–103

Walder CE, Errett CJ, Bunting S, Lindquist P, Ogez JR, Heinsohn HG, Ferrara N, Thomas GR (1996) Vascular endothelial growth factor augments muscle blood flow and function in a rabbit model of chronic hindlimb ischemia. J Cardiovasc Pharmacol 27:91–98

Waltenberger J, Claesson Welsh L, Siegbahn A, Shibuya M, Heldin CH (1994) Different signal transduction properties of KDR and Flt1, two receptors for vascular endothelial growth factor. J Biol Chem 269:26988–26995

Wang GL, Semenza GL (1995) Purification and characterization of hypoxia-inducible factor 1. J Biol Chem 270:1230–1237

Ware JA, Simons M (1997) Angiogenesis in ischemic heart disease. Nat Med 3:158–164

Warren RS, Yuan H, Matli MR, Gillett NA, Ferrara N (1995) Regulation by vascular endothelial growth factor of human colon cancer tumorigenesis in a mouse model of experimental liver metastasis. J Clin Invest 95:1789–1797

Warren RS, Yuan H, Matli MR, Ferrara N, Donner DB (1996) Induction of vascular endothelial growth factor by insulin-like growth factor 1 in colorectal carcinoma. J Biol Chem 271:29483–29488

Wizigmann Voos S, Breier G, Risau W, Plate KH (1995) Up-regulation of vascular endothelial growth factor and its receptors in von Hippel-Lindau disease-associated and sporadic hemangioblastomas. Cancer Res 55:1358–1364

Yamaguchi TP, Dumont DJ, Conlon RA, Breitman ML, Rossant J (1993) flk-1, an flt-related receptor tyrosine kinase is an early marker for endothelial cell precursors. Development 118:489–898

Yang R, Thomas GR, Bunting S, Ko A, Ferrara N, Keyt B, Ross J, Jin H (1996) Effects of vascular endothelial growth factor on hemodynamics and cardiac performance. J Cardiovasc Pharmacol 27:838–844

Yoshiji H, Gomez DE, Shibuya M, Thorgeirsson UP (1996) Expression of vascular endothelial growth factor, its receptor, and other angiogenic factors in human breast cancer. Cancer Res 56:2013–2016

Yuan F, Chen Y, Dellian M, Safabakhsh N, Ferrara N, Jain RK (1996) Time-dependent vascular regression and permeability changes in established human tumor xenografts induced by an anti-vascular endothelial growth factor/vascular permeability factor antibody. Proc Natl Acad Sci USA 93:14765–14770

# Structure, Expression and Receptor-Binding Properties of Placenta Growth Factor (PlGF)

M. G. PERSICO, V. VINCENTI, and T. DIPALMA

| | |
|---|---|
| 1 Introduction | 31 |
| 2 Human, Rat and Mouse Placenta Growth Factor (PlGF) | 32 |
| 3 Tissue/Cell Distribution of *Plgf* Gene Expression | 32 |
| 4 Post-transcriptional Regulation of the *Plgf* Gene | 35 |
| 5 In Vitro and In Vivo Activity of Homodimeric and Heterodimeric PlGF | 36 |
| 6 PlGF Receptors | 37 |
| 7 Human and Mouse *Plgf* Gene Structure and Chromosome Mapping | 38 |
| References | 39 |

## 1 Introduction

Placenta growth factor (PlGF) (MAGLIONE et al. 1991, 1992), vascular endothelial growth factor (VEGF) (FERRARA and HENZEL 1989; GOSPODAROWICZ et al. 1989; KECK et al. 1989; LEVY et al. 1989; CONN et al. 1990), VEGF-B (OLOFSSON et al. 1996), VEGF-C (JOUKOV et al. 1996) and *Fos*-induced growth factor (FIGF) (ORLANDINI et al. 1996) are members of a family of structurally related growth factors. These factors are all dimeric glycoproteins and share a number of biochemical and functional features (for review, see BUSSOLINO et al. 1997). Intra- and interchain disulphide bonds among eight characteristically spaced cysteine residues are involved in the formation of the dimeric active proteins. PlGF and VEGF can form heterodimeric molecules in cells where both genes are expressed (DISALVO et al. 1995; CAO et al. 1996a). VEGF, the first member of this family to be isolated, was identified for its ability to stimulate the proliferation of endothelial cells (ECs). VEGF exerts its action through the binding to the two receptors, VEGF receptor-1 (VEGFR-1; also denoted Flt-1) and VEGFR-2 (also denoted Flk-1/KDR), abundantly expressed on the ECs.

---

International Institute of Genetics and Biophysics, CNR, Via G. Marconi 12, 80125 Naples, Italy

An evident implication in the stimulation of proliferation of ECs and angiogenesis in vivo has only been demonstrated for VEGF and PlGF. The analysis of the expression of the genes coding for VEGFR-1 and -2 during embryo development and in the adult reveals their presence on several cell types. Therefore, these new data indicate a broad involvement of these factors, in vivo, in systems other than just the vascular system.

## 2 Human, Rat and Mouse Placenta Growth Factor (PlGF)

Alternative splicing of the PlGF primary transcript leads to two forms of the mature human PlGF protein (MAGLIONE et al. 1993a). The two forms, named PlGF1 and PlGF2 (also $PlGF_{131}$ and $PlGF_{152}$), differ only by the insertion of a highly basic 21-amino acid stretch at the carboxyl-end of the protein. This additional basic region confers to PlGF2 the ability to bind to heparin (HAUSER and WEICH 1993; MAGLIONE et al. 1993a).

Alignment of human, mouse and rat *Plgf* sequences (Fig. 1) reveals that the mouse and rat *Plgf* cDNAs encode proteins corresponding to the longer heparin-binding human form PlGF2 (DISALVO et al. 1995; DIPALMA et al. 1996). The predicted 158-amino acid murine protein shows 65% and 91% amino acid identity (78% and 95% similarity) to the human and rat *Plgf* sequences, respectively.

## 3 Tissue/Cell Distribution of *Plgf* Gene Expression

*Plgf* is highly expressed in placenta throughout all stages of human gestation, and in invasive and non-invasive hydatiform moles (Table 1; MAGLIONE et al. 1993a). Immunohistochemistry reveals the presence of PlGF in the vasculosyncytial membrane and in the media of large blood vessels of the placenta villi (KHALIQ et al. 1996). By in situ hybridization analysis, PlGF has been found to be expressed in the villous trophoblast, while VEGF is expressed in cells of mesenchymal origin within the chorionic plate (VUORELA et al. 1997). Therefore, only homodimeric PlGF is produced in the placenta cells.

Low levels of *Plgf* messenger ribonucleic acid (mRNA) are also detected by means of Northern blot analysis in human thyroid, heart, brain, lung and skeletal muscle, but not in kidney and pancreas (Table 1; MAGLIONE et al. 1993a; VIGLIETTO et al. 1995). mRNA coding for human PlGF is present in several cell lines (Table 2), such as choriocarcinoma (JAR and JEG-3), umbilical vein endothelial cells (HUVECs), hepatoma HepG2 (HAUSER and WEICH 1993; MAGLIONE et al. 1993a; CAO et al. 1996b), although it is absent in smooth muscle, astrocytoma and omental microvascular ECs (HAUSER and WEICH 1993).

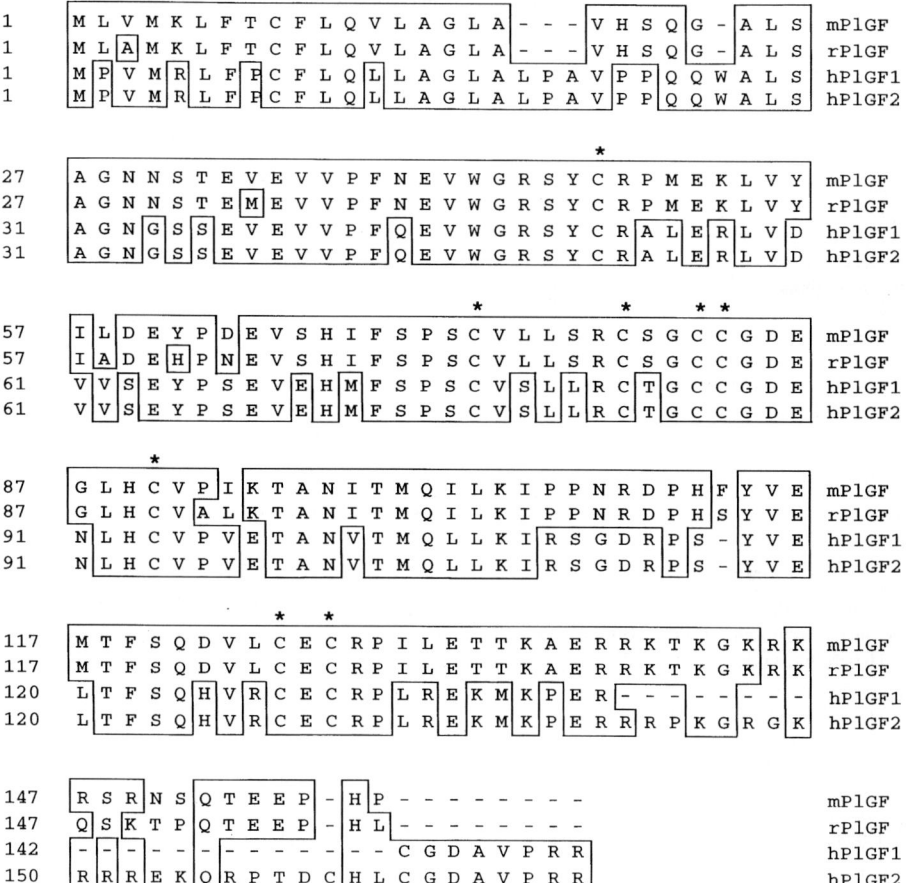

Fig. 1. Comparison of the mouse, rat and human placenta growth factors 1 and 2 (PlGF1 and PlGF2) precursor sequences. The cysteines of the platelet-derived growth factor (PDGF) motif are indicated by *asterisks*. *Dashes* indicate gaps introduced for sequence alignment

*Plgf* gene expression has been analysed in a series of different tumours (Table 2), in parallel with the analysis of *VEGF* gene expression. Human thyroid carcinoma cell lines with different tumorigenic potential and primary thyroid tumours show an elevated *VEGF* expression at the same time that *Plgf* expression is downregulated (VIGLIETTO et al. 1995). TAKAHASHI and co-workers (1994) have shown that *Plgf* is expressed in twenty-one of twenty-three hypervascular renal cell carcinomas, but not in the adjacent normal kidney tissue.

*VEGF* is dramatically upregulated in the majority of hypervascular germ cell tumours (GCTs) (VIGLIETTO et al. 1996). In contrast, the level of *Plgf* transcript in GCTs is generally very low, except for a fraction of tumours (13/44). In hemangioblastoma tumours, high levels of PlGF are only detected in a small number of cases, while VEGF is always upregulated (HATVA et al. 1996). In both thyroid

**Table 1.** Expression of placenta growth factor in human and mouse tissues

|  | Human | Mouse |
|---|---|---|
| Brain | + | − |
| Heart | + | − |
| Kidney | − | − |
| Liver | − | − |
| Lung | + | + |
| Pancreas | − | − |
| Placenta | + | + |
| Skeletal muscle | + | + |
| Smooth muscle | nd | − |
| Spleen | nd | − |
| Testis | nd | + |
| Thyroid | + | + |

+, expression of placenta growth factor; −, no expression

**Table 2.** Placenta growth factor expression

| | |
|---|---|
| Cells in culture | |
|   Human placenta cells | + |
|   Human amnion cells | + |
|   HeLa cell line | − |
|   Choriocarcinoma cell lines (JEG-3, JAR) | + |
|   Hepatoma cell line (HepG2) | + |
| Endothelial cells | |
|   Human umbilical vein (HUVEC) | + |
|   Bovine aorta (BAEC) | + |
|   Calf pulmonary artery (CPA-47) | + |
|   Omental microvascular | − |
|   Astrocytoma cells | − |
|   Human smooth muscle cells | − |
| Solid tumours | |
|   Human colon carcinoma (n) | 1/7 |
|   Human mammary carcinoma (n) | 0/39 |
|   Hypervascular renal carcinoma (n) | 21/23 |
|   Hypervascular germ cell tumours (n) | 13/44 |
|   Thyroid carcinoma (n) | 2/25 |
|   Hydatiform mole (non-invasive) | + |
|   Hydatiform mole (invasive) | + |

+, expression of placenta growth factor; −, no expression

follicular cells and NTERA2/D1 cells, the *Plgf* mRNA expression appears to be associated with the presence of differentiation markers (VIGLIETTO et al. 1995, 1996).

In rodents, *Plgf* gene expression has been studied in vivo and during embryo development (Table 1). Mouse *Plgf* mRNA is present at high levels in placenta and, at lower levels, in lung and thyroid (DIPALMA et al. 1996). *Plgf* mRNA is absent during development of the mouse embryo; at least, as analysed by in situ hybridization (our unpublished results).

In thiouracil-fed rats, development of goitres is accompanied by an increased level of thyroid-stimulating hormone (TSH) and capillary proliferation. In parallel, upregulation of mRNA and protein expression of *VEGF* and *Plgf* is observed (VIGLIETTO et al. 1997). In response to the chronic activation of the TSH receptor pathway, PlGF and VEGF are released by thyrocytes, bind to their endothelial receptors, and stimulate proliferation, permeabilization and enlargement of thyroid capillaries in a paracrine manner. At day 16 of iodide administration to goitrogenous rats, the steady-state level of *VEGF* mRNA is markedly decreased, whereas the mRNA expression of *Plgf* and its receptor Flt-1 is still elevated, suggesting that the complex vascular remodelling, occurring at later stages, may be mediated by VEGF/PlGF heterodimers and PlGF homodimers, through preferential activation of the Flt-1 receptor pathway.

## 4 Post-transcriptional Regulation of the *Plgf* Gene

Differential regulation of the expression of *VEGF* and *Plgf* genes has been observed in several systems. While *VEGF* expression appears to be regulated mostly at the transcriptional level, *Plgf* expression undergoes transcriptional and post-transcriptional regulation. Hypoxia has been shown to be a potent inducer of *VEGF* mRNA expression; however, no change in the *Plgf* mRNA level is detected in response to hypoxia in NTERA2/D1, JAR, JEG-3 and HepG2 cell lines (CAO et al. 1996b; VIGLIETTO et al. 1996). The concentration of homodimeric and heterodimeric VEGF and PlGF proteins changes in JEG-3 cell-conditioned media, as detected by means of enzyme-linked immunosorbent assay (ELISA) in hypoxic condition (CAO et al. 1996b). In particular, the concentration of VEGF and PlGF homodimers, and VEGF/PlGF heterodimers rises from 0.25, 9.3 and 0.15ng/ml to 0.8, 18 and 0.5ng/ml, respectively. These changes correspond to a fourfold increase of VEGF homodimer and VEGF/PlGF heterodimer and a twofold increase of PlGF homodimer in hypoxic conditions. The increase in PlGF protein can be explained with a post-transcriptional regulation that also controls the amount of heterodimeric and homodimeric VEGF.

An interesting post-transcriptional control mechanism of *Plgf* expression has been proposed (MAGLIONE et al. 1993b). The elimination of the 5' untranslated region (UTR) of *Plgf* mRNA increases PlGF synthesis. The *cis*-acting element responsible for this effect corresponds to a small open reading frame (ORF) positioned upstream of the *Plgf* coding sequence (MAGLIONE et al. 1993b). Mutation of the start AUG codon of this small ORF is sufficient to eliminate the post-transcriptional control. This small ORF potentially codes for a 15-amino acid long peptide and is present in human and bovine *Plgf* mRNAs, but is absent in the mouse (Fig. 2).

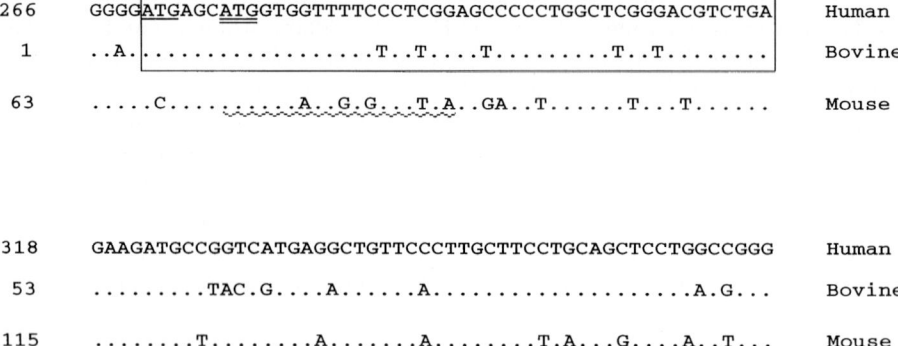

**Fig. 2.** Comparison among human, mouse and bovine placenta growth factor (PlGF) small open reading frame (ORFs).Comparison between a partial nucleotide sequence of a bovine genomic PlGF clone and the corresponding region of human and mouse PlGF cDNAs. The numbering of the sequences is as in the published sequence (MAGLIONE et al. 1991, 1993a; DIPALMA et al. 1996). The human and bovine small ORFs are *boxed-in*. The 7-aminoacid ORF of mouse is indicated by the *waved line*. *Dots* indicate nucleotide identity

## 5 In Vitro and In Vivo Activity of Homodimeric and Heterodimeric PlGF

PlGF$_{131}$ and PlGF$_{152}$ have been purified in several laboratories. Despite the contrasting results obtained, some properties of PlGF activity have now been clarified.

PlGF$_{131}$ and PlGF$_{152}$, purified from the conditioned media of transfected CEN4 cells, are unable to stimulate proliferation of adrenal cortex-derived capillary endothelial (ACCE) cells, even at concentrations as high as 530ng/ml (PARK et al. 1994).

Purified recombinant PlGF$_{152}$ has been obtained from baculovirus-infected cells. In this case, PlGF$_{152}$ is able to stimulate bovine aortic endothelial cells (BAEC) (HAUSER and WEICH 1993) and HUVECs, but not the ECs from hepatic sinusoids (SAWANO et al. 1996).

We have demonstrated that recombinant PlGF$_{131}$, purified from stably transfected human foetal kidney cells, triggers the proliferation of cultured bovine coronary post-capillary venules (CVEC) and HUVECs in a dose-dependent manner (ZICHE et al. 1997). The mitogenic effects of PlGF exerted on HUVECs (10% over control at 10ng/ml) is significantly lower than that exerted on CVEC (40% over control at 10ng/ml). However, the mitogenic effects of PlGF exerted on HUVECs is also significantly lower than that exerted by VEGF, suggesting that PlGF is a weaker mitogen for HUVECs than VEGF (see SAWANO et al. 1996). The effect exerted by PlGF$_{131}$ on CVEC is comparable to that elicited by VEGF (ZICHE et al. 1997).

These data raise several interesting queries about the heterogenic response of ECs to growth factors. The preferential target for the angiogenic effects of PlGF appears to be the endothelium of post-capillary venules, which represent the natural site for new capillary growth and which are likely to be more prone than the umbilical cord vein endothelium to respond to mitogenic stimuli. These results confirm a similar finding from SAWANO and co-workers (1996), who raised the possibility that PlGF may stimulate the growth of only certain types of ECs that have been derived from specific tissues (e.g. PlGF is able to stimulate CVEC, CPA and BAEC, less potently HUVECs, but not ECs derived from liver sinusoids).

We have demonstrated that human $PlGF_{131}$ induces a dose-dependent angiogenic response in the rabbit cornea assay and in the chorioallantoic membrane assay (ZICHE et al. 1997). The comparison of the relative potency in inducing neoangiogenesis in vivo of PlGF with the potency of VEGF and basic fibroblast growth factor (bFGF) shows that, at a dose of 100 ng/pellet, the extent of corneal vascularization induced by the three growth factors is substantially similar. At 200ng/pellet, PlGF1 is clearly more effective in promoting and supporting neovascular growth than bFGF and VEGF.

However, the kinetics of vascular growth and morphogenesis in the neoangiogenesis elicited by $PlGF_{131}$, bFGF and VEGF is slightly different, as shown by the score of vessel density. PlGF and VEGF appear to be more efficient in the early stages of cornea neovascularization than bFGF. $PlGF_{131}$ and $VEGF_{165}$ are able to induce the formation of new capillaries approximately 36 h earlier than bFGF, suggesting a faster recruitment of quiescent cells from capillary endothelium.

Two different research groups have analysed the vascular permeability activity of PlGF and shown that both $PlGF_{131}$ and $PlGF_{152}$ have very low activity, as tested in the Miles assay (PARK et al. 1994; SAWANO et al. 1996). In the Miles assay, as well as in the proliferation assay on ACCE cells, PlGF is able to potentiate the activity of minimally effective doses of VEGF (PARK et al. 1994). These data allowed PARK and co-workers (1994) to speculate on the possibility that VEGFR-1 behaves as a "decoy", having little or no transducing activity. According to this hypothesis, PlGF would release VEGF from VEGFR-1 and then increase its availability to VEGFR-2.

PlGF and VEGF can also form heterodimeric molecules in cells where both genes are expressed (DISALVO et al. 1995; CAO et al. 1996a). Comparative analysis of the mitogenic activity of the homodimeric and heterodimeric forms of these two factors on HUVECs shows that the PlGF/VEGF heterodimer is approximately 20- to 50-fold less potent that the homodimeric VEGF (CAO et al. 1996a). Comparable activity of VEGF and PlGF/VEGF is observed in the ability to induce motility of HUVECs and stimulate the phosphorylation of VEGFR-2 (CAO et al. 1996a).

## 6 PlGF Receptors

The VEGF homodimer binds to and induces autophosphorylation of both VEGFR-1 and -2 tyrosine kinase receptors; conversely, the PlGF homodimer binds only to the VEGFR-1 (TERMAN et al. 1994; CAO et al. 1996a; SAWANO et al. 1996;

LANDGREN et al. 1998). Purified heterodimeric VEGF/PlGF has been shown also to bind to VEGFR-2 (CAO et al. 1996a). The immunoglobulin (Ig)-like domain 2 of the Flt-1 is responsible for the binding specificity of PlGF (DAVIS-SMYTH et al. 1996).

Since PlGF has been shown to bind and induce autophosphorylation of only VEGFR-1 (SAWANO et al. 1996; LANDGREN et al. 1998), but not of VEGFR-2 (PARK et al. 1994; TERMAN et al. 1994), it appears that PlGF should exert its mitogenic and chemotactic effects on ECs through the activation of the VEGFR-1 intracellular signalling pathway. PlGF induces DNA synthesis, but not migration of porcine aortic ECs (PAE) overexpressing VEGFR-1 (LANDGREN et al. 1998). Our findings that PlGF is mitogenic and chemotactic for CVEC and HUVECs, in vitro, raise the problem of whether PlGF may induce VEGFR-1 directly to transduce mitogenic and chemotactic signals inside the cell or whether it may act indirectly through a mechanism of decoy, as previously proposed (PARK et al. 1994).

However, the recent observation that VEGFR-1 is able to mediate signalling in HUVECs in response to both PlGF and VEGF, leading to distinct biological responses, suggests that VEGFR-1 does not act as a decoy receptor, but is indeed able to signal intracellularly (CLAUSS et al. 1996). In addition, both PlGF and VEGF are able to induce migration of 39% and 51% monocytes, respectively, through activation of VEGFR-1 (BARLEON et al. 1996; CLAUSS et al. 1996). This suggests that PlGF may induce EC migration and proliferation through activation of VEGFR-1, although the existence of a yet unknown PlGF receptor has not been completely ruled out.

# 7 Human and Mouse *Plgf* Gene Structure and Chromosome Mapping

Sequencing of the genomic mouse and human *Plgf* genes has allowed the definition of the gene structure and the identification of the seven exons. In mouse (DIPALMA et al. 1996), the exon sizes range from 30 bp (exon 5) to 203 bp (exon 3). Exon 1, 184 nucleotides long, contains the start AUG codon and 66 bp of coding sequence. The 3′-most exon, 770 bp in length, contains 1 bp of coding sequence and all the 3′ UTR. In humans, exon 1, 388 bp in length, contains the initiator methionine at position 314 and encodes the secretory leader peptide. The other exons range in size from 30 bp to 902 bp. The 3′-most exon, 902 bp in length, contains 25 bp of coding regions as well as the whole 3′UTR (MAGLIONE et al. 1993a). The platelet-derived growth factor (PDGF)-like domain exhibited by the PlGF protein is encoded by exons 3 and 4 in both human and mouse.

The assignment of the *Plgf* gene in two syntenic regions, chromosome 14q24 in human and chromosome 12qD in mouse (DIPALMA et al. 1996), and the attempted correlation among diseases and mouse mutants located on these chromosome regions has not provided additional information on PlGF function.

*Acknowledgements.* We thank Miss A. Secondulfo for correcting and typing of the manuscript. This work was supported by a grant of the Progetto Speciale "Angiogenesi" from the Associazione Italiana Ricerca sul Cancro (AIRC).

# References

Barleon B, Sozzani S, Zhou F, Weich HA, Mantovani A, Marmé D (1996) Migration of human monocytes in response to vascular endothelial growth factor (VEGF) is mediated via the VEGF receptor flt-1. Blood 87:3336–3343

Bussolino F, Mantovani A, Persico G (1997) Molecular mechanisms of blood vessel formation. TIBS 22:251–256

Cao Y, Chen H, Zhou L, Chiang M-K, Anand-Apte B, Weatherbee JA, Wang Y, Fang F, Flanagan JG, Lik-Shing Tsang M (1996a) Heterodimers of placenta growth factor/vascular endothelial growth factor. Endothelial activity, tumor cell expression, and high affinity binding to Flk-1/KDR. J Biol Chem 271:3154–3162

Cao Y, Linden P, Shima D, Brone F, Folkman J (1996b) In vivo angiogenic activity and hypoxia induction of heterodimers of placenta growth factor/vascular endothelial growth factor. J Clin Invest 98:2507–2511

Clauss M, Weich HA, Breier G, Knies U, Röckl W, Waltenberger J, Risau W (1996) The vascular endothelial growth factor receptor Flt-1 mediates biological activities. J Biol Chem 271:17629–17634

Conn G, Bayne ML, Soderman DD, Kwok PW, Sullivan KA, Palisi TM, Hope DA, Thomas KA (1990) Amino acid and cDNA sequences of a vascular endothelial cell mitogen that is homologous to platelet-derived growth factor. Proc Natl Acad Sci USA 87:2628–2632

Davis-Smyth T, Chen H, Park J, Presta LG, Ferrara N (1996) The second immunoglobulin-like domain of the VEGF tyrosine kinase receptor Flt-1 determines ligand binding and may initiate a signal transduction cascade. EMBO J 15:4919–41927

DiPalma T, Tucci M, Russo G, Maglione D, Lago CT, Romano A, Saccone S, DellaValle G, De Gregorio L, Dragani TA, Viglietto G, Persico MG (1996) The placenta growth factor gene of the mouse. Mammalian Genome 7:6–12

DiSalvo J, Bayne ML, Conn G, Kwock PW, Trivedi PG, Soderman DD, Palisi TM, Sullivan KA, Thomas KA (1995) Purification and characterization of a naturally occurring vascular endothelial growth factor – placenta growth factor. J Biol Chem 270:7717–7723

Ferrara N, Henzel WJ (1989) Pituitary follicular cells secrete a novel heparin-binding growth factor specific for vascular endothelial cells. Biochem Biophys Res Commu. 161:851–858

Gospodarowicz D, Abraham JA, Schilling J (1989) Isolation and characterization of a vascular endothelial mitogen produced by pituitary-derived folliculo stellate cells. Proc Natl Acad Sci USA 86:7311–7315

Hatva E, Bohling T, Jaaskelainen J, Persico MG, Haltia M, Alitalo K (1996) Vascular growth factors and receptors in capillary hemangioblastomas and hemangiopericytomas. Am J Pathol 148:763–775

Hauser S, Weich HA (1993) A heparin-binding form of placenta growth factor (PlGF–) is expressed in human umbilical vein endothelial cells and placenta. Growth Factors 9:259–268

Joukov V, Pajusola K, Kaipainen A, Chilov D, Lahtinen I, Kukk E, Saksela O, Kalkkinen N, Alitalo K (1996) A novel vascular endothelial growth factor, VEGF-C, is a ligand for the Flt4 (VEGFR-3) and KDR (VEGFR-2) receptor tyrosine kinases. EMBO J 15:290–298

Keck PJ, Hauser SD, Krivi G, Sanzo K, Warren T, Feder J, Connolly DT (1989) Vascular permeability factor, an endothelial cell mitogen related to PDGF. Science 246:1309–1312

Khaliq A, Li XF, Shams M, Sisi P, Acevedo CA, Whittle MJ, Weich HA, Asmed A (1996) Localization of placenta growth factor (PlGF) in human term placenta. Growth Factors 13:243–250

Landgren E, Schiller P, Cao Y, Claesson-Welsh L (1998) Placenta growth factor stimulates MAP kinase and mitogenicity but not phospholipase C-γ and migration of endothelial cells expressing Flt 1. Oncogene 15 (in press)

Levy AP, Tamargo R, Brem H, Nathans D (1989) An endothelial cell growth factor from the mouse neuroblastoma cell line NB 41. Growth Factors 2:9–19

Maglione D, Guerriero V, Viglietto G, Delli-Bovi P, Persico MG (1991) Isolation of a human placenta cDNA coding for a protein related to the vascular permeability factor. Proc Natl Acad Sci USA 88:9267–9271

Maglione D, Guerriero V, Viglietto G, Risau W, Delli-Bovi P, Persico MG (1992) PlGF: a gene coding for a novel human vascular permeability related protein. In: Lenfant C, Paoletti R, Albertini A (eds) Growth factors of the vascular and nervous system. Karger, Basel, pp 28–33

Maglione D, Guerriero V, Viglietto G, Ferraro MG, Aprelikova O, Alitalo K, Del Vecchio S, Lei KJ, Chou JY, Persico MG (1993a) Two alternative mRNAs coding for the angiogenic factor, placenta growth factor (PlGF), are transcribed from a single gene of chromosome 14. Oncogene 8:925–931

Maglione D, Guerriero V, Rambaldi M, Russo G, Persico MG (1993b) Translation of the placenta growth factor mRNA is severely affected by a small open reading frame localized in the 5′ untranslated region. Growth Factors 8:141–152

Olofsson B, Pajusola K, von Euler G, Chilov D, Alitalo K, Eriksson U (1996) Genomic organization of the mouse and human genes for vascular endothelial growth factor B (VEGF-B) and characterization of a second splice isoform. J Biol Chem 271:19310–19317

Orlandini M, Marconcini L, Ferruzzi R, Oliviero S (1996) Identification of a c-fos-induced gene that is related to the platelet-derived growth factor/vascular endothelial growth factor family. Proc Natl Acad Sci USA 93:11675–11680

Park JE, Chen HH, Winer J, Houck KA, Ferrara N (1994) Placenta Growth Factor. Potentiation of vascular endothelial growth factor bioactivity, in vitro and in vivo, and high affinity binding to Flt-1 but not to Flk-1/KDR. J Biol Chem 269:25646–25654

Sawano A, Takahashi T, Yamaguchi S, Aonuma M, Shibuya M (1996) Flt-1 but not KDR/Flk-1 tyrosine kinase is a receptor for placenta growth factor, which is releated to vascular endothelial growth factor. Cell Growth Differ 7:213–221

Takahashi A, Sasaki H, Kim SJ, Tobisu K, Kakizoe T, Tsukamoto T, Kumamoto Y, Sugimura T, Terada M (1994) Markedly increased amount of messenger RNAs for vascular endothelial growth factor and placenta growth factor in renal cell carcinoma associated with angiogenesis. Cancer Res 54:4233–4237

Terman BI, Khandke L, Dougher-Vermazan M, Maglione D, Lassam NJ, Gospodarowicz D, Persico MG, Bohlen P, Eisinger M (1994) VEGF receptor subtypes KDR and FLT1 show different sensitivities to heparin and placenta growth factor. Growth Factors 11:187–195

Viglietto G, Maglione D, Rambaldi M, Cerutti J, Romano A, Trapasso F, Fedele M, Ippolito P, Chiappetta G, Botti G, Fusco A, Persico MG (1995) Upregulation of Vascular Endothelial Growth Factor (VEGF) and downregulation of Placenta Growth Factor (PlGF) associated with malignancy in human thyroid tumors and cell lines. Oncogene 11:1569–1579

Viglietto G, Romano A, Maglione D, Rambaldi M, Paoletti I, Lago CT, Califano D, Monaco C, Mineo A, Santelli G, Manzo G, Botti G, Chiappetta G, Persico MG (1996) Neovascularization in human germ cell tumors correlates with a marked increase in the expression of the vascular endothelial growth factor but not the placenta growth factor. Oncogene 13:577–587

Viglietto G, Romano A, Manzo G, Chiappetta G, Paoletti I, Califano D, Galati MG, Mauriello V, Bruni P, Lago CT, Fusco A, Persico MG (1997) Upregulation of the angiogenic factors PlGF, VEGF and their receptors (Flt-1, Flk-1/KDR) by TSH in cultured thyrocytes and in the thyroid gland of thiouracil-fed rats suggest a TSH-dependent paracrine mechanism for goiter hypervascularization. Oncogene 15:2687–2698

Vuorela P, Hatva E, Lymboussaki A, Kaipainen A, Joukov V, Persico MG, Alitalo K, Halmesmäki E (1997) Expression of vascular endothelial growth factor and placenta growth factor in human placenta. Biol Reprod 56:489–494

Ziche M, Maglione D, Ribatti D, Morbidelli L, Lago CT, Battisti M, Paoletti I, Barra A, Tucci M, Parise G, Vincenti V, Granger HJ, Viglietto G, Persico MG (1997) Placenta growth factor-1 is chemotactic, mitogenic and angiogenic. Lab Invest 76:517–531

# Structure, Expression and Receptor-Binding Properties of Novel Vascular Endothelial Growth Factors

U. Eriksson[1] and K. Alitalo[2]

1 Introduction . . . . . . . . . . . . . . . . . . . . . . . . . . . . . . . . . . . . 41
2 Identification and Properties of VEGF-B/VRF . . . . . . . . . . . . . . . . . 42
3 Identification and Properties of VEGF-C/VRP . . . . . . . . . . . . . . . . . 45
4 Identification and Properties of VEGF-D/FIGF . . . . . . . . . . . . . . . . . 51
5 Primary Structures of VEGF-Related Growth Factors . . . . . . . . . . . . . 52
6 Perspectives . . . . . . . . . . . . . . . . . . . . . . . . . . . . . . . . . . . 54
References . . . . . . . . . . . . . . . . . . . . . . . . . . . . . . . . . . . . . . 55

## 1 Introduction

Vascular endothelial growth factor (VEGF), an important regulator of endothelial cell physiology, was identified some 10 years ago and has, since then, been recognised as the major growth factor relatively specific for endothelial cells (reviewed in Ferrara and Davis-Smyth 1997). VEGF is a dimeric glycoprotein, closely related to placenta growth factor (PlGF). Both VEGF and PlGF are distantly related in structure to the platelet-derived growth factors A and B (PDGF A and PDGF B) (Heldin et al. 1993). Three novel growth factors belonging to the family of VEGF, PlGF and the two PDGFs were recently discovered. These growth factors, termed vascular endothelial growth factor B/VEGF-related factor (VEGF-B/VRF) (Grimmond et al. 1996; Olofsson et al. 1996a), vascular endothelial growth factor C/VEGF-related protein (VEGF-C/VRP) (Joukov et al. 1996; Lee et al. 1996)] and c-*fos*-induced growth factor (FIGF) (Orlandini et al. 1996) share structural features typical of the VEGF/PDGF growth factor family. The prominent structural similarities between VEGF-related growth factors, several of which target endothelial cells, and FIGF suggest the possibility that FIGF also targets endothelial cells, despite its identification as a fibroblast growth factor. Based on these criteria,

---

[1]Ludwig Institute for Cancer Research, Stockholm Branch, Box 240, S-171 77 Stockholm, Sweden
[2]Molecular/Cancer Biology Laboratory, Haartman Institute, PL21 (Haartmaninkatu 3), FIN-00014 University of Helsinki, Finland

we propose that the name FIGF should be changed to VEGF-D to indicate its structural and functional relatedness to the other VEGFs.

The rapidly expanding list of growth factors belonging to the VEGF-family is surprising, but underscores the complexity of regulation of endothelial cell functions and the heterogeneity among different subpopulations of endothelial cells. In this review, we will summarise known structural and functional properties of the novel VEGFs, i.e. VEGF-B, VEGF-C and VEGF-D.

## 2 Identification and Properties of VEGF-B/VRF

A serendipitously found partial, mouse complementary deoxyribonucleic acid (cDNA) clone, encoding a VEGF-related peptide, was used to isolate full-length mouse and human cDNA clones from an adult mouse-heart cDNA library and from a human tumour cell cDNA library, respectively (OLOFSSON et al. 1996a). The full-length cDNAs encoded a homologue of VEGF and, in analogy with the nomenclature of the PDGFs, the new protein was denoted VEGF-B. Independently, another group of researchers found the same gene when attempting to identify the gene for multiple endocrine neoplasia type 1 (MEN1). The protein encoded by this gene was designated VEGF-related factor (VRF, GRIMMOND et al. 1996).

The mouse and human genes for VEGF-B are almost identical, and both span about 4 kb of DNA. The genes are composed of seven exons and their exon–intron organisation resembles that of the VEGF and PlGF genes (Fig. 1) (GRIMMOND et al. 1996; OLOFSSON et al. 1996b; TOWNSON et al. 1996). Presently, two isoforms of VEGF-B, generated by alternative splicing of mRNA, have been recognised (GRIMMOND et al. 1996; OLOFSSON et al. 1996b; TOWNSON et al. 1996). These two secreted forms of VEGF-B have 167 (VEGF-$B_{167}$) and 186 (VEGF-$B_{186}$) amino acid residues, respectively. The isoforms have an identical $N$-terminal domain of 115 amino acid residues, excluding the signal sequence, while the $C$-terminal domains differ. The common $N$-terminal domain is encoded by exons 1–5. Differential use of the remaining three exons gives rise to the two splice isoforms. By the use of an alternative splice-acceptor site in exon 6, an insertion of 101 bp introduces a frame shift and a stop of the coding region of VEGF-$B_{167}$ cDNA (see Fig. 1). Thus, the two VEGF-B isoforms will have different $C$-terminal domains which are unrelated to each other. In VEGF and PlGF, several isoforms are encoded by the use of alternative splice-acceptor sites and different combinations of exons in the genes, but the corresponding transcripts are translated using the same reading frame. The use of partially overlapping, but different reading frames is fairly uncommon among higher eukaryotes.

The different $C$-terminal domains of the two splice isoforms of VEGF-B affect their biochemical and cell biological properties. The $C$-terminal domain of VEGF-$B_{167}$ is structurally related to the corresponding region in VEGF, with several conserved cysteine residues and stretches of basic amino acid residues (see Sect. 5).

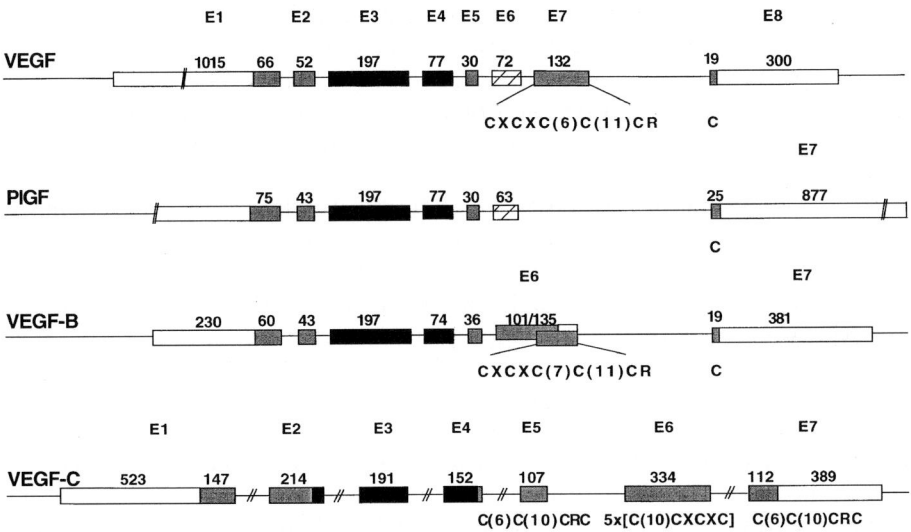

**Fig. 1.** Exon–intron organization of genes of the vascular endothelial growth factor (VEGF) family. The exons are shown as *boxes* and their lengths (bp) are indicated; the non-coding portions are *white*; *grey boxes* denote sequences encoding C- and N-terminal peptides; *black boxes* denote sequences encoding the VEGF homology domain. The *striped box* indicates the exon encoding part of the heparin-binding region. Certain cysteine motifs encoded by the different exons are shown. The structures of the genes are from TISCHER et al. 1991, OLOFSSON et al. 1996b, CHILOV et al. 1997 and DIPALMA et al. 1997. The figure was modified from CHILOV et al. 1997

Thus, this domain is highly hydrophilic and basic and, accordingly, VEGF-B$_{167}$ will remain cell-associated on secretion, unless the producing cells are treated with heparin or high salt concentrations. The cell-associated molecules binding VEGF-B$_{167}$ are likely to be cell surface or pericellular heparan sulphate proteoglycans. It is likely that the cell-association of this isoform occurs via its unique basic C-terminal region, as noted for the highly basic splice variants of VEGF. This suggestion is further supported by the observation that a fusion protein of glutathione-S-transferase and the unique C-terminal domain of VEGF-B$_{167}$ binds tightly to a heparin-Sepharose column (B. Olofsson and the authors, unpublished observation).

The C-terminal domain of the second splice isoform, VEGF-B$_{186}$, has no significant similarity with known amino acid sequences in the databases. The hydrophobic character of this domain, with several conserved alanine, proline, serine and threonine amino acid residues contrasts with the properties of the hydrophilic and basic C-terminal domain in VEGF-B$_{167}$. This is supported by the observation that VEGF-B$_{186}$ does not remain cell-associated on its secretion. Recent evidence suggests that this isoform is proteolytically processed, which regulates the biological properties of the protein (OLOFSSON et al. 1998 and unpublished data).

Isoforms of both human and mouse VEGF-B lack the consensus sequence for N-linked glycosylation (NXT/S), unlike the other growth factors of the PDGF/VEGF-family. However, VEGF-B$_{186}$ is O-glycosylated, presumably in the unique C-terminal domain rich in serine and threonine residues (OLOFSSON et al. 1996b).

The VEGF-B isoforms are produced as disulphide-linked homodimers and, under reducing conditions, the apparent molecular masses of secreted VEGF-B$_{167}$ and VEGF-B$_{186}$ isoforms are 21 kDa and 32 kDa, respectively (OLOFSSON et al. 1996a,b). The secreted 32 kDa form of VEGF-B$_{186}$ is $O$-glycosylated, while the unmodified intracellular form of VEGF-B$_{186}$ has an apparent molecular mass of 26kDa.

It is well documented that VEGF can form naturally occurring heterodimers with PlGF (DISALVO et al. 1995) and such heterodimers might display functional properties distinct from those of both VEGF and PlGF homodimers. Analysis of both isoforms of VEGF-B showed that disulphide-linked heterodimers with VEGF are generated when both are co-expressed in recipient cells (OLOFSSON et al. 1996a,b), but it has not been established whether naturally occurring VEGF–VEGF-B heterodimers exist. Homodimers of VEGF$_{165}$ are secreted from cells in a soluble form, while heterodimers of VEGF-B$_{167}$–VEGF remain cell-associated. In contrast, heterodimers of VEGF-B$_{186}$ and VEGF are freely secreted into the cell-culture medium. Thus, VEGF-B$_{167}$ appears to determine the release of the heterodimers from cells, and heterodimerization of VEGF with either of the two isoforms of VEGF-B might, therefore, control the release and bioavailability of VEGF–VEGF-B heterodimers (OLOFSSON et al. 1996a,b). It is presently unknown whether the VEGF-B polypeptides perform their in vivo function as homodimers, as heterodimers with VEGF, or as both.

The ability of VEGF-B isoforms to affect the release of VEGF–VEGF-B heterodimers from the producing cells is intriguing, since the two growth factors are co-expressed in many tissues, most prominently in the heart (OLOFSSON et al. 1996a). The almost identical patterns of expression of the two VEGF-B isoforms, predominantly in embryonic and adult muscle tissues (myocardium and skeletal muscle), makes it unlikely that differential expression of VEGF-B isoforms would contribute to a genetically controlled mechanism involved in the release of VEGF–VEGF-B heterodimers.

Conditioned medium from 293 cells transfected with an expression vector generating VEGF-B$_{167}$ stimulated thymidine incorporation into DNA in human umbilical vein endothelial cells (HUVECs) and bovine capillary endothelial (BCE) cells. This suggested that VEGF-B is an endothelial cell mitogen and that it may be angiogenic in vivo (OLOFSSON et al. 1996a). However, the possibility remains that at least part of the mitogenic activity is contributed by VEGF–VEGF-B heterodimers, as recombinant VEGF-B$_{186}$ homodimers have no detectable mitogenic activity on endothelial cells (OLOFSSON et al. 1998).

Despite their close structural similarities, the receptor-binding properties of VEGF-B differ from those of VEGF. Using soluble VEGF receptor (VEGFR) extracellular-domain fusion proteins, we have established that VEGF-B binds to VEGFR-1 with high affinity, but not to VEGFR-2 or -3 (OLOFSSON et al. 1998). The affinity for VEGFR-1 and not VEGFR-2 is not surprising, considering that certain receptor-specific epitopes defined for VEGF (KEYT et al. 1996) predict this reactivity. Thus, the acidic residues in loop 2 of VEGF, important for binding to VEGFR-1, are almost identical in VEGF-B. However, analysis of several VEGF-B mutants in which these acidic residues have been replaced by alanine residues show that the corresponding residues of VEGF-B has some effect for VEGFR-1 binding (OLOFSSON et al. 1998). Conversely, the basic residues in loop 3 of VEGF, important for VEGFR-2 binding, are not present in VEGF-B. The selective binding of VEGF-B and PlGF to VEGFR-1 suggests that the two growth factors may be differentially expressed functional homologues.

The expression of VEGF-B during most of murine development suggests that VEGF-B has a role during the establishment of the vascular system. Results from the knockout studies of VEGF have shown that VEGF-B is unable to compensate for the loss of even a single allele of VEGF (CARMELIET et al. 1996; FERRARA et al. 1996). Given that VEGF-B does not bind VEGFR-2, this is not surprising, and the functions of VEGF and VEGF-B are, thus, clearly distinct. Furthermore, the role of VEGF-B may extend beyond the vascular system as it is expressed early during the development of the central nervous system. VEGF-B expression was detected in 8-day-old embryos in structures most likely corresponding to parts of the neural tube (LAGERCRANTZ et al. 1996). On day 11.5–12.5 p.c., VEGF-B was strongly expressed in the developing heart (OLOFSSON et al. 1996a and unpublished observations), Later, on day 14 p.c., VEGF-B is expressed in most tissues of the embryo, although most prominently in heart, spinal cord and cerebral cortex. On day 17, most of the in situ hybridization signal is concentrated in the heart, brown fat and spinal cord (LAGERCRANTZ et al. 1996).

One of the unique features of VEGF expression is its upregulation under hypoxic conditions (GOLDBERG and SCHNEIDER 1994; STEIN et al. 1995) and by a variety of other stimuli, including several growth factors and cytokines (FINKENZELLER et al. 1992; GARRIDO et al. 1993; PERTOVAARA et al. 1994; FRANK et al. 1995; COHEN et al. 1996). The regulation of VEGF-B mRNA is apparently very different, as neither hypoxia nor several growth factors alter the level of expression of this gene (ENHOLM et al. 1997).

The VEGF-B gene was localised to chromosome 11q13, proximal to the cyclin D1 gene, which is amplified in a number of human carcinomas (PAAVONEN et al. 1996). The amplification of cyclin D1, however, was not accompanied by amplification of VEGF-B in several mammary carcinoma cell lines studied (PAAVONEN et al. 1996).

# 3 Identification and Properties of VEGF-C/VRP

A factor stimulating tyrosine phosphorylation of Flt4 (subsequently referred to as VEGFR-3), a receptor tyrosine kinase closely related to VEGFR-1 and VEGFR-2, was identified in conditioned medium from PC-3 prostatic adenocarcinoma cells. Receptor-affinity chromatography using the VEGFR-3 extracellular domain led to the purification of the stimulating factor. The partial amino acid sequence was obtained from the purified factor and a 5′ fragment of the cDNA encoding it was amplified by serial polymerase chain reactions (PCR) using degenerate primers. A full-length cDNA was then cloned from a library prepared from PC-3 cells, using the labelled PCR-amplified 5′ fragment as a probe (JOUKOV et al. 1996). The full-length cDNA encoded a novel homologue of VEGF and was subsequently denoted VEGF-C. Independently, an expressed sequence tag (EST) was identified in the database as being homologous with VEGF. Using the partial EST clone as the probe, a full-length VEGF-C cDNA clone was isolated. The protein encoded by this cDNA was designated VEGF-related protein (VRP) (LEE et al. 1996).

The human VEGF-C cDNA encodes a protein of 419 amino acid residues, with a predicted molecular mass of 46.9 kDa. However, the newly synthesised VEGF-C product is a pre-pro-protein, consisting of an $N$-terminal signal sequence followed by an $N$-terminal peptide, the VEGF-homology domain, and a $C$-terminal pro-peptide (JOUKOV et al. 1996). VEGF-C is secreted as a disulphide bonded homodimer, and most of it is proteolytically processed from the precursor polypeptide, which contains three putative $N$-glycosylation sites; two of these remain in mature, fully processed VEGF homology domain. Based on our results, we propose the VEGF-C proteolytic processing model schematically presented in Fig. 2. This model resembles the model for the proteolytic processing of PDGF, especially of PDGF-B (ÖSTMAN et al. 1988, 1992) in that: (1) the proteolytic cleavages occur after the formation of disulphide-bonded precursor dimers, (2) both $N$- and $C$-terminal peptides may be subject to cleavage, and (3) a variety of processed forms are secreted. However, there are several important differences between PDGF-B and VEGF-C, concerning both their processing and the structures of the mature growth factors.

The homologous part of VEGF-C is about 30% identical with $VEGF_{165}$, 27% with $VEGF-B_{167}$, 25% with PlGF-1 and 22–24% with PDGF-A and PDGF-B. Fully processed VEGF-C binds to and activates both VEGFR-3 and VEGFR-2. A single class of high-affinity sites was observed in porcine aortic endothelial (PAE)/VEGFR-3 cells ($K_d = 135$ pM) and PAE/VEGFR-2 cells ($K_d = 410$ pM). These values are of similar magnitude to the affinities reported for the VEGF–VEGFR-2 interaction (TERMAN et al. 1992; WALTENBERGER et al. 1994). VEGF-C and VEGF displace each other from VEGFR-2, indicating that the same region of this receptor is involved in the binding of both ligands. Surprisingly, none of the three basic residues reported to be critical for VEGFR-2 binding by VEGF (KEYT et al. 1996) are conserved in VEGF-C. VEGF-C also dose-dependently stimulated autophosphorylation of VEGFR-3 and VEGFR-2, but in agreement with previous reports (LEE et al. 1996), we could not detect binding to VEGFR-1 (JOUKOV et al. 1996, 1997).

The human and mouse VEGF-C genes both comprise over 40 kb of genomic DNA and consist of seven exons, all containing coding sequences (Fig. 1). The VEGF-C gene was localised to human chromosome 4q34, close to the human aspartylglucosaminidase gene (PAAVONEN et al. 1996). The VEGF homology domain of VEGF-C is encoded by exons 3 and 4. Exons 5 and 7 encode cysteine-rich motifs of the type C(6)C(10)CRC, and exon 6 encodes C(10)CXCXC motifs typical of a silk protein (CHILOV et al. 1997). The upstream promoter sequences contain

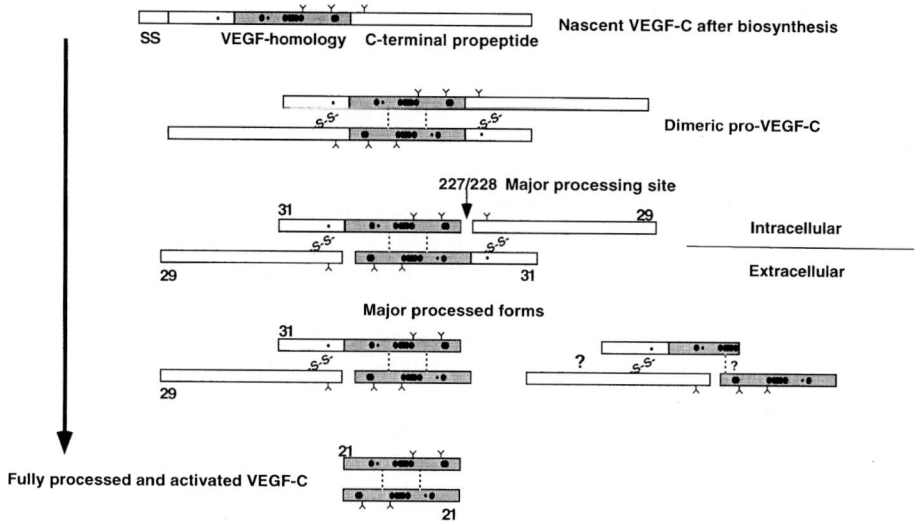

**Fig. 2.** Schematic model of the proteolytic processing of VEGF-C. The regions of VEGF-C polypeptide are marked as follows: *SS* signal sequence; *grey box* VEGF-homology domain; *open boxes* N-terminal and C-terminal peptides. Cysteine residues are shown as *black ovals*, the cysteine residues in the C-terminal pro-peptide are not marked for clarity. Putative sites of N-linked glycosylation are indicated by Y. The major proteolytic cleavage site between residues 227/228 is indicated by an *arrow*. This cleavage occurs as early as during secretion of the protein from the cellular compartment and it creates the major processed forms consisting of disulphide-linked 29/31-kDa polypeptides. Disulphide bonds are marked as -S-S-; the *dashed lines* indicate the detection of both covalent or non-covalent interactions. The proposed structure of the alternatively processed VEGF-C is indicated with a *question mark*. Several intermediate forms are omitted to simplify the scheme. The figure wad adopted from JOUKOV et al. (1997)

conserved putative binding sites for Sp-1, AP-2 and nuclear factor κB (NF-κB) transcription factors, but no TATA box, and show serum-stimulated promoter activity when transfected into cells. The VEGF-C gene structure is, thus, assembled from exons encoding pro-peptides and distinct cysteine-rich domains in addition to the VEGF homology domain, showing both similarities and distinct differences compared with other members of the VEGF/PDGF gene family.

VEGF-C mRNA was detected in Northern-blot analyses of many embryonal and adult tissues. In adult humans, the VEGF-C mRNA is expressed most prominently in heart, placenta, ovary, small intestine and the thyroid gland. Tumour cells express, almost exclusively, a 2.4-kb mRNA form, suggesting that it corresponds to the described VEGF-C cDNA clone obtained from the PC-3 tumour cell line (JOUKOV et al. 1996). The identity of another 2.0-kb mRNA, hybridizing with the VEGF-C probe in analysis of many tissues remains to be determined. Two VEGF-C cDNA clones were obtained that contained 152-bp and 557-bp deletions, corresponding to exon 2 or exons 2–4, respectively (LEE et al. 1996; CHILOV et al. 1997) Due to the shift of the reading frame, which occurs 15 amino acid residues downstream of these deletions, the predicted proteins encoded by the two deleted cDNAs contain either no or only part of the core cysteine knot region similar to that in VEGF.

Also the mouse VEGF-C cDNA was cloned and shown to encode a protein of 415 amino acid residues, which is 85% identical with human VEGF-C and similarly processed (KUKK et al. 1996). In in situ hybridization, mouse VEGF-C mRNA was detected in 8.5-day-old embryos in the cephalic mesenchyme, along the somites, in the tail region and extraembryonally in the allantois. In embryos 12.5 days p.c., VEGF-C mRNA was particularly prominent in regions where the lymphatic vessels are generated from embryonic veins, such as perimetanephric, axillary and jugular areas. The signal was also detected between the developing vertebrae, in the lung mesenchyme, in the neck region and in the developing forehead. The developing mesenterium, which is rich in lymphatic vessels, also showed strong VEGF-C expression (KUKK et al. 1996). The distribution of the VEGFR-3 mRNA follows a somewhat similar temporal and spatial pattern (KAIPAINEN et al. 1995; KUKK et al. 1996). This suggests a paracrine mode of ligand–receptor interaction, with VEGF-C expressed in mesenchymal cells adjacent to the VEGFR-3 positive endothelia. – The juxtaposed VEGFR-3 and VEGF-C expression patterns suggest that VEGF-C functions in the formation of the venous and lymphatic vascular systems during embryogenesis. Constitutive expression of VEGF-C in adult tissues further suggests that this growth factor is also involved in the maintenance of functions of, for example, differentiated lymphatic endothelium, where VEGFR-3 is expressed (KAIPAINEN et al. 1993, 1995; KUKK et al. 1996).

VEGF-C expression was detected in embryos as early as day 7 p.c. (KUKK et al. 1996). This was striking, considering the appearance of VEGFR-3 mRNA first on day 8.5 of gestation (KAIPAINEN et al. 1995). This suggests a possible role of VEGF-C during earlier stages of embryonic development. Such a function might be exercised through the ability of VEGF-C to function as a ligand for VEGFR-2, which is expressed in presumptive progenitors of yolk-sac blood islands as early as day 7 p.c. Interestingly, VEGFR-2 is essential for the development of both haematopoietic and endothelial cell lineages (CARMELIET et al. 1996; FERRARA et al. 1996). We, therefore, investigated the effect of VEGF-C on VEGFR-2 positive cells isolated from the primitive streak of gastrulating quail embryos. VEGF binding triggers endothelial differentiation of these cells, whereas haemopoietic differentiation appears to be mediated by binding of a so-far unidentified VEGFR-2 ligand. We could show that, like VEGF (EICHMANN et al. 1997), VEGF-C also triggers endothelial differentiation of these cells, presumably via VEGFR-2 (EICHMANN et al. 1998). These results indicate that VEGF and VEGF-C can act in a redundant manner via VEGFR-2.

Our results demonstrate that proteolytic processing allows VEGF-C to bind to and activate VEGFR-2 and increases its affinity and activity towards VEGFR-3 (JOUKOV et al. 1997). The biosynthesis of VEGF-C as a precursor may prevent unwanted angiogenic effects via VEGFR-2 and allow VEGF-C to signal preferentially via VEGFR-3. In certain circumstances, proteolytic processing would release mature VEGF-C, which is able to signal via both VEGFR-3 and VEGFR-2. It is also possible that activation of both VEGFR-3 and VEGFR-2, either as homo- or as heterodimers, is necessary to elicit a complete biological response to VEGF-C. Similarly, heterodimers of VEGFR-1 and VEGFR-2 could be important for the biological activities of VEGF. In the case of VEGF-C, proteolytic processing might provide a regulatory mechanism that provides the possibility for fine tuning of the biological functions of this growth factor.

The major secreted VEGF-C form contains the *C*-terminal pro-peptide, which has an unusual structure with tandemly repeated cysteine-rich motifs and is linked via disulphide bonds with the *N*-terminal peptide. The possible function of this, apparently in itself an inactive *C*-terminal half of VEGF-C, is unknown. It has domains of striking similarity to a secretory silk protein and contains short motifs homologous with the epidermal growth factor (EGF)-like domains of other secreted proteins, such as fibrillin, laminin and tenascin. All of these proteins are known to participate in protein–protein or protein–cell surface interactions. One can, thus, speculate that partially processed VEGF-C may stay associated with the extracellular matrix via its *C*-terminal pro-peptide (Fig. 2). Cleavage of the *N*-terminal pro-peptide in certain conditions by, as yet, unknown proteases would then release the active VEGF-C.

Like VEGF, VEGF-C also stimulates the migration of endothelial cells and increases vascular permeability, albeit at concentrations higher than required for VEGF (JOUKOV et al. 1996; LEE et al. 1996). About 50-fold higher concentrations of VEGF-C were required to induce the proliferation of blood vascular endothelial cells. These activities are probably mediated through VEGFR-2 activation (PARK et al. 1994; WALTENBERGER et al. 1994). The lower specific activity of VEGF-C in these assays may depend on its lower affinity for VEGFR-2 and on its inability to bind VEGFR-1, precluding the formation of VEGFR-1–VEGFR-2 heterodimers, which may be required for maximal biological responses to VEGF (WALTENBERGER et al. 1994; DISALVO et al. 1995; CAO et al. 1996; CLAUSS et al. 1996).

In order to better understand the function of VEGF-C, in vivo, its cDNA was expressed via a human keratin promoter in the basal cells of stratified squamous epithelia (JELTSCH et al. 1997). Histological examination of the transgenic mice showed that the dermis was atrophic and its connective tissue was replaced by large lymphatic vessels. In ultrastructural analysis, these vessels were shown to have overlapping endothelial junctions, anchoring filaments in the vessel wall, and a discontinuous or even partially absent basement membrane. The endothelium was also characterised by positive staining with monoclonal antibodies to desmoplakins I and II, expressed in lymphatic, but not in vascular, endothelial cells (SCHMELZ et al. 1994). VEGFR-3 and VEGFR-2, and the Tie-1 endothelial-receptor tyrosine kinase mRNAs were detected in endothelial cells lining the abnormal vessels. The VEGF-C–receptor interaction in transgenic mice apparently transduced a mitogenic signal because, in contrast to littermate controls, the lymphatic endothelium of the skin from young transgenic mice showed increased DNA synthesis. In fluorescent microlymphography, a typical honeycomb-like network with similar mesh sizes was detected in both control and transgenic mice, but the diameter of the vessels was approximately twice as large in the transgenic mice. Thus, the endothelial proliferation induced by VEGF-C led to hyperplasia of the superficial lymphatic network, but did not induce the sprouting of new vessels. Also, a relatively specific lymphangiogenic response was obtained when recombinant VEGF-

C was applied to the differentiated chick chorioallantoic membrane (OH et al. 1997).

These effects of VEGF-C overexpression were unexpectedly specific, particularly as VEGF-C is also capable of binding to and activating VEGFR-2 of blood vessel endothelial cells. In vivo, the specific effects of VEGF-C on lymphatic endothelial cells may reflect a requirement for the formation of VEGFR-3–VEGFR-2 heterodimers for endothelial cell proliferation. Such possible heterodimers may help to explain how three homologous VEGFs exert partially redundant, yet strikingly specific, biological effects. Thus, VEGF-C induces specific lymphatic endothelial proliferation and hyperplasia of the lymphatic vasculature in vivo. Further studies should establish the role of VEGF-C in lymphangiomas and in tumour metastasis via the lymphatic vasculature as well as in various other disorders involving the lymphatic system and their treatment.

Both VEGF and VEGF-C are potent vascular-permeability factors. Surprisingly, we have found that the recombinant mature VEGF-C, in which Cys156 was replaced by a Ser residue, is a selective agonist of VEGFR-3 (JOUKOV et al. 1998). This mutant, designated ΔNΔC156S, binds and activates VEGFR-3, but neither binds VEGFR-2 nor activates its autophosphorylation and downstream signalling to the ERK/MAPK pathway. Unlike VEGF-C, ΔNΔC156S neither induces vascular permeability in vivo nor stimulates migration of bovine capillary endothelial cells in culture. These data point out the critical role of VEGFR-2-mediated signal transduction for the vascular permeability activity of VEGF-C, and strongly suggest that the redundancy of biological effects of VEGF and VEGF-C is caused by their ability to bind to and activate VEGFR-2. However, the possibility exists that there are additional receptors for VEGF and VEGF-C, that are responsible for vascular permeability. The ΔNΔC156S mutant may provide a valuable tool for the analysis of VEGF-C effects mediated selectively via VEGFR-3. The ability of ΔNΔC156 S to form homodimers also emphasises differences in the structural requirements for VEGF and VEGF-C dimerization.

Serum and its component growth factors, PDGF, EGF and transforming growth factor-β (TGF-β), and tumor promoters were found to stimulate VEGF-C, but not VEGF-B, mRNA expression (ENHOLM et al. 1997). Serum induction of VEGF-C mRNA occurred independently of protein synthesis; with a slight increase of the mRNA half-life, whereas VEGF-B mRNA was very stable. Hovever, hypoxia, *Ras* oncoprotein and mutant p53 tumour suppressor, which are potent inducers of VEGF mRNA did not increase VEGF-B or VEGF-C mRNA levels. We have also studied the regulation of VEGF-C by angiogenic pro-inflammatory cytokines. Interleukin (IL)-1 induced a concentration- and a time-dependent increase in VEGF-C, but not in VEGF-B, mRNA steady-state levels in human lung fibroblasts, mainly due to increased transcription (RISTIMAKI et al. 1998). Tumour necrosis factor alpha (TNFα) and IL-1 also elevated VEGF-C mRNA steady-state levels, whereas the IL-1 receptor antagonist and dexamethasone inhibited the effect of IL-1. Hypoxia, which is an important inducer of VEGF expression, had no effect on VEGF-B or VEGF-C mRNA levels (ENHOLM et al. 1997). IL-1 and TNFα also stimulated the production of VEGF-C protein by the fibroblasts (RISTIMAKI et al.

1998). Our data suggest that in addition to VEGF, VEGF-C may also serve as a lymphangiogenic or angiogenic stimulus at sites of cytokine activation. In particular, these results raise the possibility that certain pro-inflammatory cytokines regulate the lymphatic vessels indirectly via VEGF-C.

## 4 Identification and Properties of VEGF-D/FIGF

A partial cDNA for FIGF was first isolated from a differential-display screening of murine fibroblast mRNAs from cells with or without a targeted inactivation of the c-fos locus (ORLANDINI et al. 1996). The full-length murine cDNA clone was found to encode a protein of 358 amino acid residues, including a hydrophobic putative signal sequence, with significant similarities to the PDGF/VEGF family of growth factors (see Sect. 5). FIGF was shown to stimulate mitosis of fibroblasts in a dose-dependent manner. However, based on the strong structural similarities to the PDGF/VEGF family of growth factors, we propose that FIGF should be renamed to VEGF-D to highlight this relationship and to get a rational nomenclature of the novel VEGFs. VEGF-D can be viewed as having a VEGF homology domain and long N- and C-terminal extensions. The fact that VEGF-D is most closely related to VEGF-C is apparent for two reasons (see Sect. 5): first, the VEGF homology domain of VEGF-D is much more closely related to that found in VEGF-C than to those of the other family members; second, of the other factors in the VEGF family, only VEGF-C has long N- and C-terminal extensions similar to those in VEGF-D. The presence of these extensions in VEGF-C and VEGF-D, thus, defines a new subfamily of the VEGFs.

The similarity between VEGF-D and VEGF-C exists also at the functional level, as receptor-binding studies demonstrated that VEGF-D and VEGF-C exhibit similar receptor specificities (ACHEN et al. 1998). This protein is likely to be processed in a similar fashion to VEGF-C (data not shown). A region of VEGF-D corresponding to the fully processed, mature VEGF-C can bind to the extracellular domain of VEGFR-2 and induce tyrosine phosphorylation of both VEGFR-2 and VEGFR-3. When expressed in insect cells, the full-length VEGF-D was not proteolytically processed; this protein was unable to activate VEGFR-3 and activation of VEGFR-2 was, at best, marginal. Given that VEGF-D can also activate VEGFR-3, it is possible that VEGF-D could be involved in the regulation of the growth and/or differentiation of lymphatic endothelium just like VEGF-C.

The notion that VEGF-D and VEGF-C may have similar biological functions is further supported by their similar expression patterns. For example, both genes are strongly expressed in heart, muscle and small intestine, whereas expression was undetectable in peripheral blood leucocytes, brain and liver (JOUKOV et al. 1996; LEE et al. 1996; ACHEN et al. 1998). Nevertheless, the expression patterns are not identical. A second VEGF-D transcript was detected only in skeletal muscle.

Like VEGF-C, human VEGF-D was also mitogenic for bovine aortic endothelial cells. This response is likely to involve VEGFR-2. Mouse VEGF-D has previously been reported to induce proliferation and morphological alterations of cultured fibroblasts, but the receptors responsible for mediating these effects have not been identified (ORLANDINI et al. 1996). It would be of interest to determine whether or not the cultured fibroblasts used for such studies expressed VEGFR-2 or VEGFR-3. A summary of the receptor-binding properties of known members of the VEGF family of growth factors is illustrated in Fig. 3.

## 5 Primary Structures of VEGF-Related Growth Factors

Seven polypeptides with significant similarities to VEGF and two PDGFs have been discovered so far. A multiple amino acid sequence alignment of human VEGF, PlGF, VEGF-B, VEGF-C, VEGF-D and the two PDGFs show that a central core of the proteins is well conserved during evolution (Fig. 4A). A major part of this core region is located between the eight invariant cysteine residues, shown to be involved in inter- and intramolecular disulphide bonding of VEGF and the two PDGFs. This region is encoded by the two well-conserved exons, E3 and E4, in the corresponding genes of the VEGFs (see Fig. 1). The overall amino acid sequence identity in this region varies between 20% and 56% in pairwise comparisons of available amino acid residues. Outside this central core, the overall amino acid identities are much weaker, although individual pairs of these proteins display higher sequence similarities. The receptor binding epitopes in these group of growth factors, at least for VEGF (KEYT et al. 1996) and the two PDGFs (HELDIN et al. 1993), are confined within the central core defined by the eight invariant cysteine residues. Thus, this region defines a structural and functional minimal domain. Determination of the three dimensional structures of VEGF (MULLER

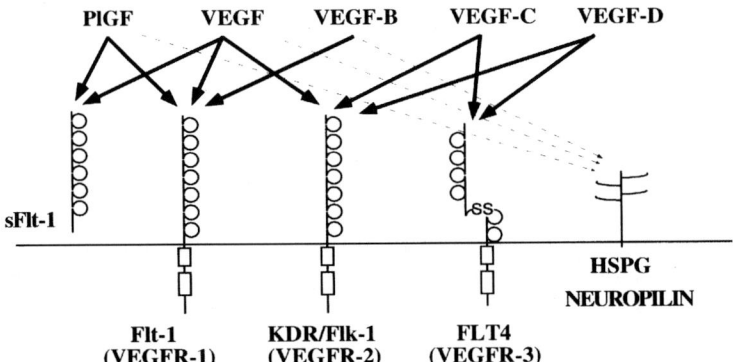

**Fig. 3.** Interactions of VEGF-related growth factor with their receptors. A schematic illustration of the receptor-binding characteristics of VEGF-related growth factors with soluble Flt-1 (sFlt-1), Flt-1 (*VEGFR-1*), KDR/Flk-1 (*VEGFR-2*), Flt-4 (*VEGFR-3*) and with heparin sulphate proteoglycans (*HSPG*) and neuropilin (SOKER et al. 1998)

## Structure, Expression and Receptor-Binding Properties of Novel VEGFs

```
VEGF-C         M H L L G F F S V A C S L L A A A L L P G P R E A P A A A A A F E S G L D L S D   40
VEGF-D         - - - - - - - - - - - M Y G E W G M G N I L M M F H V Y L V Q G F R S E H G P V   29
PDGF-A         - - - - - - - - - - - - - - - - - - - - - - - - - - - - - - - M R T L A C         6
PDGF-B         - - - - - - - - - - - - - - - - - - - - - - - - - - - - - M N R C W A L F L S    10

VEGF-C         A E P D A G E A T A Y A S K D L E E Q L R S V S S V D E L M T V L Y P E Y W K M   80
VEGF-D         K D F S F E R S S R S M L E R S E Q Q I R A A S S L E E L L - - - - Q I A H S E   65
PDGF-A         L L L L G C G Y L A H V L A E E A E I P R E V I E R L A R S Q I H S I R D L Q R   46
PDGF-B         L C C Y L R L V S A E G D P I P E E L Y E M L S D H S I R S F D D L Q R L L H G   50

VEGF 165       M N F L L S W V H W S L A L L L Y L H H A K W S Q A A P M A E G G G Q N H H E V   40
PlGF-2         M P V M R L F P C F L Q L L A G L A L P A V P P Q Q W A L S A G N G S S E V E V   40
VEGF-B167      - - - - M S P L L R R L L L A A L L Q L A P A Q A P V S Q P D A P G H Q R K V    35
Pox Orf VEGF   - - - - - - - - - - - - - M K L L V G I L V A V C L H Q Y L L N A D S N T         24
VEGF-C         Y K C Q L R K G G W Q H N R E Q A N L N S R T E E T I K F A A A H Y N T E I - L  119
VEGF-D         D W K L W R C R L K L K S L A S M D S R S A S H R S T R F A A T F Y D T E T - L  104
PDGF-A         L L E I D S V G S E D S L D T S L R A H G V H - - A T K H V P E K R P L P I R R   84
PDGF-B         D P - - - - - G E E D G A E L D L N M T R S H S G G E L E S L A R G R R S L G S   85

VEGF 165       V K F M D V Y Q R S Y C H P I E T L V D I F Q E Y P D E I E Y I F K - - P S C V    78
PlGF-2         V P F Q E V W G R S Y C R A L E R L V D V V S E Y P S E V E H M F S - - P S C V    78
VEGF-B167      V S W I D V Y T R A T C Q P R E V V V P L T V E L M G T V A K Q L V - - P S C V    73
Pox Orf VEGF   K G W S E V L K G S E C K P R P I V V P V S E T H P E L T S Q R F N - - P P C V    62
VEGF-C         K S I D N E W R K T Q C M P R E V C I D V G K E F G V A T N T F F K - - P P C V  157
VEGF-D         K V I D E E W Q R T Q C S P R E T C V E V A S E L G K T T N T F F K - - P P C V  142
PDGF-A         K R S I E E A V P A V C K T R T V I Y E I P R S Q V D P T S A N F L I W P P C V  124
PDGF-B         L T I A E P A M I A E C K T R T E V F E I S R R L I D R T N A N F L V W P P C V  125

VEGF 165       P L M R C G G C C N D E G L E C V P T E E S N I T M Q I M R I K - P H Q G Q H I  117
PlGF-2         S L L R C T G C C G D E D L H C V P V E T A N V T M Q L L K I R - S G D R P S Y  117
VEGF-B167      T V Q R C G G C C P D D G L E C V P T G Q H Q V R M Q I L M I R Y P - - S S Q L  111
Pox Orf VEGF   T L M R C G G C C N D E S L E C V P T E E V N V S M E L L G A S G S G N G M Q    102
VEGF-C         S V Y R C G G C C N S E G L Q C M N T S T S Y L S K T L F E I T V P L S Q G P K  197
VEGF-D         N V F R C G G C C N E E G V M C M N T S T S Y I S K Q L F E I S V P L T S V P E  182
PDGF-A         E V K R C T G C C N T S S V K C Q P S R V H H R S V K V A K V E Y V R K K P K L  164
PDGF-B         E V Q R C S G C C N N R N V Q C R P T Q V Q L R P V Q V R K I E I V R K K P I F  165

VEGF 165       G E M S F L Q H N K - - C E C R P K K - - - - - - - - - D R A R Q E N P C G P    145
PlGF-2         V E L T F S Q H V R - - C E C R P L R E - - - - K M K P E R R R P K G R G K R R  151
VEGF-B167      G E M S L E E H S Q - - C E C R P K K K - - - - D S A V K P D S P R P L C P R   144
Pox Orf VEGF   R L S F V E H K K - - - C D C R P R F T T T P P T T T R P P R R R R              133
VEGF-C         P V T I S F A N H T S - C R C M S K L D - - - V Y R Q V H S I I R R S L P A T -  232
VEGF-D         L V P V K I A N H T G - C K C L P T G P - - - - - R H P S I I R R S I Q T P E   216
PDGF-A         K E V Q V R L E E H L E C A C A T T S L N P D Y R E E D T G R P R E S G K K R K  204
PDGF-B         K K A T V T L E D H L A C K C E T V A A A R P V T R S P G G S Q E Q R A K T P Q  205

VEGF 165       C S S E R R K H L F V Q D P Q T C K C S C K N T D S - R C K A R Q L E L N E R T  184
PlGF-2         R E N Q R P T D C H L C G D A V P R R                                            170
VEGF-B167      C T Q H H Q R P D P R T - - - - C R C R C R R R S F L R C Q G R G L E L N P D T  180
VEGF-C         L P Q C Q A A N K T C P T N Y M W N N H I C R C L A Q E D F M F S S D A G D D S  272
VEGF-D         E D E C P H S K K L C P I D M L W D N T K C K C V L Q D E - T P L P G T E D H S  255
PDGF-A         R K R L K P T                                                                   211
PDGF-B         T R V T I R T V R V R R P P K G K H R K F K H T H D K T A L K E T L G A         241

VEGF 165       C R C D K P R R                                                                 192
VEGF-B167      C R C R K L R R                                                                 188
VEGF-C         T D G F H D I C G P N K E L D E E T C Q C V C R A G L R P A S C G P H K E L D R  312
VEGF-D         Y L Q E P T L C G P H M T F D E D R - - - - - - - - - - - - - - - - - - - - -   273

VEGF-C         N S C Q C V C K N K L F P S Q C G A N R E F D E N T C Q C V C K R T C P R N Q P  352
VEGF-D         - - C E C V C K A P C P G D L I Q H P E N - - - - - C S C F E C K E S L E S C C  306

VEGF-C         L N P G K C A C E C T E S P Q K C L L K G K K F H H Q T C S C Y R R P C T N R Q  392
VEGF-D         Q K H K I - - - - - - - - - - - - - - - F H P D T C S C E D R - C P F H T        327

VEGF-C         K A C E P G F S Y S E E V C R C V P S Y W K R P Q M S                            419
VEGF-D         R T C A S R K P A C G K H W R F P K E T R A Q G L Y S Q E N P                   358
```

**A**

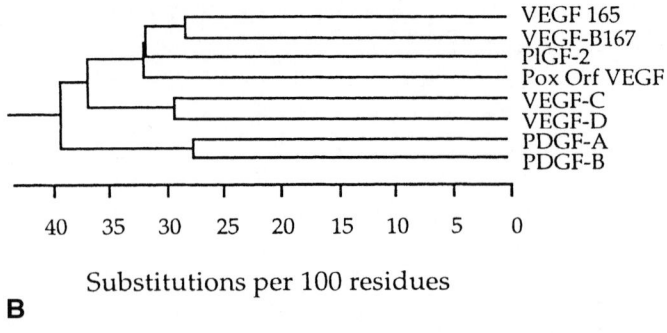

**Fig. 4A, B.** Amino acid sequence alignment of VEGF-related growth factors. **A.** Multiple amino acid sequence alignment of three novel VEGFs, e.g. VEGF-B, VEGF-C and VEGF-D and a comparison with VEGF, placenta growth factor (PlGF) and the two platelet-derived growth factors (PDGFs) as well as with the interesting viral homologue of the poxvirus *orf* virus (LYTTLE et al. 1994). The amino acid sequences were aligned using the Crystal algorithm, and the alignment was refined manually. The *boxed* residues are within two distance units using the PAM 250 matrix. **B.** An unrooted phylogenetic tree of the VEGF-related growth factors based on the amino acid sequence alignment in A

et al. 1997) and PDGF-B (OEFNER et al. 1992) by X-ray crystallography have also shown that this region forms distinct domains with remarkable structural similarities.

A phylogenetic analysis of the amino acid sequences of these growth factors shows that they can be grouped into three separate subfamilies, consisting of VEGF, PlGF, VEGF-B, and VEGF-C and D, and the two PDGFs, respectively (Fig. 4B).

# 6 Perspectives

The discovery of three novel members of VEGF family increases our understanding of the complexity of the regulatory signals for endothelial cells and promotes new areas of research in vascular biology. Many of the already-established experimental models and approaches used in VEGF studies might obviously be applied to studies of the novel VEGFs. However, not only endothelial functions should be taken into consideration here, as recent results show that VEGF might induce certain biological effects via the targeting of non-endothelial cells (MIDY and PLOUET 1994; GABRILOVICH et al. 1996).

The main questions regarding the biological roles of the novel VEGFs are not answered yet. In this regard, different molecular genetic and transgenic approaches, including gene targeting, are of great importance. Studies on VEGF-C have shown that it acts as a specific growth factor for endothelial cells of lymphatic vessels (JELTSCH et al. 1997). Studies on VEGF-B and VEGF-D are likely to provide additional information on the role of these growth factors in endothelial cell function and physiology.

Important issues also concern the analysis of tissue-specific regulation of VEGF-B, VEGF-C and VEGF-D expression by hypoxia, various growth factors and other agents or conditions known to regulate VEGF expression. Similarly, the function of the different splicing forms of VEGF-B, VEGF-C and VEGF-D should be explored. Such alternatively spliced isoforms of these growth factors might possess different functions in vivo, e.g. due to the differences in their receptor specificity/affinity, bioavailability, stability and proteolytic processing, and their ability to form heterodimers with other VEGF family members. The latter property might be of particular importance, as heterodimers of the various growth factors might express biological properties distinct from those of the corresponding homodimers. Finally, the discovery of VEGF-B, VEGF-C and VEGF-D highlights the structural similarities of VEGF family polypeptides and may simplify the search for novel homologous molecules.

*Acknowledgements.* We kindly thank all of our coworkers and collaborators for outstanding contributions to the basic findings summarized in the review and the funding agencies for continued support.

# References

Achen MG, Jeltsch M, Kukk E, Makinen T, Vitali A, Wilks AF, Alitalo K, Stacker SA (1998) Vascular endothelial growth factor-D (Vegf-D) is a ligand for the tyrosine kinases Vegf receptor 2 (Flk1) and Vegf receptor 3 (Flt4). Proc Natl Acad Sci USA 95:548–553

Cao Y, Chen H, Zhou L, Chiang MK, Anand-Apte B, Weatherbee JA, Wang Y, Fang F, Flanagan JG, Tsang ML (1996) Heterodimers of placenta growth factor/vascular endothelial growth factor. Endothelial activity, tumor cell expression, and high affinity binding to Flk-1/KDR. J Biol Chem 271:3154–3162

Carmeliet P, Ferreira V, Breier G, Pollefeyt S, Kieckens L, Gertsenstein M, Fahrig M, Vandenhoeck A, Harpal K, Eberhardt C, Declercq C, Pawling J, Moons L, Collen D, Risau W, Nagy A (1996) Abnormal blood vessel development and lethality in embryos lacking a single VEGF allele. Nature 380:435–439

Chilov D, Kukk E, Taira S, Jeltsch M, Kaukonen J, Palotie A, Joukov V, Alitalo K (1997) Genomic organization of human and mouse genes for vascular endothelial growth factor C. J Biol Chem 272:25176–25183

Clauss M, Weich H, Breier G, Knies U, Rockl W, Waltenberger J, Risau W (1996) The vascular endothelial growth factor receptor Flt-1 mediates biological activities – implications for a functional role of placenta growth factor in monocyte activation and chemotaxis. J Biol Chem 271:17629–17634

Cohen T, Nahari D, Cerem LW, Neufeld G, Levi BZ (1996) Interleukin 6 induces the expression of vascular endothelial growth factor. J Biol Chem 271:736–741

DiPalma T, Tucci M, Russo G, Maglione D, Lago C, Romano A, Saccone S, DellaValle G, De Gregorio L, Dragani T, Viglietto G, Persico M (1997) The placenta growth factor gene of the mouse. Mamm Genome 7:6–12

DiSalvo J, Bayne ML, Conn G, Kwok PW, Trivedi PG, Soderman DD, Palisi TM, Sullivan KA, Thomas KA (1995) Purification and characterization of a natural occuring vascular endothelial growth factor-placenta growth factor heterodimer. J Biol Chem 270:7717–7723

Eichmann A, Corbel C, Nataf V, Vaigot P, Breant C, Le Douarin NM (1997) Ligand-dependent development of the endothelial and hemopoietic lineages from embryonic mesodermal cells expressing vascular endothelial growth factor receptor 2. Proc Natl Acad Sci USA 94:5141–5146

Eichmann A, Corbel C, Jaffredo T, Bréant C, Joukov V, Kumar V, Alitalo K, le Douarin N (1998) Avian VEGF-C: cloning, embryonic expression pattern and stimulation of the differentiation of VEGFR2 expressing endothelial cell precursors. Dev Biol (in press)

Enholm B, Paavonen K, Ristimaki A, Kumar V, Gunji Y, Klefstrom J, Kivinen L, Laiho M, Olofsson B, Joukov V, Eriksson U, Alitalo K (1997) Comparison of VEGF, VEGF-B, VEGF-C and Ang-1 mRNA regulation by serum, growth factors, oncoproteins and hypoxia. Oncogene 14:2475–2483

Ferrara N, Davis-Smyth T (1997) The biology of vascular endothelial growth factor. Endocr Rev 18:4–25

Ferrara N, Carver-Moore K, Chen H, Dowd M, Lu L, O'Shea KS, Powell-Braxton L, Hillan KJ, Moore MW (1996) Heterozygous embryonic lethality induced by targeted inactivation of the VEGF gene. Nature 380:439–442

Finkenzeller G, Marme D, Weich HA, Hug H (1992) Platelet-derived growth factor-induced transcription of the vascular endothelial growth factor gene is mediated by protein kinase C. Cancer Res 52:4821–4823

Frank S, Hubner G, Breier G, Longaker MT, Greenhalgh DG, Werner S (1995) Regulation of vascular endothelial growth factor expression in cultured keratinocytes. Implications for normal and impaired wound healing. J Biol Chem 270:12607–12613

Gabrilovich DI, Chen HL, Girgis KR, Cunningham HT, Meny GM, Nadaf S, Kavanaugh D, Carbone DP (1996) Production of vascular endothelial growth factor by human tumors inhibits the functional maturation of dendritic cells. Nat Med 2:1096–1103

Garrido C, Saule S, Gospodarowicz D (1993) Transcriptional regulation of vascular endothelial growth factor gene expression in ovarian bovine granulosa cells. Growth Factors 8:109–117

Goldberg MA, Schneider TJ (1994) Similarities between the oxygen-sensing mechanisms regulating the expression of vascular endothelial growth factor and erythropoietin. J Biol Chem 269:4355–4359

Grimmond S, Lagercrantz J, Drinkwater C, Silins G, Townson S, Pollock P, Gotley D, Carson E, Rakar S, Nordenskjold M, Ward L, Hayward N, Weber G (1996) Cloning and characterization of a novel human gene related to vascular endothelial growth factor. Genome Res 6:124–131

Heldin C, Östman A, Westermark B (1993) Structure of platelet-derived growth factor: Implications for functional properties. Growth Factors 8:245–252

Jeltsch M, Kaipainen A, Joukov V, Meng XJ, Lakso M, Rauvala H, Swartz M, Fukumura D, Jain RK, Alitalo K (1997) Hyperplasia of lymphatic vessels in VEGF-C transgenic mice. Science 276:1423–1425

Joukov V, Pajusola K, Kaipainen A, Chilov D, Lahtinen I, Kukk E, Saksela O, Kalkkinen N, Alitalo K (1996) A novel endothelial growth factor, VEGF-C, is a ligand for the Flt4 (VEGFR-3) and KDR (VEGFR-2) receptor tyrosine kinases. EMBO J 15:290–298

Joukov V, Sorsa T, Kumar V, Jeltsch M, Claesson-Welsh L, Cao Y, Saksela O, Kalkkinen N, Alitalo K (1997) Proteolytic processing regulates receptor specificify and activity of VEGF-C. EMBO J 16:3898–3911

Joukov V, Kumar V, Sorsa T, Arighi E, Weich H, Saksela O, Alitalo K (1998) A recombinant mutant vascular endothelial growth factor C that has lost vascular endothelial growth factor receptor-2 binding, activation and vascular permeability activities. J Biol Chem 273:6599–6602

Kaipainen A, Korhonen J, Pajusola K, Aprelikova O, Persico M, Terman B, Alitalo K (1993) The related FLT4, FLT1 and KDR receptor tyrosine kinases show distinct expression patterns in human fetal endothelial cells. J Exp Med 178:2077–2088

Kaipainen A, Korhonen J, Mustonen T, van Hinsbergh VW, Fang GH, Dumont D, Breitman M, Alitalo K (1995) Expression of the fms-like tyrosine kinase 4 gene becomes restricted to lymphatic endothelium during development. Proc Natl Acad Sci USA 92:3566–3570

Keyt BA, Nguyen HV, Berleau LT, Duarte CM, Park J, Chen H, Ferrara N (1996) Identification of vascular endothelial growth factor determinants for binding KDR and FLT-1 receptors. Generation of receptor-selective VEGF variants by site-directed mutagenesis. J Biol Chem 271:5638–5646

Kukk E, Lymboussaki A, Taira S, Kaipainen A, Jeltsch M, Joukov V, Alitalo K (1996) VEGF-C receptor binding and pattern of expression with VEGFR-3 suggests a role in lymphatic vascular development. Development 122:3829–3837

Lagercrantz J, Larsson C, Grimmond S, Fredriksson M, Weber G, Piehl F (1996) Expression of the VEGF-related factor gene in pre- and postnatal mouse. Biochem Biophys Res Commun 220:147–152

Lee J, Gray A, Yuan J, Luoh S-M, Avraham H, Wood WI (1996) Vascular endothelial growth factor-related protein: a ligand and specific activator of the tyrosine kinase receptor Flt4. Proc Natl Acad Sci USA 93:1988–1992

Lyttle DJ, Fraser KM, Fleming SB, Mercer AA, Robinson AJ (1994) Homologs of vascular endothelial growth factor are encoded by the poxvirus orf virus. J Virol 68:84–92

Midy V, Plouet J (1994) Vasculotropin/vascular endothelial growth factor induces differentiation in cultured osteoblasts. Biochem Biophys Res Commun 199:380–386

Muller YA, Li B, Christinger HW, Wells JA, Cunningham BC, De Vos AM (1997) Vascular endothelial growth factor – crystal structure and functional mapping of the kinase domain receptor binding site. Proc Natl Acad Sci USA 94:7192–7197

Oefner C, D'Arcy A, Winkler FK, Eggimann B, Hosang M (1992) Crystal structure of human platelet-derived growth factor BB. EMBO J 11:3921–3926

Oh SJ, Jeltsch MM, Birkenhäger R, McCarthy JEG, Weich HA, Christ B, Alitalo K, Wilting J (1997) VEGF and VEGF-C: Specific induction of angiogenesis and lymphangiogenesis in the differentiated avian chorioallantoic membrane. Dev Biol 188:96–109

Olofsson B, Pajusola K, Kaipainen A, von Euler G, Joukov V, Saksela O, Orpana O, Pettersson R, Alitalo K, Eriksson U (1996a) Vascular endothelial growth factor B, a novel growth factor for endothelial cells. Proc Natl Acad Sci USA 93:2576–2581

Olofsson B, Pajusola K, Voneuler G, Chilov D, Alitalo K, Eriksson U (1996b) Genomic organization of the mouse and human genes for vascular endothelial growth factor B (VEGF-B) and characterization of a second splice isoform. J Biol Chem 271:19310–19317

Olofsson B, Korpeleinen E, Mandriota S, Pepper MS, Aase K, Jeltsch MM, Shibuya M, Alitalo K, Eriksson U (1998) VEGF-B binds VEGFR-1 and regulates plasminogen activator activity in endothelial cells. Proc Natl Acad Sci USA (submitted)

Orlandini M, Marconcini L, Ferruzzi R, Oliviero S (1996) Identification of a C-fos-induced gene that is related to the platelet-derived growth factor/vascular endothelial growth factor family. Proc Natl Acad Sci USA 93:11675–11680

Östman A, Rall L, Hammacher A, Wormstead MA, Coit D, Valenzuela P, Betsholtz C, Westermark B, Heldin CH (1988) Synthesis and assembly of a functionally active recombinant platelet-derived growth factor AB heterodimer. J Biol Chem 263:16202–16208

Östman A, Thyberg J, Westermark B, Heldin CH (1992) PDGF-AA and PDGF-BB biosynthesis: proprotein processing in the Golgi complex and lysosomal degradation of PDGF-BB retained intracellularly. J Cell Biol 118:509–519

Paavonen K, Horelli-Kuitunen N, Chilov D, Kukk E, Pennanen S, Kallioniemi OP, Pajusola K, Olofsson B, Eriksson U, Joukov V, Palotie A, Alitalo K (1996) Novel human vascular endothelial growth factor genes VEGF-B and VEGF-C localize to chromosomes 11q13 and 4q34, respectively. Circulation 93:1079–1082

Park J, Chen H, Winer J, Houck K, Ferrara N (1994) Placenta growth factor. J Biol Chem 269:25646–25654

Pertovaara L, Kaipainen A, Mustonen T, Orpana A, Ferrara N, Saksela O, Alitalo K (1994) Vascular endothelial growth factor is induced in response to transforming growth factor-beta in fibroblastic and epithelial cells. J Biol Chem 269:6271–6274

Ristimäki A, Narko K, Enholm B, Joukov V, Alitalo K (1998) Proinflammatory cytokines regulate expression of the lymphatic endothelial mitogen vascular endothelial growth factor-C. J Biol Chem 273:8413–8418

Schmelz M, Moll R, Kuhn C, Franke WW (1994) Complexus adhaerentes, a new group of desmoplakin-containing junctions in endothelial cells: II. Different types of lymphatic vessels. Differentiation 57:97–117

Soker S, Takashima S, Quan Miao H, Neufeld G, Klagsbrun M (1998) Neurophilin-1 is expressed by endothelial and tumor cells as an isoform-specific receptor for vascular endothelial growth factor. Cell 92:735–745

Stein I, Neeman M, Shweiki D, Itin A, Keshet E (1995) Stabilization of vascular endothelial growth factor mRNA by hypoxia and hypoglycemia and coregulation with other ischemia-induced genes. Mol Cell Biol 15:5363–5368

Terman B, Dougher-Vermazen M, Carrison M, Dimitrov D, Armellino D, Gospodarowicz D, Bölhen P (1992) Identification of the KDR tyrosine kinase as a receptor for vascular endothelial growth factor. Biochem Biophys Res Commun 187:1579–1586

Tischer E, Mitshell R, Hartman T, Silva M, Gospodarowicz D, Fiddes J, Abraham J (1991) The human gene for vascular endothelial growth factor: multiple protein forms are encoded through alternative exon splicing. J Biol Chem 266:11947–11954

Townson S, Lagercrantz J, Grimmond S, Silins G, Nordenskjold M, Weber G, Hayward N (1996) Characterization of the murine VEGF-related factor gene. Biochem Biophys Res Commun 220:922–928

Waltenberger J, Claesson-Welsh L, Siegbahn A, Shibuya M, Heldin CH (1994) Different signal transduction properties of KDR and Flt1, two receptors for vascular endothelial growth factor. J Biol Chem 269:26988–26995

# Structure and Function of Vascular Endothelial Growth Factor Receptor-1 and -2

M. Shibuya[1], N. Ito[2], and L. Claesson-Welsh[2]

| | | |
|---|---|---|
| 1 | Introduction | 60 |
| 2 | Structure | 61 |
| 2.1 | Structural Organization of Vascular Endothelial Growth Factor Receptor-1 and -2 | 61 |
| 2.2 | Binding of VEGF | 62 |
| 2.3 | Interaction with Heparin | 64 |
| 3 | Tyrosine Kinase Activity | 64 |
| 3.1 | Induction of Kinase Activity and Autophosphorylation of VEGFR-1 and -2 | 64 |
| 3.2 | Signal Transduction by VEGFR-1 and -2 | 66 |
| 4 | Regulation of Expression of VEGFR-1 and -2 | 68 |
| 4.1 | Messenger Ribonucleic Acid (mRNA) Products and Biosynthesis of VEGFR-1 and -2 | 68 |
| 4.2 | Chromosomal Localisation and Genomic Structure | 69 |
| 4.3 | Promoter/Enhancer Regions of the VEGFR-1 and -2 Genes | 70 |
| 4.4 | Regulation of VEGFR-1 and -2 Expression | 71 |
| 5 | Role of VEGFR-1 and -2 in Endothelial Cell Function | 72 |
| 5.1 | VEGF Receptors in Embryonal Development | 72 |
| 5.2 | Biological Effects of VEGFR-1 and -2 on Endothelial Cells in Vivo and in Vitro | 74 |
| 5.2.1 | Cell Proliferation | 74 |
| 5.2.2 | Migration | 75 |
| 5.2.3 | Vascular Permeability | 75 |
| 5.2.4 | Induction of Protease Activity | 75 |
| 5.2.5 | Angiogenesis in Tissue Culture | 76 |
| 5.3 | Cell Migration of Monocytes and Macrophages | 76 |
| 5.4 | VEGF Receptors and Anti-angiogenic Therapy | 76 |
| 6 | Conclusions | 78 |
| | References | 79 |

[1]Dept. of Genetics, Institute of Medical Science, University of Tokyo, 4-6-1 Shirokane-dai, Minatu-ku, Tokyo 108, Japan
[2]Dept. of Medical Biochemistry and Microbiology, Biomedical Center, Box 575, S-751 23 Uppsala, Sweden

# 1 Introduction

Receptor-type tyrosine kinases are known to be involved in a wide variety of biological processes, such as cell growth, differentiation, morphogenesis and malignant transformation. The endothelial cell-specific growth factors, denoted vascular endothelial growth factors/vascular permeability factors (VEGF/VPF), constitute a growing family of factors that bind to at least three different receptor-type tyrosine kinases (Fig. 1). A wealth of data indicates the important functions of these receptors in the normal development and for physiological and pathological angiogenesis. This review will focus on the structure and function of two of these receptors which, in many respects, are similar to other receptor-type tyrosine kinases. Thus, the VEGF receptors appear to dimerize in response to growth factor binding, become activated, tyrosine phosphorylated and couple to downstream signal transduction chains, much like other growth factor receptors. We aim to outline these similarities, but also to highlight the unique characteristics of the VEGF receptors, which are the key to our understanding of the critical roles of these receptors in fundamental biological processes.

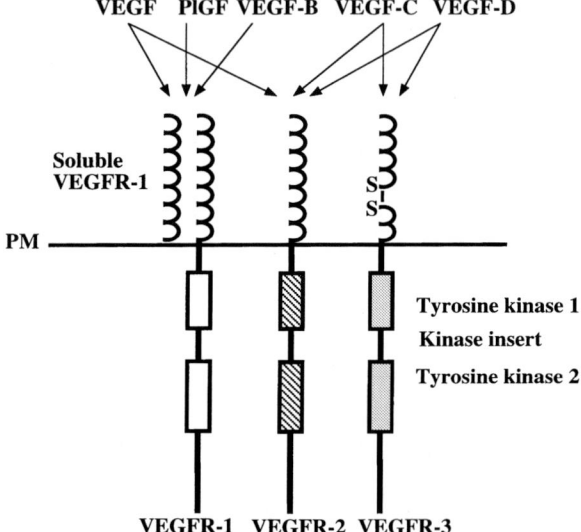

**Fig. 1.** Binding of different vascular endothelial growth factor (VEGF) family members to the three VEGF receptors. The VEGF family members, VEGF, placenta growth factor (PlGF), VEGF-B, -C and -D, bind in specific patterns to three related receptors, VEGFR-1, -2 and -3. VEGFR-1 exists also as a soluble form. The receptors have a similar general structural organization, with seven imunoglobulin-like loops extracellularly and a tyrosine-kinase domain split into two parts by insertion of the non-catalytic kinase insert

## 2 Structure

### 2.1 Structural Organization of Vascular Endothelial Growth Factor Receptor-1 and -2

The VEGF receptor family consists of three receptor-type tyrosine kinases, denoted VEGFR-1 (Flt-1), VEGFR-2 (KDR/Flk-1) and VEGFR-3 (Flt-4; see TAIPALE et al. this volume). VEGFR-1 was originally isolated from a human placenta complementary deoxyribonucleic acid (cDNA) library (SHIBUYA et al. 1990), and was referred to as "fms-like tyrosine kinase" (Flt-1) due to its structural relationship to members of the Fms family of receptor tyrosine kinases, including the colony-stimulating factor-1 (CSF-1) receptor (also denoted c-fms), the stem cell factor receptor (c-Kit) and the two related platelet-derived growth factor (PDGF) receptors (ULLRICH and SCHLESSINGER 1990). Based on similarities with the Fms family, which bind dimerized ligands, such as PDGF, a possible ligand for VEGFR-1 was suggested to be a polypeptide with a dimeric structure. In 1992, DE VRIES et al. showed that the endothelial cell-specific growth factor VEGF, a homodimer of 22-kDa polypeptide chains, bound to VEGFR-1 with high affinity, leading to transduction of VEGF signals. Thus, VEGFR-1 was found to be one of the physiological receptors for VEGF (DE VRIES et al. 1992) (Fig. 1).

A second receptor for VEGF, VEGFR-2, was subsequently identified; the human cDNA was denoted KDR (TERMAN et al. 1991) and the mouse cDNA was denoted Flk-1 (MATTHEWS et al. 1991a). MILLAUER et al. (1993) and TERMAN et al. (1992) showed that this receptor binds VEGF and responds to ligand stimulation with autophosphorylation. The overall structural organization of VEGFR-1 and -2 is highly related (Fig. 1). Their extra-cellular domains contain seven immunoglobulin (Ig)-like folds. As shown in Fig. 2, the distribution of cysteine residues in the VEGFR-1 extracellular domain indicates that the first to fifth Ig domains of this receptor correspond to the extracellular domains of other members of the Fms

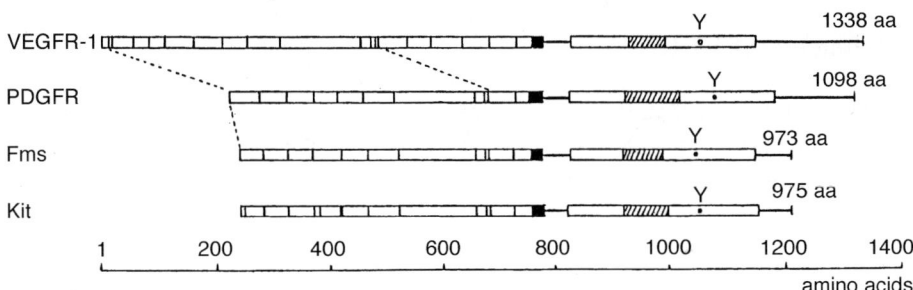

**Fig. 2.** Distribution of cysteine residues in the extracellular domains of vascular endothelial growth factor receptor 1 (VEGFR-1) and Fms/Kit/platelet-derived growth factor receptor (PDGFR). *Vertical bars* indicate cysteine residues. The amino-terminal two thirds of the VEGFR-1/VEGFR-2 extracellular domains correspond to the entire extracellular domain of the Fms family receptors (SHIBUYA 1995)

**Table 1.** Amino-acid homology of intracellular domains between vascular endothelial growth factor receptors 1 and -2 (VEGFR-1/VEGFR-2) and other receptor- or non-receptor-type tyrosine kinases

| Tyrosine kinase | Amino-half[a] % | Insert[b] % | Carboxyl-half[c] % |
| --- | --- | --- | --- |
| VEGFR-2 | 78 | 51 | 78 |
| VEGFR-3 | 80 | 11 | 81 |
| v-Fms | 57 | 5 | 53 |
| PDGFRα | 59 | 13 | 58 |
| PDGFRβ | 57 | 13 | 53 |
| v-Kit | 62 | 11 | 58 |
| Flt-3/Flk-2 | 60 | 11 | 58 |
| v-ErbB | 26 | – | 42 |
| v-Ros | 32 | – | 43 |
| v-Src | 30 | – | 38 |
| v-Fps | 32 | – | 41 |

*PDGFR* platelet-derived growth factor.
[a]VEGFR-1 residue no. 813–929.
[b]VEGFR-1 residue no. 930–994.
[c]VEGFR-1 residue no. 995–1152.

family. The fourth Ig domain of VEGFR-1 and -2, and members of the Fms family, share a unique distribution of cysteine residues. This is interesting, since the fourth Ig domain of c-Kit, a member of the Fms family, is known to contribute to receptor dimerization (BLECHMAN et al. 1995). Thus, it is possible that the same Ig domain in VEGFR-1 and -2 could play a role in dimerization and activation of the receptors.

The significance of the sixth and seventh Ig domains of VEGFR-1 and -2 is not yet clear; however, this region might also contribute to receptor dimerization (TANAKA et al. 1997). The intracellular domains of VEGFR-1 and -2 contain the catalytic tyrosine kinase sequences, which are interrupted by a non-catalytic "kinase-insert" sequence of about 70 amino-acid residues. Among tyrosine kinases, the predicted amino-acid sequence of the tyrosine-kinase domains of VEGFR-1 and -2 are most homologous (55–60% identical amino acids) to members of the Fms family (Table 1).

## 2.2 Binding of VEGF

VEGFR-1 binds VEGF, the VEGF-related growth factor PlGF (placenta growth factor) and, potentially, other novel VEGFs (see PERSICO et al. and ERIKSSON/ ALITALO at this volume) with high affinity. The amino-acid sequence of PlGF is 53% identical to that of VEGF, and PlGF is expressed in placental tissues, but not in most other tissues under normal conditions (see PERSICO et al. this volume and MAGLIONE et al. 1991, 1993). The $K_d$ value of VEGF binding to VEGFR-1 was reported to be about 10 pM, whereas PlGF binding to VEGFR-1 is tenfold weaker (100–200 pM) (KENDALL et al. 1994; PARK et al. 1994; SAWANO et al. 1996). VEGFR-2 binds both VEGF and a short form of VEGF-C (JOUKOV et al. 1997; MILLAUER et al. 1993) and the $K_d$ values were shown to be 400–900 pM and

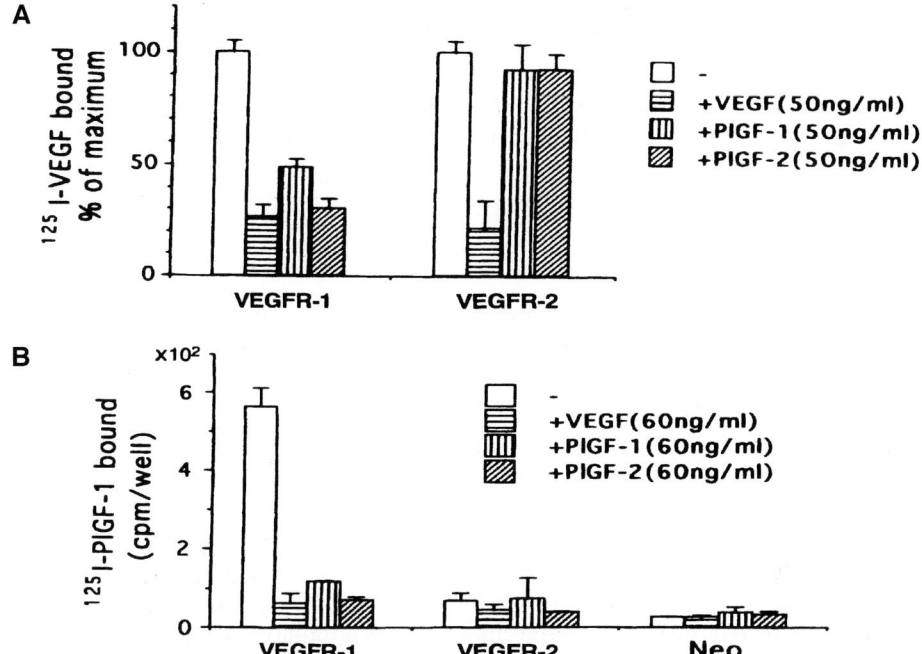

**Fig. 3A,B.** Placenta growth factor (PlGF) binds to VEGFR-1 but not to VEGFR-2. Binding assays using $^{125}$I-VEGF or $^{125}$I-PlGF in the presence or absence of unlabelled competing ligand on NIH 3T3 cells expressing VEGFR-1, VEGFR-2 or vector transfected cells (SAWANO et al. 1996)

410 pM, respectively (JOUKOV et al. 1997; SAWANO et al. 1996; WALTENBERGER et al. 1994).

Unlike VEGF, PlGF binds only to VEGFR-1, and neither to VEGFR-2 nor VEGFR-3 (Fig. 3). Thus, PlGF may be useful for characterisation of the biological and biochemical functions of VEGFR-1. In some models, the biological effects of PlGF on vascular endothelial cells in terms of cell growth and gene expression in vitro have been found to be very weak (PARK et al. 1994; SAWANO et al. 1996). In other models, PlGF has been shown to induce activation of the extracellular regulated kinase Erk2, followed by DNA synthesis (LANDGREN et al. 1998) and to stimulate angiogenic activities in vivo (ZICHE et al. 1997).

The binding sites for VEGF on the extracellular domain of VEGFR-1 have been determined using VEGFR-1/VEGFR-3 chimeric molecules, in which sequences derived from VEGFR-3 were known not to bind VEGF. In this way, the second Ig domain of VEGFR-1 was shown to contribute to the specific binding to VEGF (DAVIS-SMYTH et al. 1996). In addition, amino-acid sequences adjacent to the second Ig domain were also shown to potentiate VEGFR-1 affinity for VEGF. On the other hand, experiments using the VEGFR-1 extracellular domain as such, without fusion partner, showed that the region covering the first to third Ig domain (3 N Flt-1) bears a binding site of high affinity for both VEGF and PlGF, but that the region containing only the first to second Ig domain (2 N Flt-1) was dramatically reduced in its affinity for these ligands (BARLEON et al. 1997; TANAKA et al. 1997). These results suggest that the integrity of the structure carrying the second Ig domain is essential for the high affinity binding of VEGFR-1 to ligands, and that the surrounding sequences would support this structural integrity. The region between the fourth and seventh Ig domain appears to contribute to dimer formation of VEGFR-1 (BARLEON et al. 1997; TANAKA et al. 1997).

## 2.3 Interaction with Heparin

A heparin-binding site has been identified in the extracellular domain of VEGFR-2 (DOUGHER et al. 1997). Treatment of VEGFR-2-expressing cells with this sequence in the form of a synthetic peptide prevents binding of one of the VEGF isoforms, $VEGF_{165}$. This VEGFR-2-derived peptide showed high affinity for immobilized heparin. The full length and soluble form (see Sect. 4.1 below) of VEGFR-1, including shorter forms artificially synthesised in the baculovirus system or in other gene-expression systems, also show affinity for heparin (COHEN et al. 1995). Thus, soluble VEGFR-1 and the full-length VEGFR-1 and VEGFR-2 appear to be associated with heparan-sulphates or heparin-like molecules on the cell surface, which has implications for binding of the growth factor. Furthermore, the soluble VEGFR-1 may be retained by the extracellular matrix after having been secreted. Data indicating that binding of VEGF to the two VEGF receptors are differentially regulated by heparin have been presented; binding of VEGF to VEGFR-1 was inhibited by heparin, whereas binding to VEGFR-2 was stimulated by heparin (TERMAN et al. 1994).

# 3 Tyrosine Kinase Activity

## 3.1 Induction of Kinase Activity and Autophosphorylation of VEGFR-1 and -2

VEGFR-1 and -2 are equipped with tyrosine-kinase domains that contain all the known hallmarks of functional tyrosine kinases, including tentative ATP-binding sites. Binding of VEGF to truncated extracellular VEGFR-1 domains leads to dimerization of receptor domains (BARLEON et al. 1997; TANAKA et al. 1997) and it is conceivable that dimerization of intact VEGF receptors is critical for induction of kinase activity and transphosphorylation of receptor molecules, according to the general and well-established model for activation of receptor tyrosine kinases (for a review, see ULLRICH and SCHLESSINGER 1990). It is, therefore, unexpected, that VEGFR-1 is difficult to identify as an active kinase.

It appears that the intrinsic kinase activity of VEGFR-1 is at least one order of magnitude weaker than that of other Fms family members (SAWANO et al. 1996). However, under certain artificial conditions, VEGFR-1 autophosphorylation can be relatively easily detected. One such condition is baculovirus-mediated expression of VEGFR-1 in insect cells. Since a point mutation at the ATP-binding site completely abolishes tyrosine phosphorylation of VEGFR-1, it seems unlikely that this phosphorylation reaction is mediated by tyrosine kinases other than VEGFR-1 in the insect cells (SAWANO et al. 1996). In addition, a point mutation within the VEGFR-1 intracellular domain was found to activate the endogenous tyrosine kinase activity of VEGFR-1 (MARU et al. 1998). VEGFR-2 is more readily identified as a phosphorylated species in VEGF-stimulated cells. However, in porcine aortic endothelial cells expressing similar levels of either PDGFR-β or VEGFR-2, the PDGF receptors are tyrosine phosphorylated at considerably higher levels than VEGFR-2, in response to stimulation with the appropriate factor (CLAESSON-WELSH, unpublished observations). It is possible that the VEGF receptors form unstable dimers, are tightly regulated by tyrosine phosphatases, or carry few tyrosine phosphorylation sites.

This far, limited information is available on the position and number of tyrosine phosphorylation sites in VEGFR-1 and -2. Mapping of tyrosine phosphorylation sites has been initiated for the VEGFR-1, expressed in insect cells, which has allowed assignment of tyrosine residues 1169 and 1213 as phosphorylation sites (SAWANO et al. 1997). ITO et al. (1998) have used a similar approach to identify tyrosine residues 1213, 1242, 1327 and 1333 as phosphorylation sites in VEGFR-1. DOUGHER-VERMAZEN et al. (1994) examined tyrosine phosphorylation sites in bacterially expressed VEGFR-2 and identified phosphorylation on tyrosine residues 951, 996, 1054 and 1059 (Fig. 4).

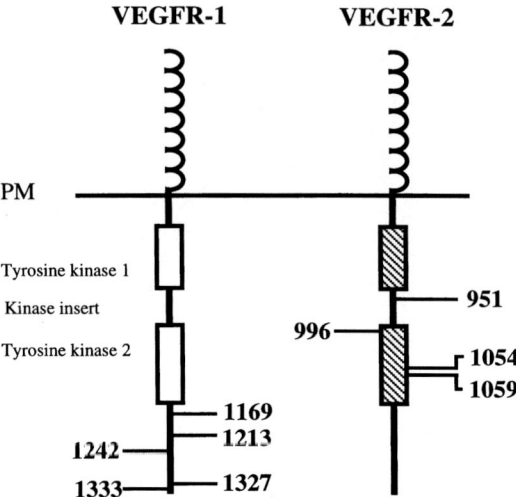

**Fig. 4.** Position of tyrosine phosphorylation sites so far described in VEGFR-1 and VEGFR-2. For references, see text

## 3.2 Signal Transduction by VEGFR-1 and -2

Signal transduction downstream of receptor tyrosine kinases depend on binding of Src Homology 2 (SH2)-domain containing signal transduction molecules to tyrosine phosphorylation sites on the receptors. The SH2 domain usually recognises a 4–7 amino-acid sequence, composed of one phosphotyrosine residue and 3–6 C-terminal amino-acid residues (for a review, see PAWSON 1995). The VEGFR-1 sequence, Tyr (794)-Leu-Ser-Ile, located in the juxtamembrane domain, and the sequences Tyr(1213)-Val-Asn-Ala and Tyr(1327)-Asn-Ser-Val, located in the C-terminal region, are examples of such conserved potential phosphotyrosine-containing sequences in the VEGFR-1 molecule (FINNERTY et al. 1993; YAMANE et al. 1994). These tyrosines, with the exception of Tyr(1327), are also present at similar positions in VEGFR-2, suggesting that VEGFR-1 and -2 might utilise these tyrosines as autophosphorylation sites and for common signal-transduction pathways under certain conditions.

The dissection of signalling events downstream of the activated VEGFR-1 and -2 have been hampered by the weak kinase activity of the receptors. A number of well-characterised signal-transduction molecules have, however, been examined for

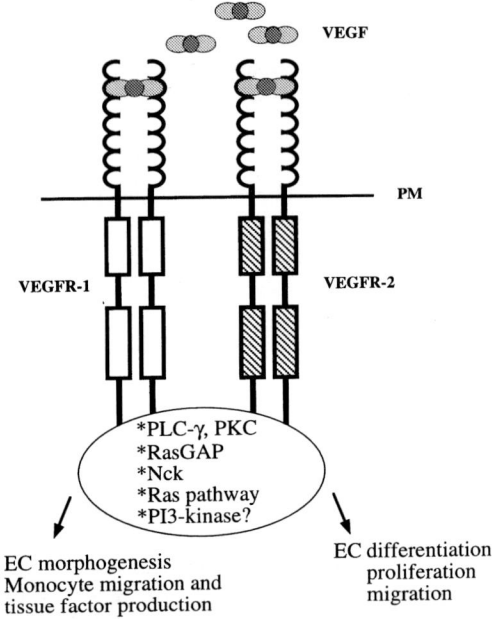

**Fig. 5.** Signal transduction downstream of VEGFR-1 and -2. Binding of VEGF family members, involving the second immunoglobulin (Ig)-like loop of the receptors, leads to dimerization and activation of receptors. The autophoshorylated receptors (see Fig. 4) can recruit signal transduction molecules, such as phospholipase C-gamma (PLC-γ)/protein kinase C (PKC), Nck, members of the Ras pathway and phosphatidyl inositol 3 (PI3)-kinase. For details, see the text. Initiation of signal transduction downstream of the receptors leads to endothelial cell (*EC*) morphogenesis, monocyte migration and tissue factor production (via VEGFR-1) and EC differentiation, proliferation and migration (via VEGFR-2)

their role in VEGF responses (Fig. 5). Thus, both receptor types appear to be able to mediate tyrosine phosphorylation of and complex-formation with phospholipase C-γ (PLC-γ). NIH 3T3 cells expressing VEGFR-1 or VEGFR-2 (SEETHARAM et al. 1995; TAKAHASHI and SHIBUYA, 1997), porcine aortic endothelial cells expressing VEGFR-2 (LANDGREN et al. 1998) and bovine aortic endothelial cells co-expressing both receptors (GUO et al. 1995) respond to VEGF stimulation with tyrosine phosphorylation of PLC-γ. In agreement, Xenopus oocytes microinjected with human VEGFR-1 messenger ribonucleic acid (mRNA) showed an increase in calcium efflux in response to VEGF, although VEGFR-1 autophosphorylation was close to undetectable (DE VRIES et al. 1992). Moreover, the GTPase-activating protein of Ras (RasGAP) is, to some extent, tyrosine phosphorylated as a consequence of activation either of VEGFR-1 and -2 (GUO et al. 1995; KROLL and WALTENBERGER 1997; SEETHARAM et al. 1995), although it remains to be shown that tyrosine phosphorylation of RasGAP is coupled to its activation.

Another signal transduction molecule that is implicated downstream of at least VEGFR-2 is the small adaptor protein Nck (GUO et al. 1995; KROLL and WALTENBERGER 1998). Recently, pathways downstream of protein kinase C (PKC) were suggested to be critically involved in VEGF signalling towards DNA synthesis in endothelial cells (XIA et al. 1996; TAKAHASHI and SHIBUYA, unpublished). The roles of PLC-γ, PKC, RasGAP and Nck in VEGF-induced biological responses remain to be clarified.

In general, the Ras pathway is utilized by most, if not all, growth stimulatory factors and there are several possible routes from an activated growth-factor receptor to Ras activation (PAWSON 1995). One such pathway is through direct binding of the small adaptor molecule Grb2 to the receptor. Both VEGFR-1 (Tyr 1213) and VEGFR-2 (Tyr 1214) are equipped with potential Grb2-binding sites (Yp-X-N; SONGYANG et al. 1993; MATTHEWS et al. 1991a; SHIBUYA et al. 1990; TERMAN et al. 1991). Another potential route from the receptors to Ras activation is through the small adaptor molecule Shc (PELICCI et al. 1992); VEGFR-2 has been shown to mediate tyrosine phosphorylation of Shc (KROLL and WALTENBERGER 1998). However, the involvement of Ras in VEGFR signaling is not yet well established.

Ras activation leads to a cascade of serine phosphorylation mediated by the Raf kinase, Mitogen-activated protein (MAP) kinase kinase (also denoted MEK) and MAP kinase (PAWSON 1995). Activation of MEK and MAP kinase can also occur independent of Ras, e.g. via PKC. It is by now well established that both VEGFR-1 and -2 mediate activation of the MAP kinases denoted Erk1 and Erk2 (KROLL and WALTENBERGER 1998; LANDGREN et al. 1998; SEETHARAM et al. 1995; TAKAHASHI and SHIBUYA 1997). SEETHARAM et al. (1995) compared activation of Erk1 and Erk2 by VEGFR-1 expressed in NIH 3T3 and sinusoidal endothelial cells and found a more efficient activation in the endothelial cells endogenously expressing VEGFR-1 and -2. ROUSSEAU et al. (1997) examined VEGF-induced activation of other members of the MAP-kinase family and found that VEGF treatment of human umbilical vein endothelial (HUVE) cells induced the stress-induced kinase p38, but not the c-Jun NH2-terminal kinase, JNK. Drug-treatment of the cells to specifically inhibit p38 led to a reduced actin reorganization and migration towards VEGF.

As described above, one characteristic of the VEGFR-1 and -2 intracellular domains is the presence of a long kinase-insert region (Fig. 4). In members of the Fms family, this region is known to contain several phosphorylation sites that present binding sites for signal transduction molecules, such as the p85 regulatory subunit of phosphatidylinositol 3-kinase (PI3-kinase). The kinase insert domains of VEGFR-1 and VEGFR-2 lack the Y(p)-X-X-M sequence for binding of p85 (SONGYANG et al. 1993); instead, a potential site is present within the tyrosine-kinase domain of each receptor (MATTHEWS et al. 1991a; MILLAUER et al. 1993; TERMAN et al. 1991, 1992). Using the yeast two-hybrid system, CUNNINGHAM et al. (1995) showed an association between the PI3-kinase p85 subunit and the phosphorylated tyrosine 1213 of the VEGFR-1 C-terminal region. Since the Tyr-Val-Asn-Ala sequence starting at amino-acid 1213 is not a typical binding motif for the p85 subunit, and since phosphorylation of Tyr (1213) has not been confirmed in vivo in endothelial cells, the strength of involvement of the PI3-kinase pathway in VEGFR-1-mediated signal transduction in endothelial cells and macrophages remains open to question.

Certain nuclear events in VEGF-stimulated cells have been identified. VEGF and members of the fibroblast growth factor (FGF) family induce a fivefold increase in mRNA and protein levels of the transcription factor Ets-1, which has been shown to regulate angiogenesis through induction of expression of urokinase-type plasminogen activator (IWASAKA et al. 1996). The molecular mechanisms underlying the effects of VEGF on Ets transcript levels are as yet unknown.

## 4 Regulation of Expression of VEGFR-1 and -2

### 4.1 Messenger Ribonucleic Acid (mRNA) Products and Biosynthesis of VEGFR-1 and -2

VEGFR-2 is encoded by a single transcript of at least 5.5 kb (MILLAUER et al. 1993). The VEGFR-1 transcripts are made up of several species of mRNA. An approximate 8.0-kb mRNA species encodes the full-length VEGFR-1, which is expressed in most normal tissues. In addition to the 8.0-kb mRNA, 3.0-kb and 2.2-kb mRNAs carrying the VEGFR-1 extracellular domain are abundantly expressed in the placenta and in several cell lines such as that of 293 E (SHIBUYA et al. 1990). The major species of these short VEGFR-1 mRNAs encodes the $NH_2$-terminal 656 amino-acid residues of the human VEGFR-1 (657 amino acids in the case of murine VEGFR-1), which consists of the extracellular domain fused to a 31 amino-acid sequence derived from the 5′ part of intron 13, as a result of premature termination and polyadenylation in this intron (KENDALL and THOMAS 1993; KONDO et al. 1998). The 31 amino-acid sequence derived from the 5′-untranslated region of intron 13 is highly homologous between man and mouse (KONDO et al.

1998). Based on the absence of the transmembrane domain, this variant short form of VEGFR-1 has been designated "soluble VEGFR-1" (Fig. 1).

Recently KENDALL et al. (1996) showed that the soluble VEGFR-1 protein does exist in vivo and is able to form heterodimers with full-length VEGFR-2. Soluble forms of VEGFR-2 have, thus far, not been identified. Some other variant species of VEGFR-1 are expected to be translated from the different species of VEGFR-1 short mRNAs, but they are likely to be minor populations among the short forms of VEGFR-1 mRNA (Subbalakshmi and Shibuya, unpublished observations).

The intracellular processing of VEGFR-1 and -2 is very similar (TAKAHASHI and SHIBUYA 1997). After synthesis of a nascent VEGFR-1 of 150 kDa, this protein is glycosylated to become an intermediate form of 170 kDa. This 170-kDa form of VEGFR-1 is further glycosylated to become the mature form of 180-kDa receptor, which is expressed on the cell surface. Autophosphorylation of VEGFR-1 occurs only on the 180-kDa, mature form, although this phosphorylation is relatively weak (SAWANO et al. 1996; SEETHARAM et al. 1995). VEGFR-2 is expressed on the cell surface as a 200-kDa to 220-kDa glycoprotein. The apparent molecular weight of VEGFR-2 appears to differ between different cell types, perhaps depending on different degrees of glycosylation of the receptor (Landgren and Claesson-Welsh, unpublished observations).

## 4.2 Chromosomal Localisation and Genomic Structure

The vegfr-1 gene is localised on human chromosome 13q12–13 (SATOH et al. 1987). This chromosomal region also carries a gene named flk-2/flt-3, which is another member of the Fms family, isolated in 1991 (MATTHEWS et al. 1991b; ROSNET et al. 1991). These two genes are located close to each other and are separated by DNA sequences of only about 600 kb (ROSNET et al. 1993). The vegfr-2 gene, however, is located in a gene cluster on the human chromosome 4q11–12. This cluster also contains the genes encoding the PDGF receptor-α and c-Kit (SAIT et al. 1995; SPRITZ et al. 1994). Interestingly, the human chromosome 5q33-q35 carries still another cluster of genes encoding members of the VEGFR-Fms family; VEGFR-3 (Flt-4), PDGFR-β and the CSF-1 receptor (c-fms). Therefore, a putative ancestral gene for the VEGFR-Fms family may have generated a set of genes containing one vegfr-family gene and two fms-family genes as a result of a *cis* gene duplication. Following this initial event, this set of three genes might have been triplicated into three clusters on different chromosomes in a *trans*-acting manner (Fig. 6).

The vegfr-1 gene in mice consists of 30 exons distributed over more than 140 kb of chromosomal DNA (KONDO et al. 1998). A clear difference in the genomic organization between the vegfr-1 gene and the fms family genes was found in the region corresponding to the first and the second Ig domains of the extracellular domain. Each of these Ig domains in the fms family is encoded by a single exon, whereas each of the corresponding domains in the vegfr-1 gene is encoded by two exons. The organization of the vegfr-2 gene has not yet been described.

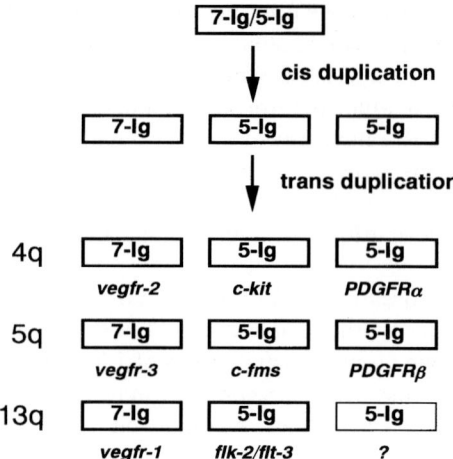

**Fig. 6.** Possible models for phylogenetic development of 7-immunoglobulin (Ig) (VEGF receptors) and 5-Ig (Fms family) receptor genes. (ROSNET et al. 1993)

## 4.3 Promoter/Enhancer Regions of the vegfr-1 and -2 Genes

VEGF and its receptors have specific functions during vasculogenesis and angiogenesis, as demonstrated by the phenotypes of mice with targeted inactivation of the vegf and vegf receptor genes (see Sect. 5.1 below and CARMELIET/COLLEN this volume). The critical function of VEGF and its receptors in the vasculature is, furthermore, underscored by the almost exclusive expression of VEGFR-1 and -2 on vascular endothelial cells and the possible presence of specific transduction pathway(s) in endothelial cells coupling to these receptors.

The promoter/enhancer region of the human vegfr-1 gene was examined by use of the chloramphenicol acetyltransferase assay, with or without the introduction of point mutations (IKEDA et al. 1996; MORISHITA et al. 1995; WAKIYA et al. 1996). A single transcription initiation site was identified 281 bp upstream of the initiation codon ATG (A as +1) and 29 bp downstream of a TATA box. In a region of about 200 bp (−200 to +1), which is essential for significant gene expression of vegfr-1, four typical ETS motifs and one CRE (cyclic AMP response element)/ATF (activating transcription factor) motif were identified. Point mutation analysis revealed that the fourth ETS motif at −54 to −51 and the CRE/ATF at −83 to −76 positively regulate the vegfr-1 gene expression in a co-operative manner. These findings are consistent with the report by WERNERT et al. (1992), showing that expression of the transcription factor Ets-1 is upregulated in vascular endothelial cells under conditions of embryonal and pathological angiogenesis (WERNERT et al. 1992).

The promoter region of the vegfr-2 gene has been characterised both in human and mouse. In the human promoter (PATTERSON et al. 1995), the transcription initiation site was identified to a nucleotide 303 bp upstream of the initiating methionine. Five Sp1 elements were identified, but there was no TATA consensus sequence. An approximate 4-kb sequence of the vegfr-2 5′-flanking region mediated high levels of transcriptional activity in endothelial cells, but not in other cell types. A number of positive regulatory sites were identified within this 4-kb region. Characterisation of the murine vegfr-2 promoter (RONICKE et al. 1996) revealed an endothelium-specific positive regulatory element in the 5'-untranslated region of the first exon of the vegfr-2 gene. In addition, two endothelium-specific negative regions were identified between nucleotides −4100 and −623 and generally activating elements in the region between nucleotides −96 and −37.

## 4.4 Regulation of VEGFR-1 and -2 Expression

Using isotope-labelled VEGF, JAKEMAN et al. (1992, 1993) examined the localisation of VEGF-binding sites in various tissues of adult and embryonic rats, and showed that expression of the VEGF receptors is highly specific for vascular endothelial cells located throughout the body. VEGF binding sites detected with $^{125}$I-VEGF could be presented by either of VEGFR-1 or -2. More specific analyses, using in situ hybridization clearly showed that both VEGFR-1 and -2 mRNA are expressed in an endothelial cell-specific manner in most, if not all, of the tissues in mouse and human embryos (PETERS et al. 1993; KAIPAINEN et al. 1993).

During the early stages of the embryonal development, VEGFR-2 is first expressed in presumptive mesodermal yolk-sac blood-island progenitors at 7 days postcoitum and, then, in the primitive endothelial cells surrounding the blood islands (SHALABY et al. 1995). This is in contrast to VEGFR-1, which is not expressed in the mesodermal area at day 7, but is expressed in the primitive endothelial cells outside the blood islands at 8.5 days and in vascular endothelial cells of the embryo (FONG et al. 1995; PETERS et al. 1993). Both VEGFR-1 and -2 are expressed at relatively high levels in vascular endothelial cells at this stage of embryonic development. This is followed by a significant decrease in expression of VEGFR-1 during the later stages at 14.5–16.5 days (KAIPAINEN et al. 1993; PETERS et al. 1993). The expression of VEGFR-1 is upregulated in newborn mice and this receptor continues to be expressed at a relatively high level during the adult stages (PETERS et al. 1993; YAMANE et al. 1994). VEGFR-2 is expressed in endothelial cells during embryonal development after day 7, but the expression level declines towards the end of gestation (MILLAUER et al. 1993). Thus, the receptors are differentially regulated and the differences in their expression levels during embryonal development reinforces the notion that they have different functions during vasculogenesis and angiogenesis.

Whether all endothelial cells in different organs express VEGFR-1 and -2 in the adult stage is relatively unexplored. In some organs, such as the brain, it has been shown that endothelial cells lack both VEGFR-1 and -2 expression (MILLAUER et al. 1993). Furthermore, it appears that the expression levels of the VEGF receptors become downregulated in quiescent endothelial cells (MILLAUER et al. 1993). Under conditions of pathological angiogenesis, such as tumour angiogenesis, VEGFR-1 and VEGFR-2 mRNA are significantly upregulated compared with the levels of these mRNAs in the vascular endothelial cells of the normal tissues surrounding the tumours (BOOCOCK et al. 1995; BROWN et al. 1993; HATVA et al. 1996; PLATE et al. 1992; WIZIGMANN-VOOS et al. 1995; YOSHIJI et al. 1996). A similar pattern appears to exist in non-tumorigenic diseases; SATO et al. (1995) reported upregulation of VEGFR-1 and -2 in hyperthyroidism, which is a disease characterised by increased angiogenesis.

It is well established that VEGF mRNA is strongly induced during hypoxic conditions (SHWEIKI et al. 1992). Interestingly, the vegfr-1 gene has been reported to be transiently expressed in pericytes under hypoxic culture conditions (NOMURA et al. 1995). Induction of VEGFR-1 mRNA has also been demonstrated in hypoxic microvessels in skin explant cultures (DETMAR et al. 1997). Under these conditions, VEGFR-2 mRNA was not induced, in agreement with the fact that the receptors are differentially regulated. Similar results, showing upregulation of VEGFR-1, but not VEGFR-2, under hypoxic conditions were demonstrated for HUVE cells (GERBER et al. 1997), and a region in the vegfr-1 promoter involved in binding the hypoxia-inducible transcription factor was identified. In endothelial cells from postnatal brain, hypoxia does induce expression of VEGFR-2 (KREMER et al. 1997), which could be an indirect effect due to a stimulation by the increased levels of VEGF, rather than the hypoxic atmosphere per se.

Expression of the VEGFR-1 and -2 have been demonstrated in certain non-endothelial cells. Thus, monocytes express VEGFR-1 and migrate towards VEGF (CLAUSS et al. 1996). VEGF also induces activation of mononuclear phagocytes, based on expression of tissue factor pro-coagulant activity (CLAUSS et al. 1990; SHEN et al. 1993). Pancreatic duct cells express VEGFR-2 (ÖBERG et al. 1994). Both receptor types appear to be expressed in normal human testicular tissue (ERGUN et al. 1997).

The expression patterns of the VEGF receptors and their ligand VEGF in the liver during the adult stage in rats are demonstrated in Fig. 7. The transcripts of both VEGFR-1 and -2 are specifically expressed at significant levels on the sinusoidal endothelial cell fraction, whereas VEGF transcripts are detectable only in the parenchymal cell (hepatocyte) fraction. These expression patterns indicate that, under normal conditions, VEGF and its receptor system are basically used in a paracrine manner.

The data outlined above indicate three major characteristics of VEGFR-1 and -2 expression: (i) endothelial cell-specific expression with certain exceptions; (ii) regulated expression during embryogenesis; and (iii) upregulation in pathological angiogenesis.

## 5 Role of VEGFR-1 and -2 in Endothelial Cell Function

### 5.1 VEGF Receptors in Embryonal Development

Targeted inactivation of the vegfr-1 and -2 genes revealed unique and interesting properties for these receptors in vasculogenesis during embryonic development (FONG et al. 1995; SHALABY et al. 1995) (Fig. 8). To generate the VEGFR-1 null

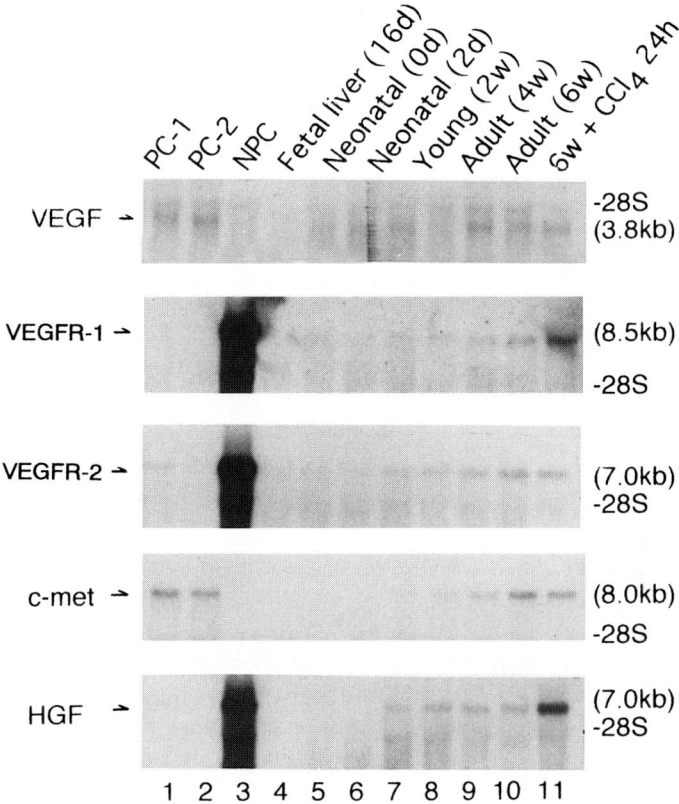

**Fig. 7.** Differential expression of vascular endothelial growth factor (VEGF), VEGFR-1 and VEGFR-2 genes in parenchymal and non-parenchymal cells of rat liver. *PC* parenchymal cells (more than 95% are hepatocytes); *NPC* non-parenchymal cells (about 85% are sinusoidak endothelial cells. Other cell types are macrophage-like Kupffer cells, epithelial cells and Ito cells; YAMANE et al. 1994)

mice, the first exon of the mouse vegfr-1 gene, which carries the initiation codon in addition to the downstream sequence for the signal peptide, was replaced by the *lacZ* gene, so as to completely disrupt vegfr-1 gene expression. The vegfr-1 (+/−) heterozygote mice were essentially normal compared with the wild-type mice; however, the vegfr-1 (−/−) homozygotes did not survive beyond 8.5–9.5 days after gestation due to disorganization of the vascular system. Histopathological analysis revealed that, although the cardiovascular system had been almost completely established within the embryo, monolayers of endothelial cells on a variety of vascular walls were not well organized and abnormal cells were found to have accumulated frequently within the blood vessels. These abnormal cells were positive for the cell surface markers of endothelial cells, such as vegfr-2 and PECAM-1, and also positive for the mRNAs of *vegfr-3*, Tie-1 and Tie-2. Therefore, these cells were most likely derived from endothelial cell lineages, but lacked important cues for the proper spatial organization into lumen-containing vessels.

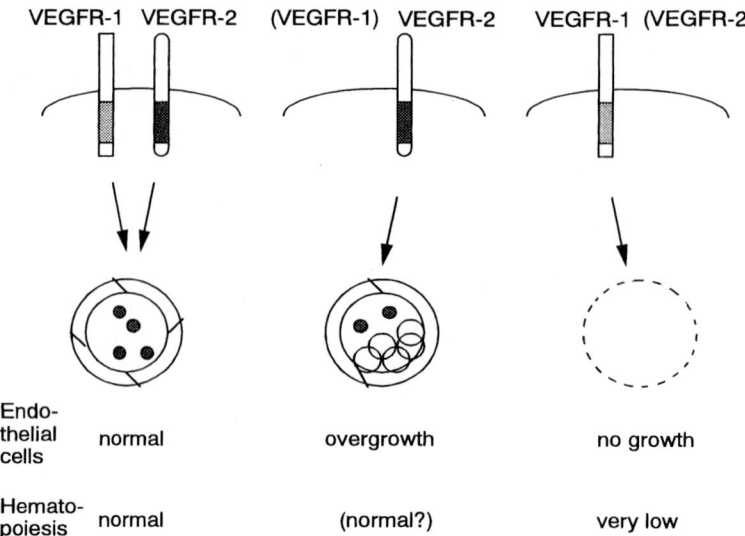

**Fig. 8.** Effect of targeted inactivation of genes encoding *vegfr*-1 and *vegfr*-2 in mouse (FONG et al. 1995; SHALABY et al. 1995)

The homozygous vegfr-2 knockout mice show a completely different phenotype (SHALABY et al. 1995). In these animals, differentiated endothelial cells are lacking and hematopoietic development is severely impaired. It is interesting that the phenotype of vegf (−/−) mice is very similar to that of the vegfr-2 knockout (CARMELIET et al. 1996; FERRARA et al. 1996; see FERRARA and CARMELIET/COLLEN this volume). Thus, it seems clear that VEGFR-2 is involved in the commitment to endothelial-cell lineages and to cell proliferation, whereas VEGFR-1 is responsible for guiding endothelial cells for the precise morphogenesis of blood vessels (Fig. 8). Therefore, these two structurally related VEGF receptors appear to play complementary roles during vasculogenesis at the early stages of embryogenesis.

## 5.2 Biological Effects of VEGFR-1 and -2 on Endothelial Cells in Vivo and in Vitro

### 5.2.1 Cell Proliferation

Essentially all endothelial cells in tissue culture will respond to VEGF stimulation with increased DNA synthesis. Some endothelial cells will also increase in cell number when stimulated with VEGF. It is probable that the growth stimulatory signals are transduced, to a major extent, via VEGFR-2. Thus, VEGF treatment of NIH 3T3 cells over-expressing VEGFR-1 led to only weak activation of Erk1/2 (SEETHARAM et al. 1995) and no induction of DNA synthesis. It therefore appears that when ectopically expressed in cell types other than capillary endothelial cells, VEGFR-1 fails to mediate increased DNA synthesis, whereas cells ectopically expressing VEGFR-2 are able to increase their DNA synthesis when stimulated with VEGF (WALTENBERGER et al. 1994; SEETHARAM et al. 1995). To properly judge the contribution of the two receptors to endothelial cell function, it may be important to first consider the possibility that endothelial cells may harbour specifically expressed signal-transduction components, which operate downstream of the VEGF receptors. Second, since both receptors bind VEGF with high affinity, it is possible that co-expression of VEGFR-1 and -2 in endothelial cells leads to formation of heterodimers in response to VEGF. Heterodimers may respond differently than homodimeric receptor complexes to VEGF treatment in biological assays, such as DNA synthesis.

### 5.2.2 Migration

It is well established that endothelial cells are able to migrate directionally towards VEGF. Porcine aortic endothelial cells, individually expressing VEGFR-1 or -2 after transfection, differed in their abilities to migrate towards VEGF (WALTENBERGER et al. 1994). The VEGFR-1-expressing cells failed to migrate, whereas the VEGFR-2-expressing cells migrated efficiently with a maximal response at 10 ng/ml VEGF. It is, thus, likely that VEGFR-2 expressed in endothelial cells mediate chemotaxis. However, monocytes endogenously expressing VEGFR-1, but not VEGFR-2, will migrate directionally towards VEGF and PlGF (see Sect. 5.3 and BARLEON et al. 1996; CLAUSS et al. 1996).

### 5.2.3 Vascular Permeability

VEGF/VPF is a potent inducer of vascular permeability (see DVORAK et al. this volume). In the Miles assay, PlGF-1 and PlGF-2 showed 20 to 40-fold less activity (PARK et al. 1994; SAWANO et al. 1996). These results indicate that the activation of VEGFR-1 is not sufficient or at least not directly involved in the regulation of vascular permeability, and it is likely that VEGFR-2 alone can mediate this response.

### 5.2.4 Induction of Protease Activity

For formation of new vessels, the basement membrane has to be degraded to allow migration of endothelial cells into the surrounding tissue. VEGF has been shown to induce plasminogen activators (tPA and uPA; PEPPER et al. 1991). Using the porcine aortic endothelial cells individually expressing VEGFR-1 and -2 as a model system, VEGF and PlGF were shown to stimulate uPA production in VEGFR-1-expressing cells, but not in VEGFR-2-expressing cells (LANDGREN et al. 1998). Further studies are necessary to rule out the possibility that the results represent cell-specific phenomena.

### 5.2.5 Angiogenesis in Tissue Culture

Some endothelial cell types in tissue culture are able to form vessel-like structures in three-dimensional collagen gels, when appropriately stimulated. It is known that FGF-2 is a potent inducer of vessel-like structures in vitro (KANDA et al. 1996; MONTESANO et al. 1986) and growth factor-induced tube formation in three-dimensional matrixes has been employed as a model for endothelial cell differentiation. PEPPER et al. (1992) have shown that VEGF treatment will also induce the vessel-like structures; it is noteworthy that there is a synergistic effect of the two growth factors, indicating that different signal transduction pathways are engaged by the FGF-2 and VEGF receptors. It is likely that VEGFR-2 is responsible for the cellular response to VEGF; at least, PlGF fails to induce tube formation of a murine brain endothelial cell line, which responds to FGF-2 with tube formation in three-dimensional collagen gels (LANDGREN et al. 1998).

## 5.3 Cell Migration of Monocytes and Macrophages

Although the VEGFR-1 gene is preferentially expressed on vascular endothelial cells, human peripheral blood monocytes also express this gene at relatively high levels (BARLEON et al. 1996; CLAUSS et al. 1996). Cell migration and tissue factor production are induced by treatment of these monocytes with VEGF. The two biological responses appear to be mediated through VEGFR-1, since the VEGFR-1-specific ligand, PlGF, was able to induce these responses almost to the same extent as VEGF. In addition, an antibody against VEGFR-2, which blocks VEGF binding to VEGFR-2, failed to inhibit these processes. These findings represent the first case of transduction of intracellular signals by endogenously expressed VEGFR-1 (Fig. 9).

VEGF is known to be induced and secreted from a variety of cells under conditions of hypoxia, hypoglycaemia, inflammation and malignancy. Macrophages, lymphocytes and polymorphonuclear leucocytes migrate and infiltrate pathological tissues, such as those resulting from inflammation, atherosclerosis and cancer. Therefore, it seems likely that a portion of macrophage migration in vivo is mediated by the VEGF/VEGFR-1 system. Since a number of cytokines are also known to stimulate macrophage migration, it seems important to measure the extent of VEGF/VEGFR-1 involvement in total macrophage migration under each condition, and to determine the unique features of VEGF-induced macrophage migration, if any, with respect to other cytokine-induced motility.

## 5.4 VEGF Receptors and Anti-angiogenic Therapy

The establishment or expansion of a number of diseases, such as cancer, rheumatoid arthritis and diabetic retinopathy depend on deregulated angiogenesis (see FOLKMAN 1995). The critical role for the VEGF receptors in endothelial cell

|  | Expression of receptors | | VEGF treatment | |
|---|---|---|---|---|
|  | VEGFR-1 | VEGFR-2 | migration | TF production |
| Human peripheral Monocyte | + | - | + (+) | + (+) |
| HUVEC | + | + | ++ (-) | ++ (±) |

( ); results in the presence of VEGFR-2-blocking antibody

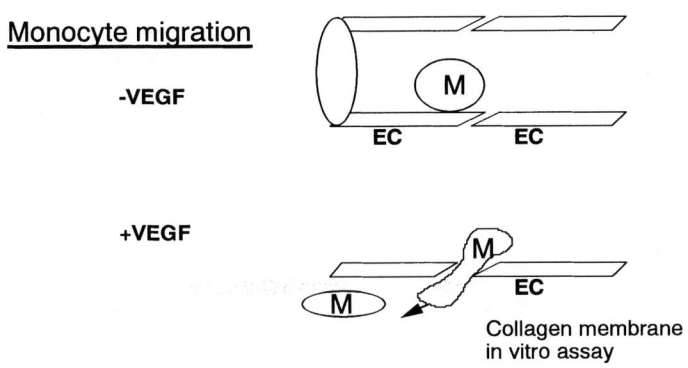

**Fig. 9.** Involvement of vascular endothelial growth factor receptor-1 (VEGFR-1) in cell migration and gene expression of monocytes/macrophages (CLAUSS et al. 1996). TF, tissue factor

function make them ideal drug targets for treatment of deregulated angiogenesis. Inhibition of receptor function may be achieved by different strategies. One possibility is to treat cells with ATP analogues that will, more or less, specifically interact with the VEGF receptors by binding to the ATP-binding lysine in the first kinase domain. Such drugs do not yet exist on the market, but it is likely that their therapeutic potential will encourage intense research for this purpose.

Alternatively, VEGF receptor function can be blocked by expression of truncated soluble receptors to create a dominant-negative inhibition of receptor function. It is noteworthy that soluble VEGFR-1 exists in vivo and may have a role in fine tuning of the biological effects of VEGF. Soluble VEGFR-1 bears strong affinity for VEGF at a level almost equal to that of the native VEGFR-1. Human soluble VEGFR-1 has already been demonstrated to have efficient inhibitory effect in pathological angiogenesis in both an experimental solid-tumour system and in hypoxia-induced retinopathy in mice, where it results in the suppression of abnormal angiogenesis in vivo (AIELLO et al. 1995; KENDALL and THOMAS 1993).

Furthermore, VEGFR-2 lacking the intracellular domain was introduced in a rat glioblastoma model through retroviral-mediated production in situ (MILLAUER et al. 1994). Compared with the control, the glioblastomas treated with dominant-negative VEGFR-2 grew less efficiently (MILLAUER et al. 1994). A wide range of solid tumours in rats were later shown to be inhibited in their growth by treatment with dominant-negative VEGFR-2 (MILLAUER et al. 1996). Thus, dominant-negative VEGFR-1 or VEGFR-2 may be useful tools in inhibiting VEGF activity through sequestration during pathological angiogenesis and may help to reduce the increased vascular permeability induced by overexpression of VEGF.

# 6 Conclusions

Based on the information presented above, it appears that VEGFR-1 and -2 have different roles during vasculogenesis and angiogenesis. Moreover, most of the biological effects that can be induced by VEGF on cultured cells are mediated by VEGFR-2. Why are the effects of VEGF binding to VEGFR-1 so limited? Three models can be outlined:

1. The limited kinase activity and weak signal transduction of VEGFR-1 is important for both morphogenesis of vessels during vasculogenesis and angiogenesis and VEGFR-1-mediated macrophage migration
2. Under as yet unidentified conditions, the VEGFR-1 tyrosine kinase is strongly activated and elicits a variety of signals, including those leading to proliferation of endothelial cells
3. The tyrosine kinase activity of VEGFR-1 is not required and is dispensable for its biological role in vascular formation.

These models are not necessarily mutually exclusive and certain elements may function together under different conditions. Recent studies suggest that model 3 is relevant in the case of mouse embryogenesis (HIRATSUKA et al., 1998)

Future efforts will involve further delineation of the signal transduction pathways downstream of VEGFR-1 and -2 and will show whether there are unique signal transduction molecules in endothelial cells. With increased understanding of the consequences of the different signal transduction pathways in endothelial cell responses, it is likely that suitable drug targets for anti-angiogenesis therapy will be found. The interplay between different receptor tyrosine kinases on endothelial cells, such as the VEGF receptors and the angiopoietin/Tie receptors is considered important for the development and maintenance of the vascular system (MAISONPIERRE et al. 1997 and PARTANEN/DUMONT and DAVIS/YANCOPOULOS this volume) and will be the focus of intense studies.

Another interesting aspect of vascular development and maintenance is the interplay among different cell types in and around the vessels, such as endothelial cells, monocytes, pericytes and smooth muscle cells, which is likely to involve a multitude of paracrine loops of different growth factors. Thus, although our knowledge of the VEGF/VEGF receptor system has developed at an immense rate during the last few years, a number of key questions remain to be answered.

# References

Aliello LP, Pierce EA, Foley ED, Takagi H, Chen H, Riddle L, Ferrara N, King GL, Smith LE (1995) Suppression of retinal neovascularization in vivo by inhibition of vascular endothelial growth factor (VEGF) using soluble VEGF-receptor chimeric proteins. Proc Natl Acad Sci USA 92:10457–10461

Barleon B, Sozzani S, Zhou D, Weich HA, Martovani A, Marme D (1996) Migration of human monocytes in response to Vascular Endothelilal Growth Factor (VEGF) is mediated via the VEGF receptor flt-1. Blood 87:3336–3343

Barleon B, Totzke F, Herzog C, Blanke S, Kremmer E, Siemeister G, Marme D, Martiny-Baron G (1997) Mapping of the sites for ligand binding and receptor dimerization at the extracellular domain of the vascular endothelial growth factor receptor FLT-1. J Biol Chem 272:10382–10388

Blechman JM, Lev S, Barg J, Eisenstein M, Vaks B, Vogel Z, Givol D, Yarden Y (1995) The fourth immunoglobulin domain of the Stem Cell Factor receptor couples ligand binding to signal transduction. Cell 80:103–113

Boocock CA, Charnock-Jones DS, Sharkey AM, McLaren J, Varker PJ, Wright KA, Twentyman PR, Smith SK (1995) Expression of vascular endothelial growth factor and its receptors flt and KDR in ovarian carcinoma. J Natl Cancer Inst 87:506–516

Brown LF, Berse B, Jackman RW, Tognazzi K, Manseau EJ, Dvorak HF, Senger DR (1993) Increased expression of vascular permeability factor (vascular endothelial growth factor) and its receptors in kidney and bladder carcinomas. Am J Pathol 143:1255–1262

Carmeliet P, Ferreira V, Breier G, Pollefeyt S, Kieckens L, Gertsenstein M, Fahrig M, Vandenhoeck A, Harpal K, Eberhardt C, Declercq C, Pawling J, Moons L, Collen D, Risau W, Nagy A (1996) Abnormal blood vessel development and lethality in embryos lacking a single VEGF allele. Nature 380:435–439

Clauss M, Gerlach M, Gerlach H, Brett J, Wang F, Familletti PC, Pan YC, Olander JV, Connolly DT, Stern D (1990) Vascular permeability factor: a tumor-derived polypeptide that induces endothelial cell and monocyte procoagulant activity, and promotes monocyte migration. J Exp Med 172:1535–1545

Clauss M, Weicht H, Breier G, Knies U, Rockl W, Waltenberger J, Risau W (1996) The vascular endothelial growth factor receptor Flt-1 mediates biological activities. J Biol Chem 271:17629–17634

Cohen T, Gitay-Goren H, Sharon R, Shibuya M, Halaban R, Levi BZ, Neufeld G (1995) VEGF121, a VEGF isoform lacking heparin binding ability, requires cell surface heparin-sulfates for efficient binding to the VEGF receptors of human melanoma cells. J Biol Chem 270:11322–11326

Cunningham SA, Waxham MN, Arrate PM, Brock TA (1995) Interaction of the Flt-1 tyrosine kinase receptor with the p85 subunit of phosphatidylinositol 3-kinase. J Biol Chem 270:20254–20257

Davis-Smyth T, Chen H, Park J, Presta LG, Ferrara N (1996) The second immunoglobulin-like domain of the VEGF tyrosine kinase receptor Flt-1 determines ligand binding and may initiate a signal transduction cascade. EMBO J 15:4919–4927

Detmar M, Brown LF, Berse B, Jackman RW, Elicker BM, Dvorak HF, Claffey KP (1997) Hypoxia regulates the expression of vascular permeability factor/vascular endothelial growth factor (VPF/VEGF) and its receptors in human skin. J Invest Dermatol 108:263–268

De Vries C, Escobedo JA, Ueno H, Houck K, Ferrara N, Williams LT (1992) The fms-like tyrosine kinase, a receptor for vascular endothelial growth factor. Science 255:989–991

Dougher AM, Wasserstrom H, Torley L, Shridaran L, Westdock P, Hileman RE, Fromm JR, Anderberg R, Lyman S, Linhardt RJ, Kaplan J, Terman BI (1997) Identification of a heparin binding peptide on the extracellular domain of the KDR VEGF receptor. Growth Factors 4:257–268

Dougher-Vermazen M, Hulmes JD, Bohlen P, Terman BI (1994) Biological activity and phosphorylation sites of the bacterially expressed cytosolic domain of the KDR VEGF-receptor. Biochem Biophys Res Commun 205:728–738

Ergun S, Kilic N, Fiedler W, Mukhopadhyay AK (1997) Vascular endothelial growth factor and its receptors in normal human testicular tissue. Mol Cell Endocrinol 131:9–20

Ferrara N, Carver-Moore K, Chen H, Dowd M, Lu L, O'Shea KS, Powell-Braxton L, Hillan KJ, Moore MW (1996) Heterozygous embryonic lethality induced by targeted inactivation of the VEGF gene. Nature 380:439–442

Finnerty H, Kelleher K, Morris GE, Bean K, Merberg DM, Kriz R, Morris JC, Sookdeo H, Turner KJ, Wood CR (1993) Molecular cloning of murine FLT and FLT4. Oncogene 8:2293–2298

Folkman J (1995) Angiogenesis in cancer, vascular, rheumatoid and other disease. Nat Med 1:27–31
Fong GH, Rossant J, Gertsentein M, Breitman ML (1995) Role of the Flt-1 receptor tyrosine kinase in regulating the assembly of vascular endothelium. Nature 376:66–70
Gerber HP, Condorelli F, Park J, Ferrara N (1997) Differential transcriptional regulation of the two vascular endothelial growth factor receptor genes. Flt-1, but not Flk-1/KDR, is up-regulated by hypoxia. J Biol Chem 272:23659–23667
Guo D, Jia Q, Song HY, Warren RS, Donner DB (1995) Vascular endothelial cell growth factor promotes tyrosine phosphorylation of mediators of signal transduction that contain SH2 domains. Association with endothelial cell proliferation. J Biol Chem 270:6729–6733
Hatva E, Böhling T, Jääskeläinen J, Persico MG, Haltia M, Alitalo K (1996) Vascular growth factors and receptors in capillary hemangioblastomas and hemangiopericytomas. Am J Pathol 148:763–775
Hiratsuka S, Minowa O, Kuno J, Noda T, Shibuya M (1998) Flt-1 lacking the tyrosine kinase domain is sufficient for normal development and angiogenesis in mice. Proc. Natl. Acad. Sci. USA, in press
Ikeda T, Wakiya K, Shibuya M (1996) Characterization of the promoter region for flt-1 tyrosine kinase gene, a receptor for vascular endothelial growth factor. Growth Factors 13:151–162
Ito N, Wernstedt C, Engström U, Claesson-Welsh L (1998) Identification of VEGF receptor-1 tyrosine phosphorylation sites and binding of SH2 domain-containing molecules. J Biol Chem, in press
Iwasaka C, Tanaka K, Abe M, Sato Y (1996) Ets-1 regulates angiogenesis by inducing the expression of urokinase-type plasminogen activator and matrix metalloproteinase-1 and the migration of vascular endothelial cells. J Cell Physiol 169:522–531
Jakeman L, Winer J, Bennett GL, Alter CA, Ferrara N (1992) Binding sites for vascular endothelial growth factor are localized on endothelial cells in adult rat tissues. J Clin Invest 89:244–253
Jakeman LB, Armanini M, Phillips HS, Ferrara N (1993) Developmental expression of binding sites and messenger ribonucleic acid for vascular endothelial growth factor suggests a role for this protein in vasculogenesis and angiogenesis. Endocrinology 133:848–859
Joukov V, Sorsa T, Kumar V, Jeltsch M, Claesson-Welsh L, Cao Y, Saksela O, Kalkkinen N, Alitalo K (1997) Proteolytic processing regulates receptor specificity and activity of VEGF-C. EMBO J 16:3898–3911
Kaipainen A, Korhonen J, Pajusola K, Aprelikova O, Persico MG, Terman BI, Alitalo K (1993) The related FLT4, FLT1, and KDR receptor tyrosine kinases show distinct expression patterns in human fetal endothelial cells. J Exp Med 178:2077–2088
Kanda S, Landgren E, Ljungstrom M, Claesson-Welsh L (1996) Fibroblast growth factor receptor 1-induced differentiation of endothelial cell line established from tsA58 large T transgenic mice. Cell Growth Differ 7:383–395
Kendall RL, Thomas KA (1993) Inhibition of vascular endothelial cell growth factor activity by an endogenously encoded soluble receptor. Proc Natl Acad Sci USA 90:10705–10709
Kendall RL, Wang G, DiSalvo J, Thomas KA (1994) Specificity of vascular endothelial cell growth factor receptor ligand binding domains. Biochem Biophys Res Commun 201:326–330
Kendall RL, Wang G, Thomas KA (1996) Identification of a natural soluble form of the vascular endothelial growth factor receptor, FLT-1, and its heterodimerization with KDR. Biochem Biophys Res Commun 226:324–328
Kondo K, Hiratsuka S, Subbalakshmi E, Matsushime H, Shibuya M (1998) Genomic organization of the flt-1 gene encoding for vascular endothelial growth factor (VEGF) receptor-1 suggests an intimate evolutionary relationship between 7-Ig and the 5-Ig tyrosine kinase receptors. Gene (in press)
Kremer C, Breier G, Risau W, Plate KH (1997) Up-regulation of flk-1/vascular endothelial growth factor receptor 2 by its ligand in a cerebral slice culture system. Cancer Res 57:3852–3859
Kroll J, Waltenberger J (1997) The vascular endothelial growth factor receptor KDR activates multiple signal transduction pathways in porcine aortic endothelial cells. J Biol Chem 272:32521–32527
Landgren E, Schiller P, Cao Y, Claesson-Welsh L (1998) Placenta growth factor stimulates MAP kinase and mitogenicity but not phospholipase C-gamma and migration of endothelial cells expressing Flt 1. Oncogene 16:359–367
Maglione D, Guerriero V, Viglietto G, Delli-Bovi P, Persico MG (1991) Isolation of a human placenta cDNA coding for a protein related to the vascular permeability factor. Proc Natl Acad Sci USA 88:9267–9271
Maglione D, Guerriero V, Viglietto G, Ferraro MG, Aprelikova O, Alitalo K, Vecchio SD, Lei KJ, Chou JY, Persico MG (1993) Two alternative mRNAs coding for the angiogenic factor, placenta growth factor (PlGF), are transcribed from a single gene of chromosome 14. Oncogene 8:925–931
Maru Y, Yamaguchi S, Shibuya M (1998) Flt-1, a receptor for vascular endothelial growth factor, has transforming and morphogenic potentials. Oncogene (in press)

Maisonpierre PC, Suri C, Jones PF, Bartunkova S, Wiegand SJ, Radziejewski C, Compton D, McClain J, Aldrich TH, Papadopoulos N, Daly TJ, Davis S, Sato TN, Yancopoulos GD (1997) Angiopoietin-2, a natural antagonist for Tie2 that disrupts in vivo angiogenesis. Science 277:55–60

Matthews W, Jordan CT, Gavin M, Jenkins NA, Copeland NG, Lemischka IR (1991a) A receptor tyrosine kinase cDNA isolated from a population of enriched primitive hematopoietic cells and exhibiting close genetic linkage to c-kit. Proc Natl Acad Sci USA 88:9026–9030

Matthews W, Jordan CT, Wiegand GW, Pardoll D, Lemischka IR (1991b) A receptor tyrosine kinase specific to hematopoietic stem and progenitor cell-enriched populations. Cell 65:1143–1152

Millauer B, Wizigmann-Voos S, Schnurch H, Martinez R, Moller NPH, Risau W, Ullrich A (1993) High affinity VEGF binding and developmental expression suggest flk-1 as a major regulator of vasculogenesis and angiogenesis. Cell 72:835–846

Millauer B, Shawver LK, Plate KH, Risau W, Ullrich A (1994) Glioblastoma growth inhibited in vivo by a dominant-negative Flk-1 mutant. Nature 367:576–579

Millauer B, Longhi MP, Plate KH, Shawver LK, Risau W, Ullrich A, Strawn LM (1996) Dominant-negative inhibition of Flk-1 suppresses the growth of many tumor types in vivo. Cancer Res 56:1615–1620

Montesano R, Vassalli JD, Baird A, Guillemin R, Orci L (1986) Basic fibroblast growth factor induces angiogenesis in vitro. Proc Natl Acad Sci USA 83:7297–7301

Morishita K, Johnson DE, Williams LT (1995) A novel promoter for vascular endothelial growth factor receptor (flt-1) that confers endothelial-specific gene expression. J Biol Chem 270:27948–27953

Nomura M, Yamanashi SI, Harada SI, Hayashi Y, Yamashima T, Yamashita J, Yamamoto H (1995) Possible participation of autocrine and paracrine vascular endothelial growth factors in hypoxia-induced proliferation of endothelial cells and pericytes. J Biol Chem 270:28316–28324

Oeberg C, Waltenberger J, Claesson-Welsh L, Welsh M (1994) Expression of protein tyrosine kinases in islet cells: possible role of the Flk-1 receptor for b-cell maturation from duct cells. Growth Factors 10:115–126

Park JE, Chen HH, Winer J, Houck KA, Ferrara N (1994) Placenta growth factor: potentiation of vascular endothelial growth factor bioactivity, in vitro and in vivo, and high affinity binding to Flt-1 but not to Flk-1/KDR. J Biol Chem 269:25646–25654

Patterson C, Perrella MA, Hsieh CM, Yoshizumi M, Lee ME, Haber E (1995) Cloning and functional analysis of the promoter for KDR/flk-1, a receptor for vascular endothelial growth factor. J Biol Chem 270:23111–23118

Pawson T (1995) Protein modules and signalling networks. Nature 373:573–580

Pelicci G, Lanfrancone L, Grignani F, McGlade J, Cavallo F, Forni G, Nicoletti I, Grignani F, Pawson T, Pelicci PG (1992) A novel transforming protein (SHC) with an SH2 domain is implicated in mitogenic signal transduction. Cell 70:93–104

Pepper MS, Ferrara N, Orci L, Montesano R (1991) Vascular endothelial growth factor (VEGF) induces plasminogen activators and plasminogen activator inhibitor-1 in microvascular endothelial cells. Biochem Biophys Res Commun 181:902–906

Pepper MS, Ferrara N, Orci L, Montesano R (1992) Potent synergism between vascular endothelial growth factor and basic fibroblast growth factor in the induction of angiogenesis in vitro. Biochem Biophys Res Commun 189:824–831

Peters KG, De Vries C, Williams LT (1993) Vascular endothelial growth factor receptor expression during embryogenesis and tissue repair suggests a role in endothelial differentiation and blood vessel growth. Proc Natl Acad Sci USA 90:8915–8919

Plate KH, Breier G, Weich HA, Risau W (1992) Vascular endothelial growth factor is a potential tumour angiogenesis factor in human gliomas in vivo. Nature 359:845–848

Ronicke V, Risau W, Breier G (1996) Characterization of the endothelium-specific murine vascular endothelial growth factor receptor-2 (Flk-1) promoter. Circ Res 79:277–285

Rosnet O, Mattei MG, Marchetto S, Birnbaum D (1991) Isolation and chromosomal localization of a novel FMS-like tyrosine kinase gene. Genomics 9:1–6

Rosnet O, Stephenson D, Mattei MG, Marchetto S, Shibuya M, Chapman VM, Birnbaum D (1993) Close physical linkage of the FLT1 and FLT3 genes on chromosome 13 in man and chromosome 5 in mouse. Oncogene 8:173–179

Rousseau S, Houle F, Landry J, Huot J (1997) p38 MAP kinase activation by vascular endothelial growth factor mediates actin reorganization and cell migration in human endothelial cells. Oncogene 15:2169–2177

Sait SN, Dougher-Vermazen M, Shows TB, Terman BI (1995) The kinase insert domain receptor gene (KDR) has been relocated to chromosome 4q11→q12. Cytogenet Cell Genet 70:145–146

Sato K, Yamazaki K, Shizume K, Kanaji Y, Obara T, Ohsumi K, Demura H, Yamaguchi S, Shibuya M (1995) Stimulation by TSH and Graves' IgG of vascular endothelial growth factor mRNA expression in human thyroid follicles in vitro and flt mRNA expression in the rat thyroid in vivo. J Clin Invest 96:1295–1302

Satoh H, Yoshida MC, Matsushime H, Shibuya M, Sasaki M (1987) Regional localization of the human c-ros-1 on 6q22 and flt on 13q12. Jpn J Cancer Res (Gann) 78:772–775

Sawano A, Takahashi T, Yamaguchi S, Aonuma T, Shibuya M (1996) Flt-1 but not KDR/Flk-1 tyrosine kinase is a receptor for placenta growth factor (PlGF), which is related to vascular endothelial growth factor (VEGF). Cell Growth Differ 7:213–221

Sawano A, Takahashi T, Yamaguchi S, Shibuya M (1997) The phosphorylated 1169-tyrosine containing region of flt-1 kinase (VEGFR-1) is a major binding site for PLC-gamma. Biochem Biophys Res Commun 238:487–491

Seetharam L, Gotoh N, Maru Y, Neufeld G, Yamaguchi S, Shibuya M (1995) A unique signal transduction from FLT tyrosine kinase, a receptor for vascular endothelial growth factor VEGF. Oncogene 10:135–147

Shalaby F, Rossant J, Yamaguchi TP, Gertsenstein M, Wu XF, Breitman ML, Schuh AC (1995) Failure of blood-island formation and vasculogenesis in Flk-1-deficient mice. Nature 376:62–66

Shen H, Clauss M, Ryan J, Schmidt AM, Tijburg P, Borden L, Connolly D, Stern D, Kao J (1993) Characterization of vascular permeability factor/vascular endothelial growth factor receptors on mononuclear phagocytes. Blood 81:2767–2773

Shibuya M (1995) Role of VEGF-Flt receptor system in normal and tumor angiogenesis. Adv Cancer Res 67:281–316

Shibuya M, Yamaguchi S, Yamane A, Ikeda T, Tojo A, Matsushime H, Sato M (1990) Nucleotide sequence and expression of a novel human receptor-type tyrosine kinase gene (flt) closely related to the fms family. Oncogene 5:519–524

Shweiki D, Itin A, Soffer D, Keshet E (1992) Vascular endothelial growth factor induced by hypoxia may mediate hypoxia-initiated angiogenesis. Nature 359:843–845

Songyang Z, Shoelson SE, Chaudhuri M, Gish G, Pawson T, Haser WG, King F, Roberts T, Ratnofsky S, Lechleider RJ et al (1993) SH2 domains recognize specific phosphopeptide sequences. Cell 72:767–778

Spritz RA, Strunk KM, Lee ST, Lu-Kuo JM, Ward DC, Le Paslier D, Altherr MR, Dorman TE, Moir DT (1994) A YAC contig spanning a cluster of human type III receptor protein tyrosine kinase genes (PDGFRA-KIT-KDR) in chromosome segment 4q12. Genomics 22:431–436

Takahashi T, Shibuya M (1997) The 230 kDa mature form of KDR/Flk-1 (VEGF receptor-2) activates the PLCg pathway and partially induces mitotic signals in NIH3T3 fibroblasts. Oncogene 14:2079–2089

Tanaka K, Yamaguchi S, Sawano A, Shibuya M (1997) Characterization of the extracellular domain in vascular endothelial growth factor receptor-1 (Flt-1 tyrosine kinase). Jpn J Cancer Res 88:867–876

Terman BI, Carrion ME, Kovacs E, Rasmussen BA, Eddy RL, Shows TB (1991) Identification of a new endothelial cell growth factor receptor tyrosine kinase. Oncogene 6:1677–1683

Terman BI, Dougher-Vermazen M, Carrion ME, Dimitrov D, Armellino DC, Gospodarowicz D, Bohlen P (1992) Identification of the KDR tyrosine kinase as a receptor for vascular endothelial growth factor. Biochem Biophys Res Commun 187:1579–1586

Terman B, Khandke L, Dougher-Vermazan M, Maglione D, Lassam NJ, Gospodarowicz D, Persico MG, Bohlen P, Eisinger M (1994) VEGF receptor subtypes KDR and FLT1 show different sensitivities to heparin and placenta growth factor. Growth Factors 11:187–195

Ullrich A, Schlessinger J (1990) Signal transduction by receptors with tyrosine kinase activity. Cell 61:203–212

Wakiya K, Begue A, Stehelin D, Shibuya M (1996) A cAMP response element and an Ets motif are involved in the transcriptional regulation of flt-1 tyrosine kinase (Vascular Endothelial Growth Factor Receptor 1) gene. J Biol Chem 271:30823–30828

Waltenberger J, Claesson-Welsh L, Siegbahn A, Shibuya M, Heldin CH (1994) Different signal transduction properties of KDR and Flt1, two receptors for Vascular Endothelial Growth Factor. J Biol Chem 269:26988–26995

Wernert N, Raes MB, Lassalle P, Dehouck MP, Gosselin B, Vandenbunder B, Stehelin D (1992) c-ets1 proto-oncogene is a transcription factor expressed in endothelial cells during tumor vascularization and other forms of angiogenesis in humans. Am J Pathol 140:119–127

Wizigmann-Voos S, Breier G, Risau W, Plate KH (1995) Up-regulation of vascular endothelial growth factor and its receptors in von Hippel-Lindau disease-associated and sporadic hemangioblastomas. Cancer Res 55:1358–1364

Xia P, Aiello LP, Ishii H, Jiang ZY, Park DJ, Robinson GS, Takagi H, Newsome WP, Jirousek MR, King GL (1996) Characterization of vascular endothelial growth factor's effect on the activation of protein kinase C, its isoforms, and endothelial cell growth. J Clin Invest 98:2018–2026

Yamane A, Seetharam L, Yamaguchi S, Gotoh N, Takahashi T, Neufeld G, Shibuya M (1994) A new communication system between hepatocytes and sinusoidal endothelial cells in liver through vascular endothelial growth factor and Flt tyrosine kinase receptor family (Flt-1 and KDR/Flk-1). Oncogene 9:2683–2690

Yoshiji H, Gomez DE, Shibuya M, Thorgeirsson UP (1996) Expression of Vascular Endothelial Growth Factor, its receptor and other angiogenic factors in human breast cancer. Cancer Res 56:2013–2016

Ziche M, Maglione D, Ribatt D, Morbidelli L, Lago CT, Battisti M, Paoletti I, Barra A, Tucci M, Parise G, Vincenti V, Granger HJ, Viglietto G, Persico MG (1997) Placenta growth factor-1 is chemotactic, mitogenic, and angiogenic. Lab Invest 76:517–531

# Vascular Endothelial Growth Factor Receptor-3

J. TAIPALE, T. MAKINEN, E. ARIGHI, E. KUKK, M. KARKKAINEN, and K. ALITALO

| | |
|---|---|
| 1 Introduction | 85 |
| 2 VEGFR-3 Structure | 86 |
| 3 VEGFR-3 Ligands | 86 |
| 4 Regulation of VEGFR-3 Expression | 87 |
| 5 VEGFR-3 Regulates Lymphangiogenesis and Endothelial Cell Growth and Differentiation | 89 |
| 6 VEGFR-3 Signalling | 90 |
| 7 VEGFR-3 in Vascular Endothelium and in Tumours | 92 |
| 8 Perspective | 93 |
| References | 94 |

## 1 Introduction

Cell–cell communication during vascular development and tumour angiogenesis seems to involve at least five endothelial cell-specific tyrosine kinase receptors belonging to two distinct subclasses: two receptors of the Tie family, and three vascular endothelial cell growth factor receptors, VEGFR-1, -2 and -3, originally named Flt1 (Fms-like tyrosine kinase), KDR/Flk-1 (Kinase insert-domain containing receptor or fetal-liver kinase-1) and Flt4, respectively. VEGFRs are subclass-III receptor tyrosine kinases, homologous to the platelet-derived growth factor (PDGF)-receptor family, having seven immunoglobulin homology domains in the extracellular domain, and a tyrosine kinase intracellular domain split by a kinase insert sequence (for recent reviews, see KLAGSBRUN and D'AMORE 1996; FOLKMAN and D'AMORE 1996; MUSTONEN and ALITALO 1995; KORPELAINEN and ALITALO, 1998, CLAESSON-WELSH, this book).

---

Molecular/Cancer Biology Laboratory, Haartman Institute, University of Helsinki, P.O.B. 21 (Haartmaninkatu 3), FIN-00014 Helsinki, Finland

## 2 VEGFR-3 Structure

VEGFR-3 was cloned from human placental and erythroleukaemia cell complementary deoxyribonucleic acid (cDNA) libraries (APRELIKOVA et al. 1992; PAJUSOLA et al. 1992, 1993; GALLAND et al. 1992, 1993). The protein shows approximately 35% amino acid identity with VEGFR-1 and -2 in the extracellular domain and about 80% in the tyrosine-kinase domain (PAJUSOLA et al. 1992). Mouse and quail (designated Quek2) homologues of VEGFR-3 have also been cloned (FINNERTY et al. 1993; EICHMANN et al. 1996, 1998). They are relatively conserved in evolution, the quail homologue showing 70% amino acid identity with the human receptor and having similar ligand-binding characteristics.

The human VEGFR-3 gene is located in chromosome 5q35, and the mouse gene in a syntenic region of mouse chromosome 11 (APRELIKOVA et al. 1992; ARMSTRONG et al. 1993; ROSNET et al. 1993; WATKINS-CHOW et al. 1997). A theory has been suggested for the evolution of the class-III receptor tyrosine kinases (RTKs), in which an ancestral chromosome accommodating one seven- and one five-immunoglobulin homology domain-containing receptor (of subclass III) underwent a *cis*- and subsequent *trans*-duplication, resulting in the present arrangement of the genes in mammals (ROSNET et al. 1993).

The major human VEGFR-3 mRNA transcript is 5.8 kb in size, and an alternative 3′ polyadenylation signal results in a minor 4.5 kb transcript. The shorter transcript codes for a protein that has 65 residue truncation at the C-terminus. However, the longer form of VEGFR-3 is the major form detected in tissues. This form is synthesised as a 195-kDa precursor that is glycosylated and proteolytically cleaved after Arg472 to yield a disulphide linked two-chain form (PAJUSOLA et al. 1993, 1994; FOURNIER et al. 1995; BORG et al. 1995; Fig. 1). The C-terminus contains three tyrosine residues that are not encoded in the shorter transcript, Tyr1333, Tyr1337 and Tyr1363.

## 3 VEGFR-3 Ligands

The ligands for VEGFR-3, VEGF-C and VEGF-D are secreted as dimeric forms containing N- and C-terminal pro-peptides (JOUKOV et al. 1996; LEE et al. 1996; ORLANDINI et al. 1996; ACHEN et al. 1998; YAMADA et al. 1997). At least in the case of VEGF-C, the pro-form is proteolytically processed in the secretory pathway between its C-terminal pro-peptide and the VEGF-homology domain. This partially processed form of VEGF-C can already bind to VEGFR-3 (Fig. 2). Subsequent extracellular proteolytic cleavage between the N-terminal pro-peptide and the VEGF-homology domain leads to the generation of a short form of VEGF-C that interacts with VEGFR-3 with high affinity ($K_d$ 135 pM). This fully processed form can also bind to VEGFR-2 ($K_d$ 410 pM; JOUKOV et al. 1996, 1997, 1998; ACHEN et al. 1998). It is not yet clear whether these receptors can heterodimerize in

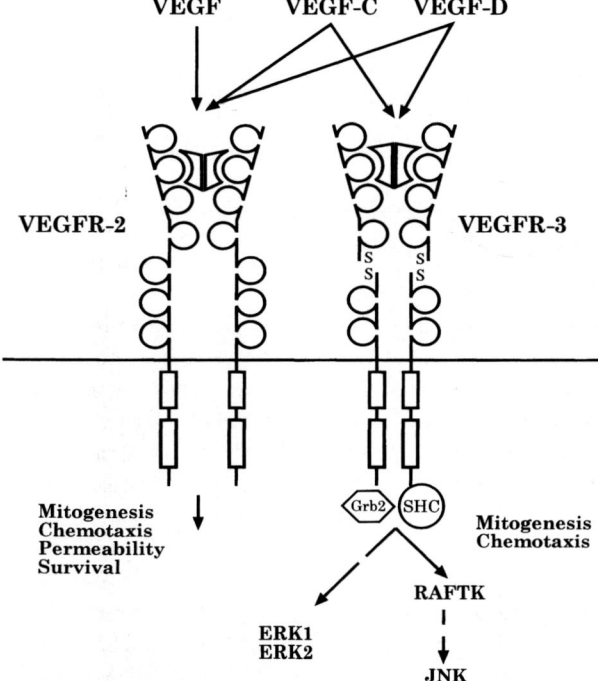

**Fig. 1.** Schematic view of vascular endothelial growth factor receptor-3 (VEGFR-3) signalling. VEGF-C and VEGF-D bind to the second immunoglobulin-homology domain of VEGFR-3, and induce receptor dimerization and transphosphorylation. Subsequently, VEGFR-3 phosphorylates and recruits several intracellular substrates, including Shc and Grb2. Downstream effectors include related adhesion focal tyrosine kinase (*RAFTK*) which is a novel focal adhesion kinase-like protein, the mitogen-activated protein kinases *ERK1* and *ERK2*, and the *c-Jun* N-terminal kinase, *JNK*. The binding of VEGF to VEGF-R2 is also shown, but VEGFR-1 and its other ligands are not illustrated

response to the fully processed ligands. The VEGFR-3 specific mutant, VEGF-C156S should help in defining some of the receptor-specific signaling functions (Joukov et al., 1998).

## 4 Regulation of VEGFR-3 Expression

The VEGFR-3 ribonucleic acid (RNA) signals are detected in 8.5-day-old mouse embryos, in the angioblasts of the head mesenchyme and the cardinal vein (KAIPAINEN et al. 1995). In 12.5-day-old embryos, VEGFR-3 is expressed in venous and presumptive lymphatic endothelia, but expression is gradually lost from the developing arteries. During later stages of development, VEGFR-3 expression becomes largely confined to endothelial cells lining the lymphatic vessels and some high endothelial venules (KAIPAINEN et al. 1995). A similar pattern of VEGFR-3 expression is found in developing quail embryos, where the messenger RNA

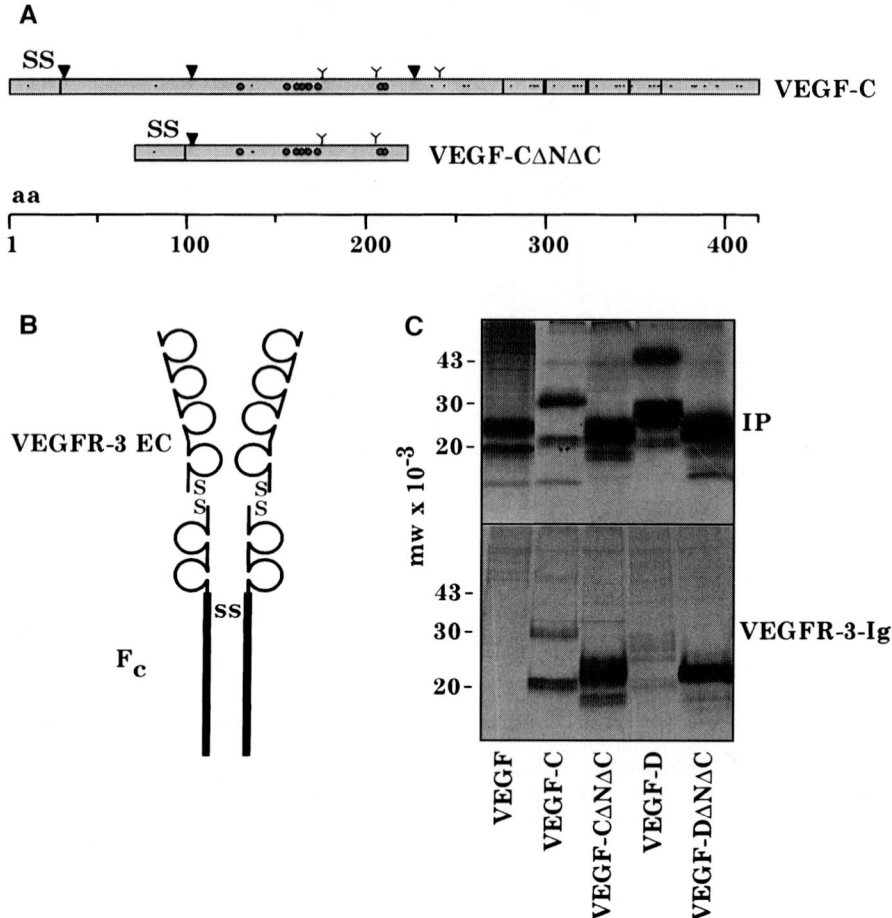

**Fig. 2A–C.** Vascular endothelial growth factor receptor-3 (VEGFR-3) binding to full-length and processed forms of VEGF-C and VEGF-D. **A** Schematic structure of the wild-type VEGF-C protein (*VEGF-C*) and recombinant short form of VEGF-C which lacks the N- and C-terminal pro-peptides (*VEGF-CΔNΔC*). Cys residues (·) non-conserved and (●)-conserved among platelet-derived growth factor (PDGF)/VEGF family members; *Y* N-linked glycosylation sites; ∀ proteolytic cleavage sites; *SS* signal sequence; *BR3P* motifs are outlined. **B** Structure of the recombinant VEGFR-3-Immunoglobulin (Ig) fusion protein used for ligand-affinity precipitation. **C** Precipitation of metabolically labelled VEGF-C and VEGF-D by specific antibodies or VEGFR-3-Ig. Supernatants of metabolically labelled cells expressing VEGF or different forms of VEGF-C and VEGF-D were analysed by immunoprecipitation using antibodies specific to the expressed protein (*upper panel, IP*) or affinity precipitation using VEGFR-3-Ig (*lower panel*) see Achen et al. (1998)

(mRNA) is first found in the intraembryonic vascular plexus. During further development, the expression is lost from endothelial cells that form arteries and veins, but retained in the lymphatic vasculature. In the quail, certain non-endothelial cells have also been reported to express VEGFR-3. These include cells of the notochord, podocytes of the kidney and some epithelial cells of the gallbladder and extra-hepatic bile ducts (WILTING et al. 1997). It is not clear which transcription factors regulate VEGFR-3 expression or what molecular mechanisms are responsible for the restriction of its expression to the lymphatic endothelium. Apart from VEGFR-3, no other lymphatic endothelial-specific genes have, thus far, been isolated.

The pattern of expression of VEGFR-3, in relation to its ligand VEGF-C during the development of the lymphatic vessels in mouse embryos, suggests that VEGF-C acts on VEGFR-3-expressing endothelial cells in a paracrine fashion. Thus, the expression pattern is consistent with the hypothesis that VEGFR-3 plays a role in the genesis of the lymphatic system (KUKK et al. 1996). VEGF-C RNA expression is detected in mesenchymal cells of post-implantation mouse embryos (KUKK et al. 1996). Strong expression is seen in the perimesonephric, axillary and jugular regions during embryonic day 12, when, according to the theory of SABIN (1902), lymphatic vessels are forming by centrifugal sprouting from embryonic veins. In addition, the developing mesenterium, which is rich in lymphatic vessels, shows strong VEGF-C expression.

## 5 VEGFR-3 Regulates Lymphangiogenesis and Endothelial Cell Growth and Differentiation

VEGF, the major ligand for VEGFR-2, is known to have pleiotrophic effects on blood vessel morphogenesis and function. It increases vascular permeability and induces angiogenesis and growth of existing vessels (KLAGSBRUN and D'AMORE 1996). Similarly, VEGF-C, upon its interaction with VEGFR-3, has multiple roles in regulating the lymphatic and blood vascular systems.

Mice devoid of VEGFR-3 function die in utero after embryonic day 10 (L. JUSSILA, D. DUMONT, unpublished data). The phenotype of the VEGFR-3 –/– embryos indicates that VEGFR-3 has a general function of promoting the formation of major blood vessels during early development since, at this stage, the development of the lymphatic vessels has not yet started, and VEGFR-3 is still expressed in most endothelial cells.

Lymphatic vessels are very different in structure from arteriae, veins and capillaries. The lymphatics are characterised by an extremely permeable, thin endothelial lining devoid of a basal lamina. Simple end-to-end cell junctions and interdigitating and especially overlapping junctions between endothelial cells are a characteristic feature of lymphatic vessels. In addition, the lymphatics typically lack supporting cells, such as pericytes or smooth muscle cells (LEAK 1970, 1971; LEAK and JAMUAR 1983).

Interestingly, overexpression of VEGF-C in the basal keratinocytes of transgenic mice resulted in lymphatic, but not vascular, endothelial proliferation and vessel enlargement (JELTSCH et al. 1997). However, the number of lymphatic vessels and their hexagonal organisation in the skin of the animals was preserved. Similarly, expression of VEGF-C in the pancreatic endocrine β-cell islands under the insulin promoter resulted in hyperplasia of the surrounding lymphatic vessels

(S. MANDRIOTTA, M. JELTSCH, K. ALITALO, M. Pepper, unpublished data). Since lymphatic endothelial cells express both VEGFR-2 and VEGFR-3, it is unclear whether the lymphatic endothelial growth signal is transduced by VEGFR-3 alone or requires signals from both receptors.

The role of VEGFR-3 signalling in the induction of lymphangiogenesis has been studied most extensively in the chorioallantoic membrane (CAM) of 13-day-old chick embryos (OH et al. 1997). At this stage, the vasculature of the CAM has been formed, and very little angiogenesis or lymphangiogenesis is occurring. The CAM is drained by lymphatic vessels that are regularly arranged around all arteries and arterioles, and major veins. Pairs of lymphatic vessels are interconnected by capillaries which, in turn, contain blind-ending extensions. The lymphatic endothelial cells of the CAM express both VEGFR-2 and VEGFR-3, whereas in the blood vasculature, VEGFR-3 is not expressed. The capability of different factors to induce angiogenesis or lymphangiogenesis can be studied by adding exogenous growth factors to the CAM. VEGF-C was the first lymphangiogenic factor identified in this system, and it seems to be highly chemoattractive for lymphatic endothelial cells. It induces proliferation of lymphatic endothelial cells and development of new lymphatic sinuses which are located immediately beneath the chorionic epithelium, on top of which the growth factor is placed. The effects of VEGF-C on blood vasculature of the CAM are minor, suggesting that the (lymph) angiogenic signal is transduced by VEGFR-3, or by a heterodimer between VEGFR-3 and VEGFR-2. This is in keeping with the fact that VEGF-C is 50-fold less potent than VEGF in stimulating the DNA synthesis of capillary endothelial cells (JOUKOV et al. 1996; LEE et al. 1996).

The differences in the structure and cellular composition of lymphatic and non-lymphatic vessels suggest that the lymphatic endothelial cells may represent a very differentiated form of endothelial cells. VEGF-C could act as a differentiation factor for the lymphatic endothelial cells, selecting the population of endothelial cells that differentiates and forms the lymphatic vessels, which then maintain VEGFR-3 expression. Since VEGFR-3 has a role in the vascular system prior to the development of the lymphatic vessels, it is likely that the loss of VEGFR-3 expression from arteriae and veins prevents them from receiving the VEGF-C growth factor signals during the development of the lymphatic system around them.

## 6 VEGFR-3 Signaling

Prior to the identification of the natural ligands of VEGFR-3, the signal transduction of VEGFR-3 was studied using CSF-1R/VEGFR-3 chimeras. When expressed in fibroblasts, the chimeric receptor was tyrosine phosphorylated and elicited a mitogenic response upon ligand stimulation (PAJUSOLA et al. 1994; BORG et al. 1995; FOURNIER et al. 1995). Interestingly, only the chimera containing the cytoplasmic tail of the long isoform of VEGFR-3 was able to mediate anchorage-independent growth in soft agar and tumorigenicity in nude mice (BORG et al. 1995; FOURNIER et al. 1995). The transforming activity of VEGFR-3 is probably related to the capacity of a tyrosine 1337-containing peptide to interact with the N-terminal phosphotyrosine binding (PTB) domain of the Shc (Src homologous and collagen) protein (FOURNIER et al. 1996).

After identification of the VEGFR-3 ligands, it was shown that VEGFR-3 is strongly tyrosine phosphorylated when stimulated with either VEGF-C or VEGF-D (JOUKOV et al. 1996; ACHEN et al. 1998). Stimulation of VEGFR-3 expressing cells with either VEGF-C or VEGF-D induced rapid tyrosine phosphorylation of the Shc protein, and activation of the mitogen-activated protein kinases ERK1 and ERK2 (Fig. 3). In addition, in a human erythroleukaemia cell line that expresses high levels of the VEGFR-3 protein, VEGF-C stimulation induces cell growth and recruitment of the signalling molecules Shc, Grb2, and hSOS (human son of sevenless) to the activated VEGFR-3 (WANG et al. 1997). In these cells, VEGF-C stimulation also induced tyrosine phosphorylation of the cytoskeletal protein paxillin, resulting in an increased association of paxillin with related adhesion focal tyrosine kinase (RAFTK), a recently identified member of the focal adhesion kinase family. The c-Jun $NH_2$-terminal kinase (JNK) is also activated following VEGF-C stimulation of these cells or Kaposi's sarcoma cells (LIU et al. 1997).

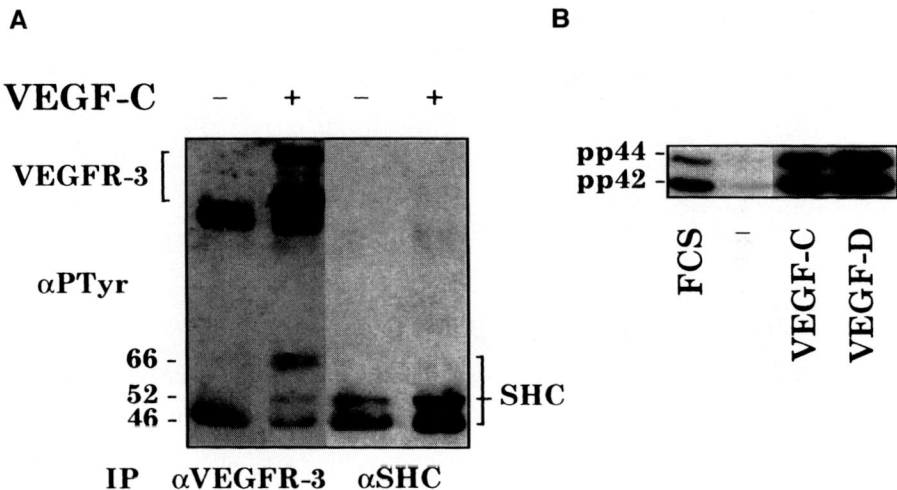

Fig. 3A,B. Signalling by vascular endothelial growth factor receptor-3 (VEGFR-3). A Shc is phosphorylated and recruited by VEGFR-3. PAE/VEGFR-3 cells were treated with VEGF-C, the cells were subsequently lysed and the lysates immunoprecipitated with the indicated antibodies, followed by Western blotting with anti-phosphotyrosine antibodies. Phosphorylated VEGFR-3, and the 46, 52 and 66 kDa Shc polypeptides are indicated by brackets on the left and right, respectively. B VEGF-C stimulation induces the activation of ERK1 and ERK2 mitogen-activated protein kinases (MAPKs) through VEGFR-3. Serum starved PAE/VEGFR-3 cells were treated with serum (10%) or the factors indicated. Cells were lysed after 5 min, and ERK1 and ERK2 phosphorylation was analysed by means of Western blotting using phospho-specific p44/42 MAPK antibodies

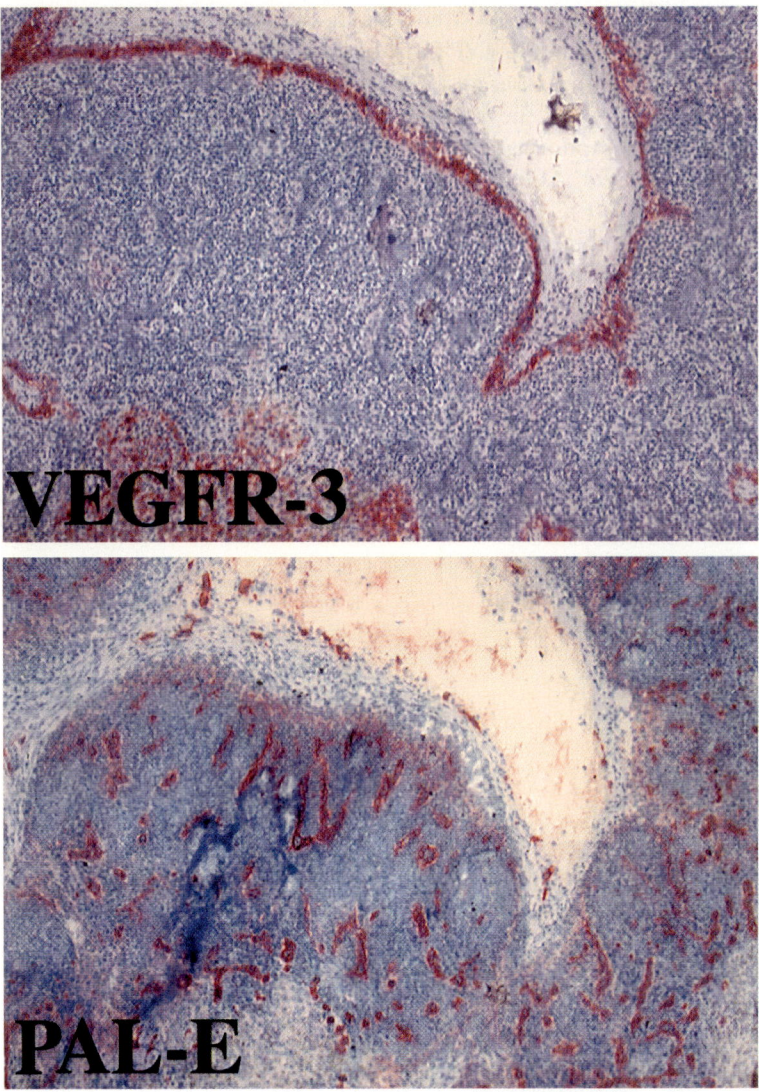

**Fig. 4.** Imunoperoxidase staining of VEGFR-3 and PAL-E antigens in adjacent sections of a lymph node, revealing the location of lymphatic sinuses and blood vessels, respectively

## 7  VEGFR-3 in Vascular Endothelium and in Tumours

Lymphatic spread of cancer cells by invasion is an important prognostic indicator of tumour aggressiveness. The involvement of lymph nodes forms the basis for staging of many cancer types and for treatment directed at the draining regional lymph nodes. Despite an increasing interest in tumour angiogenesis involving blood vessels (reviewed in FOLKMAN 1996), the lymphatic vessels have been, generally, neglected and the possible existence of lymphangiogenesis in tumour growth and metastasis is, thus, unclear. Based on morphological characterisation, solid tumours have, generally, been thought to lack lymphatic vessels. Indeed, the high interstitial pressure of solid tumours is likely to induce a collapse of lymphatic vessels, thus preventing their ingrowth into the tumour interstitium.

Lymphatic vessels have been difficult to study in detail in both normal and tumour tissues because of the lack of molecular markers. However, according to our studies, VEGFR-3 is a specific marker for the lymphatic vessels in most normal tissues. VEGFR-3 was also strongly expressed in the endothelium of cutaneous lymphangiomatosis, but staining of endothelial cells in cutaneous haemangiomas was weaker (LYMBOUSSAKI et al. 1998). Indeed, the monoclonal antibodies developed against the VEGFR-3 allowed us to distinguish VEGFR-3-positive vessels in tissues surrounding lymphomas. Intraductal breast carcinomas (JUSSILA et al. 1998) and several solid tumours and cell lines established from them, including the human prostate cancer cell line PC-3 and human fibrosarcoma cell line HT-1080, were found to express abundant VEGF-C (SALVEN et al. 1998). However, it seems that also tumour blood vessels can exhibit VEGFR-3 expression as do the embryonic vessels and only some of the VEGFR-3-positive vessels around tumours represent true lymphatic vessels (Reija Valtola, unpublished data).

Interestingly, the spindle cells of several cutaneous nodular acquired immunodeficiency syndrome (AIDS)-associated Kaposi's sarcomas (KS) and the endothelium around the nodules were VEGFR-3 positive (JUSSILA et al. 1998). Our results are compatible with the concept that the KS cells originate from an undifferentiated mesenchymal stem cell, which shows aspects of endothelial differentiation, or that their origin is in the lymphatic endothelium (DICTOR and ANDERSSON 1988; BECKSTEAD et al. 1985). Also, cultured KS cells have been reported to express VEGFR-3 (LIU et al. 1997).

VEGFR-3 has also been detected in some haematopoietic and leukaemia cells, such as the erythroleukaemia cell line HEL (Aprelikova et al. 1992) and promonocytic leukaemic cell line U937. FIELDER et al. (1997) reported that VEGFR-3 mRNA was detected by reverse-transcription polymerase chain reaction (RT-PCR) in leukaemic cells from a subset patients with de novo acute myeloid leukaemia (AML). In some of the patients, the leukaemic cells also expressed VEGF-C, suggesting that autocrine stimulation of the VEGFR-3 might be involved in driving the proliferation of the tumour cells. The possible role of the VEGF-C–VEGFR-3 system in normal haematopoiesis is, as yet, unclear.

## 8  Perspective

VEGFR-3 may play a role in disorders involving the lymphatic system and angiogenesis and it may be of potential use in drug targeting, in vivo imaging of the lymphatic vessels, in therapeutic lymphangiogenesis, e.g. after breast cancer, and in the treatment of ascites and oedemas.

*Acknowledgements.* We kindly thank all our coworkers and collaborators for outstanding contributions to the findings summarized in the review. The original contributions were supported by the Finnish Academy of Sciences, the University of Helsinki, the Finnish Cancer Research Foundation, the Helsinki University Central Hospital Research Funds (TYH 8105), the State Technology Development Centre and the European Union (Biomedicine grant PL 963380).

# References

Achen MG, Jeltsch M, Kukk E, Mäkinen T, Vitali A, Wilks AF, Alitalo K, Stacker SA (1998) Vascular endothelial growth factor-D (VEGF-D) is a ligand for the tyrosine kinase VEGF receptor-2 (Flk1) and VEGF receptor-3 (Flt4). Proc Natl Acad Sci USA 95:548–553

Aprelikova O, Pajusola K, Partanen J, Armstrong E, Alitalo R, Bailey SK, McMahon J, Wasmuth J, Huebner K, Alitalo K (1992) FLT4, a novel class III receptor tyrosine kinase in chromosome 5q33-qter. Cancer Res 52:746–748

Armstrong E, Kastury K, Aprelikova O, Bullrich F, Nezelof C, Gogusev J, Wasmuth JJ, Alitalo K, Morris S, Huebner K (1993) FLT4 receptor tyrosine kinase gene mapping to chromosome band 5q35 in relation to the t(2;5), t(5;6), and t(3;5) translocations. Genes Chromosom Cancer 7:144–151

Beckstead JH, Wood GS, Fletcher V (1985) Evidence for the origin of Kaposi's sarcoma from lymphatic endothelium. Am J Pathol 119:294–300

Borg J-P, deLapeyrière O, Noguchi T, Rottapel R, Dubreuil P, Birnbaum D (1995) Biochemical characterization of two isoforms of FLT4, a VEGF receptor-related tyrosine kinase. Oncogene 10:973–984

Dictor N, Andersson C (1988) Lymphaticovenous differentiation in Kaposi's sarcoma. Am J Pathol 130:411–417

Eichmann A, Corbel C, Jaffredo T, Bréant C, Joukov V, Kumar V, Alitalo K, le Douarin N (1998) Avian VEGF-C: cloning embryonic expression pattern and stimulation of the differentiation of VEGFR2 expressing endothelial cell precursors. Development (in press)

Eichmann A, Marcelle C, Breant C, Le Douarin NM (1996) Molecular cloning of Quek 1 and 2, two quail vascular endothelial growth factor (VEGF) receptor-like molecules. Gene 174:3–8

Fielder W, Graeven U, Ergun S, Verago S, Kilic N, Stcokschlader M, Hossfeld DK (1997) Expression of FLT4 and its ligand VEGF-C in acute myeloid leukemia. Leukemia 11:1234–1237

Finnerty H, Kelleher K, Morris GE, Bean K, Merberg DM, Kriz R, Morris J C, Sookdeo H, Turner KJ, Wood CR (1993) Molecular cloning of murine FLT and FLT4. Oncogene 8:2293–2298

Folkman J (1996) Angiogenesis and tumor growth. N Engl J Med 334:921

Folkman J, D'Amore P (1996) Blood vessel formation: what is its molecular basis? Cell 87:1153–1155

Fournier E, Dubreuil P, Birnbaum D, Borg JP (1995) Mutation at tyrosine residue 1337 abrogates ligand-dependent transforming capacity of the FLT4 receptor. Oncogene 11:921–931

Fournier E, Rosnet O, Marchetto S, Turck CW, Rottapel R, Pelicci PG, Birnbaum D, Borg JP (1996) Interaction with the phosphotyrosine binding domain/phosphotyrosine interacting domain of Shc is required for the transforming activity of the FLT4/VEGFR3 receptor tyrosine kinase. J Biol Chem 271:12956–12963

Galland F, Karamysheva A, Mattei M-G, Rosnet O, Marchetto S, Birnbaum D (1992) Chromosomal localization of FLT4, a novel receptor-type tyrosine kinase gene. Genomics. 13:475–478

Galland F, Karamysheva A, Pebusque M-J, Borg J-P, Rottapel R, Dubreuil P, Rosnet O, Birnbaum D (1993) The FLT4 gene encodes a transmembrane tyrosine kinase related to the vascular endothelial growth factor receptor. Oncogene 8:1233–1240

Jeltsch M, Kaipainen A, Joukov V, Meng X, Lakso M, Rauvala H, Swartz M, Fukumura D, Jain RK, Alitalo K (1997) Hyperplasia of lymphatic vessels in VEGF-C transgenic mice. Science 276:1423–1425

Joukov V, Pajusola K, Kaipainen A, Chilov D, Lahtinen I, Kukk E, Saksela O, Kalkkinen N, Alitalo K (1996) A novel vascular endothelial growth factor, VEGF-C, is a ligand for the Flt4 (VEGFR-3) and KDR (VEGFR-2) receptor tyrosine kinases. EMBO J 15:290–298

Joukov V, Sorsa T, Kumar V, Jeltsch M, Claesson-Welsh L, Cao Y, Saksela O, Kalkkinen N, Alitalo K (1997) Proteolytic processing regulates receptor specificity and activity of VEGF-C. EMBO J 16:3898–3911

Joukov V, Kumar V, Sorsa T, Arighi E, Weich H, Saksela O, Alitalo K (1998) A Recombinant Mutant Vascular Endothelial Growth Factor-C that Has Lost Vascular Endothelial Growth Factor Receptor-2 Binding, Activation, and Vascular Permeability Activities*

Jussila L, Valtola R, Partanen TA, Salven P, Heikkilä P, Matikainen M-T, Renkonen R, Kaipainen A, Detmar M, Tschachler E, Alitalo R, Alitalo K (1998) Lymphatic endothelium and Kaposi's sarcoma spindle cells detected by antibodies against the vascular endothelial growth factor receptor-3. Cancer Res (in press)

Kaipainen A, Korhonen J, Mustonen T, van Hinsbergh VM, Fang G-H, Dumont D, Breitman M, Alitalo K (1995) Expression of the fms-like tyrosine kinase FLT4 gene becomes restricted to endothelium of lymphatic vessels during development. Proc Natl Acad Sci USA 92:3566–3570

Klagsbrun M, D'Amore P (1996) Vascular endothelial growth factor and its receptors. Cytokine Growth Factor Rev 7:259–270

Korpelainen, E. and Alitalo, K. 1998 Signaling angiogenesis and lymphangiogenesis. Curr. Opin. Cell Biol. 10:159–164

Kukk E, Lymboussaki A, Taira S, Kaipainen A, Jeltsch M, Joukov V, Alitalo K (1996) VEGF-C receptor binding and pattern of expression with VEGFR-3 suggests a role in lymphatic vascular development. Development 122:3829–3837

Leak LV (1970) Electron microscopic observations on lymphatic capillaries and the structural components of the connective tissue-lymph interface. Microvasc Res 2:361–391

Leak LV (1971) Studies on the permeability of lymphatic capillaries. J Cell Biol 50:300–323

Leak LV, Jamuar MP (1983) Ultrastructure of pulmonary lymphatic vessels. Am Rev Respir Dis 128:S59–S65

Lee J, Gray A, Yuan J, Louth S-M, Avraham H, Wood W (1996) Vascular endothelial growth factor-related protein: a ligand and specific activator of the tyrosine kinase receptor Flt4. Proc Natl Acad Sci USA 93:1988–1992

Liu Z-Y, Ganju RK, Wang J-F, Ona MA, Hatch WC, Zheng T, Avraham S, Gill P, Groopman JE (1997) Cytokine signaling through the novel tyrosine kinase RAFTK in Kaposi's sarcoma cells. J Clin Invest 99:1798–1804

Lymboussakis A, Partanen TA, Olofsson, B, Thomas-Crusells J, Fletcher CDM, de Waal RMW, Kaipainen A, Alitalo K (1998) Expression of the vascular endothelial growth factor C receptor VEGFR-3 in cutaneous lymphatic vascular endothelium and in vascular skin lesions Am J Pathol 153:395–403

Mustonen, T and Alitalo, K (1995) Endothelial receptor tyrosine kinases involved in angiogenesis. J. Cell Biol. 129:895–898

Oh SJ, Jeltsch MM, Birkenhager R, McCarthy JE, Weich HA, Christ B, Alitalo K, Wilting J (1997) VEGF and VEGF-C: specific induction of angiogenesis and lymphangiogenesis in the differentiated avian chorioallantoic membrane. Dev Biol 188:96–109

Orlandini M, Marconcini L, Ferruzzi R, Oliviero S (1996) Identification of a c-fos-induced gene that is related to the platelet-derived growth factor/vascular endothelial growth factor family. Proc Natl Acad Sci USA 93:11675–11680

Pajusola K, Aprelikova O, Korhonen J, Kaipainen A, Pertovaara L, Alitalo R, Alitalo K (1992) FLT4 receptor tyrosine kinase contains seven immunoglobulin-like loops and is expressed in multiple human tissues and cell lines. Cancer Res 52:5738–5743

Pajusola K, Aprelikova O, Armstrong E, Morris S, Alitalo K (1993) Two human FLT4 receptor tyrosine kinase isoforms with distinct carboxyterminal tails are produced by alternative processing of primary transcripts. Oncogene 8:2931–2937

Pajusola K, Aprelikova O, Pelicci G, Weich H, Claesson-Welsh L, Alitalo K (1994) Signalling properties of FLT4, a proteolytically processed receptor tyrosine kinase related to two VEGF receptors. Oncogene 9:3545–3555

Rosnet O, Stephenson D, Mattei M-G, Marchetto S, Shibuya M, Chapman VM, Birnbaum D (1993) Close physical linkage of the FLT1 and FLT3 genes on chromosome 13 in man and chromosome 5 in mouse. Oncogene 8:173–179

Sabin FR (1902) On the origin of the lymphatic system from the veins and the development of the lymph hearts and thoracic duct in the pig. Am J Anat 1:367–391

Salven P, Lymboussaki A, Heikkilä P, Jääskelä-Saari H, Aase K, von Euler G, Eriksson U, Joensuu H, Alitalo K (1998) Expression of the novel vascular endothelial growth factors VEGF-B and VEGF-C (submitted)

Wang JF, Ganju RK, Liu ZY, Avraham H, Avraham S, Groopman JE (1997) Signal transduction in human hematopoietic cells by vascular endothelial growth factor related protein, a novel ligand for the FLT4 receptor. Blood 90:3507–3515

Watkins-Chow DE, Douglas KR, Buckwalter MS, Probst FJ, Camper SA (1997) Construction of a 3-Mb contig and partial transcript map of mouse chromosome 11. Genomics 45:147–157

Wilting J, Eichmann A, Christ B (1997) Expression of the avian VEGF receptor homologues quek1 and quek2 in blood-vascular and lymphatic endothelial and non-endothelial cells during quail embryonic development. Cell Tissue Res 288:207–223

Yamada Y, Nezu J, Shimane M, Hirata Y (1997) Molecular cloning of a novel vascular endothelial growth factor, VEGF-D. Genomics 42:483–488

# Vascular Permeability Factor/Vascular Endothelial Growth Factor and the Significance of Microvascular Hyperpermeability in Angiogenesis

H. F. Dvorak, J. A. Nagy, D. Feng, L. F. Brown, and A. M. Dvorak

| | | |
|---|---|---|
| 1 | Introduction | 98 |
| 2 | VPF/VEGF as a Vascular Permeabilizing Cytokine | 99 |
| 3 | Overexpression of VPF/VEGF and Its Receptors in Tumors | 101 |
| 4 | Microvascular Hyperpermeability – A Characteristic Feature of Tumor Blood Vessels | 101 |
| 5 | Increased VPF/VEGF Expression and Microvascular Hyperpermeability in Non-neoplastic States Associated with Angiogenesis and New Stroma Generation | 104 |
| 5.1 | Wound Healing | 105 |
| 5.2 | Retinopathies | 106 |
| 5.3 | Delayed Hypersensitivity | 107 |
| 5.4 | Rheumatoid Arthritis | 107 |
| 5.5 | Psoriasis and Inflammatory Skin Disorders | 108 |
| 5.6 | Increased VPF/VEGF Expression and Microvascular Hyperpermeability in Ovarian Follicle Development and Corpus Luteum Formation | 108 |
| 5.7 | Expression of VPF/VEGF in Normal Adult Tissues. Possible Relationship to Intrinsic, Low-Level Permeability of Normal Microvessels | 109 |
| 6 | Consequences of Increased Microvascular Permeability: Deposition of a Pro-angiogenic Provisional Matrix | 109 |
| 7 | Signal Transduction Pathways by which VPF/VEGF Renders Microvascular Endothelium Hyperpermeable | 110 |
| 8 | The Structural Basis of VPF/VEGF-Induced Microvascular Hyperpermabililty | 111 |
| 8.1 | Extravasation of Plasma and Its Solutes from Normal Blood Vessels | 111 |
| 8.2 | Extravasation of Plasma and Its Solutes From Hyperpermeable Tumor Blood Vessels | 113 |
| 8.3 | Vesiculo-vacuolar Organelles (VVOs) | 114 |
| 8.4 | Structural Properties of VVOs | 117 |
| 8.5 | Trans-Endothelial Openings. Are They Inter-Endothelial Cell Gaps or Trans-Endothelial Cell Pores? | 118 |
| 8.6 | The Controversy Regarding Trans-Endothelial Openings: Gaps, Pores and VVOs | 123 |
| 8.7 | Possible Relationships Between VVOs and Trans-Endothelial Cell Pores | 123 |
| 9 | Summary | 124 |
| References | | 125 |

Departments of Pathology, Beth Israel Deaconess Medical Center, 330 Brookline Ave, Boston, MA 02215 USA and Harvard Medical School, Boston, MA 02215, USA

# 1 Introduction

Vascular permeability factor/vascular endothelial growth factor (VPF/VEGF) was originally discovered in the late 1970s because of its capacity to increase the permeability of microvessels to plasma and plasma proteins (DVORAK et al. 1979a,b). Using plastic-embedded, light microscopic sections and, subsequently, immunohistochemistry, we noted that transplantable tumors growing in guinea pigs and rodents exhibit substantial deposits of fibrin in their stroma. Fibrin results from the clotting of fibrinogen, a 340kDa plasma protein which, under normal circumstances, is retained almost quantitatively within the blood vasculature. For fibrin to be deposited outside of blood vessels in tumor stroma, it was necessary that two requirements be met; namely, (1) that microvessels be abnormally hyperpermeable to permit the escape of fibrinogen and other plasma proteins necessary for blood clotting and (2) that there be a mechanism in place for activating the clotting system. In fact, both requirements were found to be met in tumors. The microvessels supplying tumors were hyperpermeable to fibrinogen and other plasma proteins, and both tumor cells and host stromal cells were capable of initiating extravascular coagulation via the tissue-factor pathway. Encouraged by these findings, we initiated a search for a tumor product that could account for tumor-vessel hyperpermeability. A potent vascular permeabilizing protein was soon found in serum-free tumor culture supernatants (DVORAK et al. 1979a,b) and was subsequently purified to homogeneity and given the name vascular permeability factor (VPF) (SENGER et al. 1983, 1986, 1987, 1990).

Several years later, our collaborators at the Monsanto Company found that, in addition to enhancing vascular permeability, VPF was also a selective mitogen for cultured vascular endothelial cells (CONNOLLY et al. 1989). Independently, investigators in California isolated an endothelial cell mitogen from pituitary cell cultures and gave this protein the name vascular endothelial growth factor (VEGF) (FERRARA and HENZEL 1989; GOSPODAROWICZ et al. 1989; LEUNG et al. 1989). Subsequent experiments determined that the vascular permeabilizing and the endothelial cell mitogenic activities were mediated by the same molecule, hence the designation of this molecule as VPF/VEGF (KECK et al. 1989; CLAUSS et al. 1990; CONN et al. 1990a,b; FERRARA et al. 1992; KOCH et al. 1994).

In addition to its vascular permeabilizing activity, VPF/VEGF exerts a number of other important actions on vascular endothelium. Many of these are discussed elsewhere in this volume and in several reviews (DVORAK et al. 1992, 1995; FERRARA et al. 1992; SENGER et al. 1993; ALON et al. 1995; BROWN et al. 1997a; WATANABE and DVORAK 1997; WATANABE et al. 1997). Increased microvascular permeability is the earliest biological activity detected following interaction of tissues with VPF/VEGF, becoming evident within a matter of seconds to a few minutes. In contrast, changes in endothelial cell shape, adhesion, migration, altered messenger ribonucleic acid (mRNA) and protein expression, cell division and protection from apoptosis and senescence develop more slowly over a period of hours to days to weeks.

Recent work has led to the discovery of several additional proteins that are structurally related to VPF/VEGF. VPF/VEGF is therefore the founding member of a family of proteins that presently includes placenta growth factor (PlGF), VEGF-B and VEGF-C (CONN et al. 1990a,b; HOUCK et al. 1991; FERRARA et al. 1992; KIM et al. 1992; PARK et al. 1993; JOUKOV et al. 1996; OLOFSSON et al. 1996; JELTSCH et al. 1997). Of these, only VPF/VEGF (and to a lesser extent VEGF-C) has been found to increase microvascular permeability.

In this chapter, we review the significance of VPF/VEGF-induced microvascular hyperpermeability for angiogenesis and stroma generation in both pathology and normal physiology. We also consider the mechanisms by which VPF/VEGF induces microvascular hyperpermeability and the structural basis for macromolecular extravasation from VPF/VEGF-hyperpermeabilized microvessels.

## 2 VPF/VEGF as a Vascular Permeabilizing Cytokine

VPF/VEGF's capacity to increase microvascular permeability is arguably its most potent activity and one that is unique among known cytokines. VPF/VEGF is among the most potent vascular permeabilizing agents known. With a potency some 50,000 times that of histamine (DVORAK et al. 1992; SENGER et al. 1993; BROWN et al. 1997), it is effective at concentrations well below 1 nM in the Miles assay (Fig. 1). VPF/VEGF selectively permeabilizes venular endothelium to plasma and plasma proteins in a number of vascular beds, including those of skin, subcutaneous tissue, peritoneum, pleura, mesentery, diaphragm, retina and skeletal muscle. Vascular hyperpermeability becomes evident within a minute or two of VPF/VEGF injection into normal skin or other tissues and persists for ~20 min. The hyperpermeability induced is therefore time-limited, reversible and is not associated with detectable injury to endothelial cells or other microvascular components (SENGER et al. 1983; DVORAK 1986, 1990; DVORAK et al. 1988, 1992, 1995).

Before proceeding further, it will be good to define with greater precision what is meant by "vascular permeability". Physiologists note that extravasation of a particular solute, such as albumin from plasma into tissues, is dependent on a number of different variables (RIPPE and HARALDSSON 1994; BATES and CURRY 1996; JAIN 1996). These include: properties of the solute molecule (i.e., size, shape, charge); surface area of the microvasculature available for flux; the relative concentrations of solute in plasma, relative to tissues; the net filtration rate; the ratio of convective to diffusive flux; the solute permeability coefficient; and the solute reflection coefficient. In addition, the net filtration rate is dependent on the net driving force for water flux (the balance of hydrostatic and colloid osmotic pressures) and hydraulic conductivity (Lp), a measure of the resistance of the microvessel wall to water flow that depends on pore size, density and thickness. Obviously, therefore, increased solute extravasation can reflect changes in a number of different variables besides intrinsic microvascular permeability.

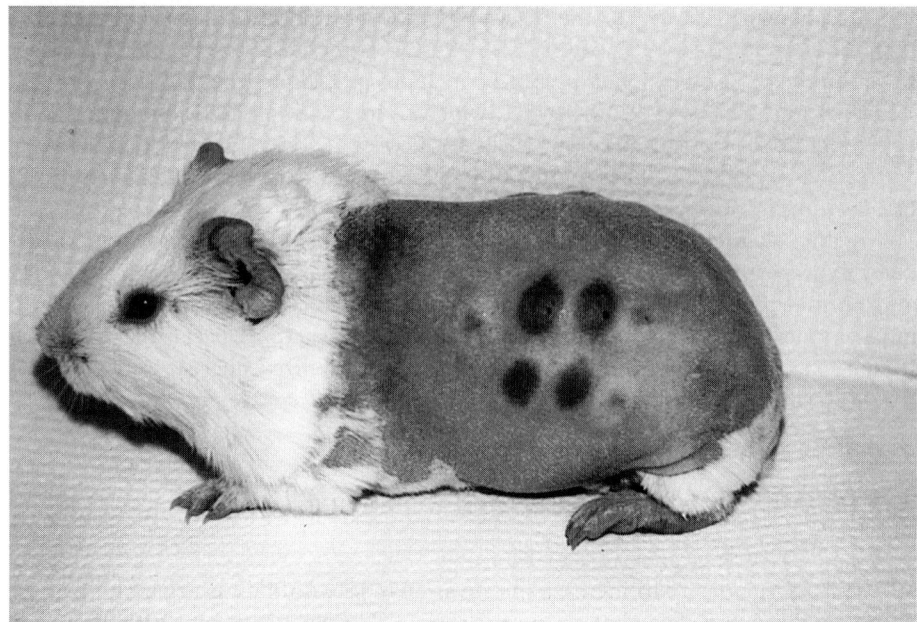

**Fig. 1.** The Miles assay provides a sensitive measure of increased microvascular permeabililty to plasma proteins (MILES and MILES 1952; DVORAK et al. 1979). Evans-blue dye is injected intravenously into a guinea pig (or other laboratory animal), whose flank skin has been shaved and depilated. Evans-blue dye binds to albumin and other plasma proteins and, therefore, is normally retained within the circulation. Putative vascular permeabilizing factors, such as vascular permeability factor/vascular endothelial growth factor (VPF/VEGF) and control substances that do not enhance permeability are then injected intradermally. Albumin, along with bound Evans-blue dye, extravasates at sites of increased microvascular permeability, generating a visible blue spot, the dimensions and intensity of which, compared with control sites, provide a measure of microvascular permeability. *Top row, left to right*: neutralizing anti-VPF/VEGF antibody, line-10 tumor ascites fluid, mix of line-10 tumor ascites fluid and control immunoglobulin (Ig) G, mix of line-10 tumor ascites fluid and anti-VPF/VEGF antibody. *Bottom row, left to right*: line-1 tumor ascites fluid, mix of line-1 ascites fluid with control IgG, mix of line-1 ascites fluid with anti-VPF/VEGF antibody. (Photograph courtesy of Dr. Donald R. Senger)

So, what is being measured when VPF/VEGF leads to increased extravasation of plasma proteins in the Miles assay? VPF/VEGF is reported to relax coronary arteries (KU et al. 1993) and, if applicable to the smooth muscle cells of cutaneous venules, would be expected to result in increased vessel diameter and surface area and perhaps reduced endothelial cell (and therefore pore) thickness. However, recent studies have shown that VPF/VEGF dramatically increases Lp (BATES and CURRY 1996). Thus, studies with isolated, perfused, frog mesenteric vessels have demonstrated a nearly eightfold acute increase in Lp in response to VPF/VEGF, with a return to baseline within 2min. In addition, these studies showed that 24h after a 10min perfusion with 1nM VPF/VEGF, baseline hydraulic conductivity was increased ~fivefold above the starting baseline; thus, VPF/VEGF exerts both acute and chronic effects on vascular permeability (BATES and CURRY 1996).

VPF/VEGF does not itself provoke mast cell degranulation or induce a significant inflammatory cell infiltrate. The vascular permeabilizing action of VPF/VEGF is not blocked by inhibitors of inflammation, including those that block histamine, thrombin and platelet activating factor (DVORAK et al. 1979; SENGER et al. 1993).

All of VPF/VEGF's actions, including that of permeabilizing venules, are thought to be mediated by interaction with two high affinity receptors, VEGF receptor-1 (VEGFR-1; also denoted Flt-1) and VEGFR-2 (also denoted KDR in man and Flk-1 in rodents) (MATTHEWS et al. 1991; TERMAN et al. 1991, 1992; DE VRIES et al. 1992; MILLAUER et al. 1993; QUINN et al. 1993). Both receptors are expressed predominantly, though not exclusively, on vascular endothelium (HATVA et al. 1995; BROWN et al. 1997), and this distribution accounts for the selective action of VPF/VEGF on endothelial cells.

## 3 Overexpression of VPF/VEGF and Its Receptors in Tumors

VPF/VEGF is substantially overexpressed at both the mRNA and protein levels by many of the most important human and animal tumors and by many immortalized and transformed cell lines as determined by Northern analysis, in situ hybridization, Western blotting, "sandwich" or other immunoassays and immunohistochemistry (DVORAK et al. 1979a,b; SENGER et al. 1983, 1986; PLATE et al. 1992; BROWN et al. 1993a,b, 1995b, 1996; GUIDI et al. 1995; ABU-JAWDEH et al. 1996; CHOW et al. 1997). However, important exceptions exist even among tumors that are able to elicit new blood vessels, indicating that some tumors induce angiogenesis by VPF/VEGF-independent mechanisms (YOSHIJI et al. 1997). Of interest, the endothelial cells lining microvessels that supply VPF/VEGF-secreting tumors themselves overexpress both VEGFR-1 and -2. Finally, at least some malignant tumors overexpress other members of the VPF/VEGF family as well as VPF/VEGF itself (WEINDEL et al. 1994; JOUKOV et al. 1996).

## 4 Microvascular Hyperpermeability – A Characteristic Feature of Tumor Blood Vessels

The blood vessels that supply tumors differ from those supplying normal tissues in a number of important respects. Thus, tumor vessels are distributed unevenly, often assume a serpentine pattern and are relatively undifferentiated so that they do not correspond closely to the arterioles, capillaries and venules of normal tissues (WARREN 1979). In addition, tumor vessels are hyperpermeable to plasma proteins and to other circulating macromolecules (reviewed in DVORAK 1986; NAGY et al. 1988; DVORAK et al. 1992, 1995; SENGER et al. 1993; BROWN et al. 1997); this is not unexpected in that most tumors overexpress VPF/VEGF.

Tumor vessel hyperpermeability can be readily demonstrated with radiolabeled plasma proteins or with other macromolecular tracers, such as fluoresceinated dextrans (FITC-D) in tumor-bearing animals. Leakage of 150-kDa FITC-D from tumor microvessels is noticeable within a few minutes of intravenous injection, and tracer accumulates progressively in the tumor interstitium (Fig. 2); in contrast, this and other macromolecular tracers extravasate much more slowly from the vessels that supply most normal tissues (GERLOWSKI and JAIN 1986; DVORAK et al. 1988; LIN et al. 1994; NAGY et al. 1995a, b, c). For obvious reasons, tracer studies of this type cannot be performed ethically in cancer patients. Nonetheless, the microvessels supplying human tumors must be hyperpermeable because, like their animal counterparts, their stroma also contain fibrin and other extravasated plasma proteins (A.M. Dvorak, unpublished data).

Quantitative measurements of tumor-vessel hyperpermeability have generally been made by following the time-dependent extravasation and accumulation of radiolabeled macromolecular tracers. Fibrinogen has proved to be particularly sensitive for such studies in that it rapidly undergoes clotting and covalent cross-linking upon extravasation (DVORAK et al. 1984, 1985). Thus, unlike other extravasated plasma-protein tracers, insoluble fibrin cannot re-enter the blood or host lymphatics without first being re-solubilized by proteolytic degradation, a time-dependent process even in tumors with increased fibrinolytic capacity.

Using a variety of macromolecular tracers, a number of investigators have determined that the vessels supplying different solid tumors are approximately four to ten or more times more permeable to macromolecules than the microvessels supplying normal tissues (DEWEY 1959; SPAR et al. 1967; SONG and LEVITT 1971; DVORAK et al. 1979, 1984, 1988; O'CONNOR and BALE 1984; GERLOWSKI and JAIN 1986; BROWN et al. 1987, 1988a,b; NAGY et al. 1988, 1993, 1995). However, there are limitations to this type of measurement because, as noted above, a strict comparison requires that blood flow, vessel surface area and intra- and extravascular-pressure relationships are the same in tumors as in the normal tissues used as controls. These assumptions are frequently invalid. Nonetheless, the experiments that have been performed clearly indicate that tumor vessels are hyperpermeable to circulating macromolecules and provide a rough measure of the extent of that hyperpermeability.

Particularly careful measurements of tracer influx and efflux have been performed in mice bearing ascites tumors.[1] Using a series of FITC-D tracers varying in size from Mr 3kDa to 5000kDa, NAGY et al. demonstrated that the rate constant for tracer influx into the peritoneal cavity was up to 40-fold greater in mice with established tumor ascites than in normal mice. No absolute size barrier to macromolecular influx was found with tracers, up to the largest studied (5000kDa, Stokes' radius ~40nm). The rate constants for influx increased 10- to 40-fold for FITC-D tracers of 20–150kDa; however, influx of 5000kDa FITC-D was also

---

[1] As used here, influx refers to the passage of tracer from the blood to the peritoneal cavities of normal or tumor-bearing mice. Efflux refers to the reverse process, that of tracer passage out of the peritoneal cavity. Influx and efflux are complex processes, modulated by properties of the tracer, by the vasculature (vascular area, hydraulic conductivity and reflectivity and intra-and extravascular-pressure relationships) and by rates of convection and diffusion that determine entry of extravasated tracers from the immediate extravascular space into the peritoneal cavity (RIPPE and HARALDSSON 1994; BATES and CURRY 1996; JAIN 1996). While a complete understanding of macromolecular extravasation requires detailed measurement of each of these parameters individually, it is useful, as a first approximation, to consider influx and efflux as composite entities, i.e., as a summation of the several variables listed (NAGY et al. 1988, 1993, 1995a–c).

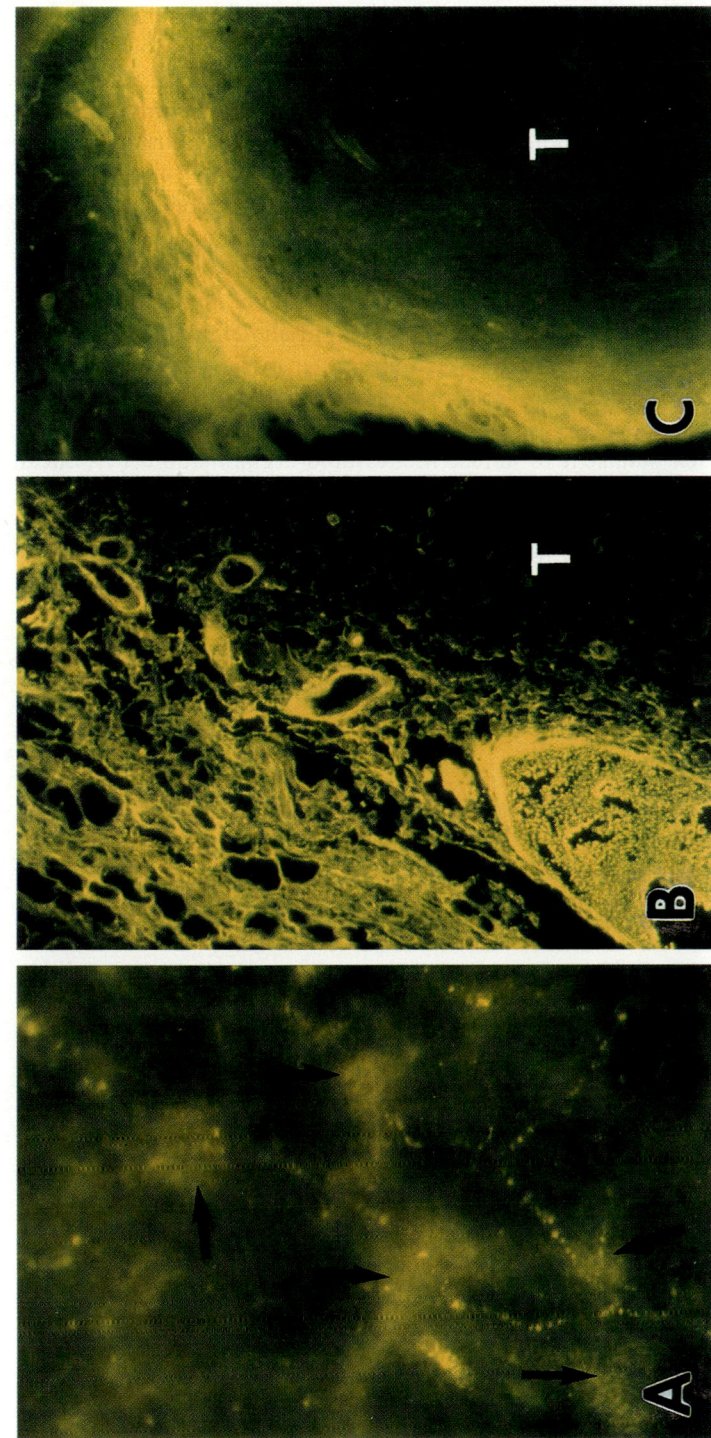

**Fig. 2A–C.** Hyperpermeability of tumor blood vessels to high molecular weight (150 kDa) fluoresceinated dextran (FITC-D). **A** "Puffs" of fluorescence (*arrows*) identify sites of leakage from surface vessels of line-1 bile-duct carcinoma, growing in a syngeneic strain-2 guinea pig 15 min after intravenous injection of FITC-D. **B, C** Paraffin sections illustrating extravasation of FITC-D from the microvessels enveloping nodules of Lewis lung carcinomas growing in syngeneic mice; tissue was harvested at 5 min (**B**) and 15 min (**C**) after intravenous injection of FITC-D. Extravasated tracer surrounds individual leaky vessels (**B**) and forms a circumferential rim at the tumor–host interface (**C**) that does not significantly penetrate into the tumor parenchyma (*T*). **A** ×30; **C, D** ×130 (Modified from Yeo and Dvorak 1995 with permission)

enhanced, but only two to fourfold above that of normal vessels. Immunofluorescence microscopy suggests that the lower rate constant for influx for 5000kDa FITC-D is attributable, at least in part, to impaired diffusion of this large molecule after it has extravasated from vessels, i.e., the rate-limiting step may not be at the level of microvascular endothelium, but likely reflects impaired diffusion of extravasated tracer into the peritoneal cavity.

Impaired efflux of tracers from the peritoneal cavity to the blood plasma was also a feature of ascites tumors (NAGY et al. 1988, 1993). Within a day of intraperitoneal tumor cell injection, and prior to the onset of angiogenesis or of detectable increases in peritoneal vessel hyperpermeability, efflux of plasma albumin from the peritoneal cavity was reduced three- to fivefold and efflux of particulate, $^{51}$Cr-labeled erythrocytes was reduced even more extensively. The mechanisms responsible for retarded tracer efflux from the peritoneal cavity have not yet been discovered, although lymphatic obstruction is a possibility. However, impaired efflux was not by itself sufficient to induce ascites fluid accumulation. Peritoneal fluid only began to accumulate 5–7 days after tumor cell injection, when tumor cells had increased greatly in number and were secreting sufficient amounts of VPF/VEGF to increase influx, i.e., to render peritoneal-lining blood vessels hyperpermeable to plasma (Fig. 3). Accumulation of peritoneal fluid closely paralleled the appearance of VPF/VEGF in ascites (NAGY et al. 1995).

Besides the close correlation that exists between VPF/VEGF overexpression and tumor vessel hyperpermeability, there is additional evidence that points strongly to a causal relationship. Following transfection with VPF/VEGF cDNA, human melanoma cells that otherwise expressed only low levels of VPF/VEGF induced strikingly hyperpermeable vessels when transplanted into immunodeficient mice (CLAFFEY et al. 1996). Moreover, antibodies to VPF/VEGF inhibit the leakage of plasma proteins induced by both VPF/VEGF and tumors (SENGER et al. 1983; YUAN et al. 1996). Finally, although some investigators have postulated that kinins (GREENBAUM et al. 1978; MATSUMURA et al. 1988) and perhaps other vasoactive mediators may be involved, VPF/VEGF is the only vascular permeabilizing agent that has, thus far, been found in tumor culture supernatants or in tumor ascites fluids isolated and maintained under physiological conditions. The evidence, therefore, is strong that tumor cell expression and secretion of VPF/VEGF is responsible for the microvascular hyperpermeability characteristic of tumor microvessels.

## 5 Increased VPF/VEGF Expression and Microvascular Hyperpermeability in Non-neoplastic States Associated with Angiogenesis and New Stroma Generation

VPF/VEGF and its endothelial cell receptors are also overexpressed in a number of non-neoplastic pathologies and in certain examples of normal physiology in which angiogenesis and stroma generation figure prominently.

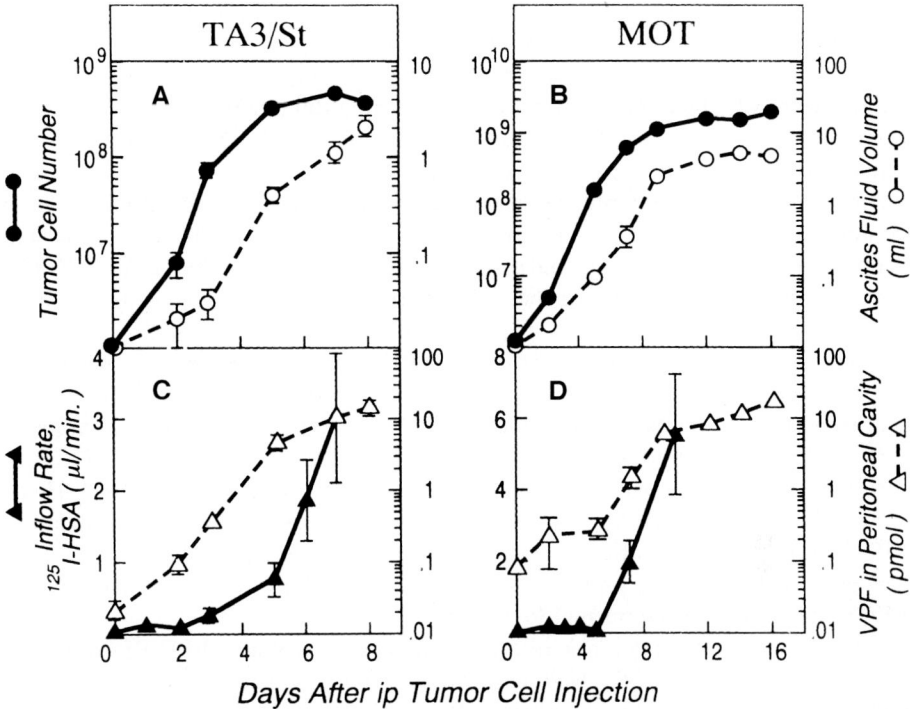

**Fig. 3A–D.** Kinetics of tumor cell proliferation, ascites fluid accumulation, vascular permeability factor/vascular endothelial growth factor (VPF/VEGF) accumulation and $^{125}$I human serum albumin (HSA) inflow rate into the peritoneal cavities of syngeneic mice bearing TA3/St (**A, C**) or MOT (**B, D**) ascites tumor cells. Modified from NAGY et al. 1995 with permission

## 5.1 Wound Healing

In healing skin wounds, VPF/VEGF mRNA expression increases dramatically in epidermal keratinocytes at the wound edge and in residual hair follicles in the wound base within 24h of injury. VPF/VEGF overexpression reaches a peak at 2–3 days and persists at an elevated level for ~1 week – the time required for granulation tissue to form and migrating keratinocytes to cover the wound defect (BROWN et al. 1992). Certain mononuclear cells infiltrating the dermis, most likely a sub-population of macrophages, also overexpress VPF/VEGF. Consistent with these findings, elicited (but not native) peritoneal macrophages express readily detectable levels of VPF/VEGF mRNA (BERSE et al. 1992). Increased expression of at least one VPF/VEGF receptor mRNA, VEGFR-1 (VEGFR-2 has not, to our knowledge, been investigated), has been documented in the endothelial cells lining the new blood vessels that form as part of the developing granulation tissue accompanying wound healing (PETERS et al. 1993).

Like VPF/VEGF overexpression, microvascular hyperpermeability is an early and persistent feature of wound healing (BROWN et al. 1988b, 1992; BREUING et al.

1992). Following split thickness excisional wounding in guinea-pig or rodent skin, overt bleeding stops within a few minutes. However, measurements with circulating protein tracers indicate that local microvascular hyperpermeability persists for a number of days. Thus, for the first 4 days after wounding, permeability of the local vasculature to circulating $^{125}$I fibrinogen was 7- to 15-fold greater than in normal skin and only, thereafter, fell slowly to control levels over the course of a week. The extent of vascular hyperpermeability in wounds is, thus, comparable to that found in tumors and corresponds temporally to the overexpression of VPF/VEGF by skin keratinocytes and infiltrating macrophages. In contrast to tumors, VPF/VEGF expression was downregulated as healing progressed and, in parallel with decreased expression of VPF/VEGF, vascular permeability returned to normal.

The factors responsible for regulating VPF/VEGF expression in wound healing have not been carefully investigated, but lowered tissue oxygen levels are characteristic of wounds (KNIGHTON et al. 1983) and hypoxia is a potent stimulus to VPF/VEGF upregulation (KOOS and OLSON 1991; SHWEIKI et al. 1992; LADOUX and FRELIN 1993; GOLDBERG and SCHNEIDER 1994; MINCHENKO et al. 1994; HATA et al. 1995). Local tissue hypoxia, therefore, may be responsible for upregulating VPF/VEGF expression after wounding; likewise, restoration of blood flow with healing restores tissue oxygen levels to normal and likely serves to reduce VPF/VEGF expression to the normal low levels.

In addition to having a role in the healing of skin wounds, VPF/VEGF is also likely to be important in the healing of other forms of tissue injury. For example, normal adult cardiac myocytes express detectable levels of VPF/VEGF and these levels, and the levels of both VPF/VEGF receptors on cardiac microvessels, are strikingly upregulated within hours of ischemic injury. Thus, VPF/VEGF probably plays an important role in the healing of myocardial infarcts (LADOUX and FRELIN 1993; BANAI et al. 1994; HASHIMOTO et al. 1994; MINCHENKO et al. 1994; LI et al. 1996) and presumably also in the repair of tissue injury in other organs.

## 5.2 Retinopathies

Regulated overexpression of VPF/VEGF and its receptors has also been demonstrated in the retinas of patients and animals subjected to various forms of ocular ischemia, e.g., diabetes mellitus, central vein occlusion, fetal prematurity, etc. (ADAMIS et al. 1993, 1994; AIELLO et al. 1994; MILLER et al. 1994; PE'ER et al. 1995; PIERCE et al. 1995). Several types and layers of retinal cells express VPF/VEGF, and, in various disease states, upregulated VPF/VEGF mRNA expression is selectively localized to those layers of retina that are expected to be ischemic, e.g., the outer retinal layer in chronic retinal detachment (PE'ER et al. 1995). Immunoassays performed on aqueous or vitreous humor have demonstrated a strong correlation between VPF/VEGF content and proliferative retinopathy in patients with diabetes, central retinal-vein occlusion and fetal prematurity. However, VPF/VEGF levels are very low in the eyes of patients affected by disorders that do not involve vascular proliferation (e.g., diabetes without retinopathy) and in the eyes of patients with quiescent vascular proliferations. Moreover, all layers of the retinas of diabetic rats display increased amounts of VPF/VEGF by immunohistochemistry, and retinal vessels are hyperpermeable and stain with antibodies to VPF/VEGF (MURATA et al. 1995), much as has been described for tumor blood vessels (DVORAK et al. 1991a,b; PLATE et al. 1992; QU-HONG et al. 1995).

## 5.3 Delayed Hypersensitivity

Overexpression of VPF/VEGF mRNA by keratinocytes and infiltrating mononuclear cells is observed in delayed hypersensitivity reactions of both the tuberculin and contact allergy type and in both rodent and human skin (BROWN et al. 1995). In addition, VEGFR-1 and -2 mRNAs are overexpressed in the microvascular endothelium supplying these reactions.

The microvessels supplying delayed hypersensitivity reactions have long been known to be hyperpermeable to plasma and circulating macromolecules (COLVIN and DVORAK 1975). At one time a lymphocyte product, designated "skin-reactive factor" (SRF), was reported to increase microvascular permeability in normal skin vessels (MAILLARD et al. 1972); however, unlike VPF/VEGF, SRF acted with delayed kinetics, such that increased permeability did not begin for at least 20min after intracutaneous injection and then persisted for up to 4 h. Another factor, also named "vascular permeability factor" (VPF), was reported to be released by mitogen-stimulated human blood lymphocytes, taken from patients with nephrotic syndrome (SOBEL and LAGRUE 1980); comparable culture supernatants prepared with lymphocytes from normal control subjects exhibited little or no permeability enhancing activity. The lymphocyte VPF was never well-characterized, but was thought to have a molecular weight of $\sim$12,000Da, well below that of VPF/VEGF. Attempts in our laboratory to find VPF/VEGF or a permeability-enhancing activity in cultures of sensitized guinea-pig or rat lymphocytes, with or without antigen or lectin, have proved unsuccessful. However, cultured macrophages can be stimulated to produce VPF/VEGF (BERSE et al. 1992); this finding is consistent with results obtained from in situ hybridization, which indicate that a sub-population of infiltrating macrophages synthesize the VPF/VEGF found in delayed hypersensitivity reactions as well as in healing skin wounds (BROWN et al. 1992).

## 5.4 Rheumatoid Arthritis

Overexpression of VPF/VEGF and its receptors is a prominent feature of rheumatoid arthritis (FAVA et al. 1994; KOCH et al. 1994), an inflammatory disorder which is often classified as an example of cell-mediated or delayed-type hypersensitivity. Both VPF/VEGF mRNA and protein are overexpressed by the synovial cells that line actively inflamed joints, and both VEGFR-1 and -2 are overexpressed by microvessels that supply the pannus. Accumulation of plasma protein-rich joint fluid is a characteristic feature of rheumatoid arthritis and provides decisive evidence of local microvascular hyperpermeability. In addition, VPF/VEGF is present in rheumatoid joint fluids in amounts comparable to those found in tumor ascites (YEO et al. 1993; FAVA et al. 1994; KOCH et al. 1994; NAGY et al. 1995).

## 5.5 Psoriasis and Inflammatory Skin Disorders

Overexpression of VPF/VEGF and its receptors characterizes a variety of inflammatory skin disorders. Prominent among these are such important clinical entities as psoriasis, bullous pemphigoid, dermatitis herpetiformis and erythema multiforme (DETMAR et al. 1994; BROWN et al. 1995). In all of these examples, epidermal keratinocytes overexpress VPF/VEGF and dermal microvascular endothelial cells overexpress both VEGFR-1 and -2. Also, papillary dermal edema is characteristic of these disorders and high levels of VPF/VEGF are detected in the blister fluid elicited in several of these entities.

## 5.6 Increased VPF/VEGF Expression and Microvascular Hyperpermeability in Ovarian Follicle Development and Corpus Luteum Formation

VPF/VEGF overexpression correlates closely with the vascular hyperpermeability characteristic of developing ovarian follicles and corpora lutea (PHILLIPS et al. 1990; RAVINDRANATH et al. 1992; SHWEIKI et al. 1993; KAMAT et al. 1995). As graafian follicles develop in the ovary and accumulate fluid, the lining granulosa cells become increasingly positive for VPF/VEGF by immunohistochemistry, and the surrounding theca cells also become weakly positive. After ovulation, follicular granulosa cells differentiate into granulosa lutein cells. In early corpora lutea, granulosa lutein cells exhibit strong cytoplasmic immunoperoxidase staining for VPF/VEGF protein and also express substantial VPF/VEGF mRNA as determined by in situ hybridization.

Accompanying VPF/VEGF expression by follicular cells, adjacent blood vessels become hyperpermeable, and it is thought that at least a portion of the proteinaceous fluid (liquor folliculi) that accumulates within developing follicles is derived from plasma. Fluid accumulation accelerates after the pre-ovulatory gonadotropin surge when capillaries of the thecal layer of the follicle become hyperpermeable (MORRIS and SASS 1966; PAYER 1975; OKUDA et al. 1980; MOOR and SEAMARK 1986; CULLINAN-BOVE and KOOS 1993). It is at this time, shortly before ovulation, that both granulosa and theca cells of the human ovary stain strongly for VPF/VEGF. Thus, it is likely that local secretion of VPF/VEGF, under hormonal regulation, renders thecal microvessels hyperpermeable, leading to increased extravasation of plasma and accumulation of antral fluid in the maturing graafian follicles.

Careful studies in the ewe and rat indicate that microvessels associated with developing corpora lutea are strikingly hyperpermeable to plasma and plasma proteins (MORRIS and SASS 1966); in fact, the microvessels that supply the corpus luteum are reported to be among the leakiest in the body (OKUDA et al. 1980; CULLINAN-BOVE and KOOS 1993). The close temporal and spatial correlation between VPF/VEGF expression by luteal cells and local vessel hyperpermeability suggests that VPF/VEGF is the responsible mediator. As a further point of evidence, VPF/VEGF has been localized by immunohistochemistry to the new blood vessels supplying human corpora lutea; by analogy with other species, it is likely that these new vessels are also hyperpermeable (KAMAT et al. 1995).

## 5.7 Expression of VPF/VEGF in Normal Adult Tissues. Possible Relationship to Intrinsic, Low-Level Permeability of Normal Microvessels

In the adult, strong expression of VPF/VEGF mRNA is detected in several tissues by in situ hybridization in the absence of angiogenesis, e.g., visceral glomerular epithelium (podocytes) of normal adult kidneys, epithelial cells of lung alveoli and adrenal cortex, and cardiac myocytes (BERSE et al. 1992; BREIER et al. 1992; MONACCI et al. 1993; LI et al. 1995). Of possible relevance, two of these tissues (renal glomeruli, adrenal cortex) have fenestrated endothelium, and an association between tissue VPF/VEGF expression and endothelial fenestration has been noted (BERSE et al. 1992; BREIER et al. 1992; ROBERTS and HASAN 1993; SIOUSSAT et al. 1993; ROBERTS and PALADE 1995; SIMON et al. 1995).

Albumin, γ globulin and other plasma proteins have important physiological functions in tissues and it is, therefore, necessary that they be able to cross normal microvessels in limited amounts. It is quite possible that the low level extravasation of plasma proteins that occurs in normal tissues is mediated by VPF/VEGF. Certain microvessels, such as those found in renal glomeruli, are relatively more permeable to plasma proteins than others and, as noted, glomerular podocytes express large amounts of VPF/VEGF (BERSE et al. 1992).

## 6 Consequences of Increased Microvascular Permeability: Deposition of a Pro-angiogenic Provisional Matrix

The work reviewed above has established that VPF/VEGF is responsible for the increased microvascular permeability that is consistently associated with angiogenesis and new stroma generation as they occur in tumors, in many non-neoplastic pathologies and in certain physiological states, such as ovarian cycling. The regular association among overexpression of VPF/VEGF and its receptors, microvascular hyperpermeability and angiogenesis suggests a causal relationship. Indeed, such a relationship has been established and has been reviewed elsewhere in detail (DVORAK et al. 1983, 1991b, 1992, 1995; DVORAK 1986; SENGER et al. 1993; YEO and DVORAK 1995; BROWN et al. 1997).

In brief, overexpression of VPF/VEGF renders local microvessels hyperpermeable to circulating macromolecules; as a consequence, plasma proteins extravasate into the tissue interstices, depositing a new, provisional extravascular matrix that is rich in fibrin and other adhesive proteins. In contrast to the pre-existing mature extracellular matrix of adult tissues, which is not supportive of tissue remodeling, this new provisional fibrin matrix is pro-angiogenic and pro-stromagenic. Fibrin deposited in the extravascular space contributes to angiogenesis and new stroma formation in at least three important ways: (1) it provides a provisional matrix that affords structure and tissue organization; 2) it provides a favorable matrix that encourages and supports the inward migration of mesenchymal cells (fibroblasts and new blood vessels) to form granulation tissue; and (3) it induces granulation tissue maturation. Taken together, the microvascular hyperpermeability induced by VPF/VEGF overexpression initiates angiogenesis and new stroma formation as these occur in a variety of different pathological and physiological settings. Of course, VPF/VEGF contributes to angiogenesis and stroma generation at later stages as well, by inducing endothelial cell division and by altering integrin expression so that endothelial cells can migrate effectively on the newly deposited provisional matrix proteins.

## 7 Signal Transduction Pathways by which VPF/VEGF Renders Microvascular Endothelium Hyperpermeable

Interaction of VPF/VEGF with endothelial cells leads to the phosphorylation of its two high-affinity receptors (VEGFR-1 and -2) and a number of proteins including phospholipase C$\gamma$; phosphatidylinositol 3-kinase; GAP, the RAS GTPase-activating protein; two GAP-associated proteins, p190 and p62; and the oncogenic adaptor protein NcK (DOUGHER-VERMAZEN et al. 1994; WALTENBERGER et al. 1994; GUO et al. 1995; SEETHARAM et al. 1995). However, it is not certain whether all or only some of these phosphorylations are associated with the permeability enhancing function of VPF/VEGF. Another early event following interaction of VPF/VEGF with its tyrosine kinase receptors is a severalfold increase in cytoplasmic calcium, beginning after a 10–15s delay (BROCK et al. 1991). Other agonists that increase vascular hyperpermeability and intracellular calcium concentration ($[Ca^{2+}]_i$) do so through activation of phospholipase C, which leads to the generation of two intracellular messengers, inositol 1, 4, 5-triphosphate ($IP_3$) and 1, 2-diacylglycerol (DAG). $IP_3$, in turn, increases $[Ca^{2+}]_i$. It is likely that VPF/VEGF acts in similar fashion in that it leads to activation of phospholipase C$\gamma$ and endothelial cell accumulation of $IP_3$ (BROCK et al. 1991).

Recent work has implicated nitric oxide (NO) in the regulation of microvascular hyperpermeability (KUBES 1995). However, the results have been somewhat contradictory (FUKUMURA et al. 1997). Some reports indicate that the NO synthesis inhibitor $N^w$-nitro-L-arginine methylester (L-NAME) increases microvascular leakage (KUBES and GRANGER 1992; FILEP and FOLDES-FILEP

1993), whereas other authors, studying different tissues, obtained the opposite result (YUAN et al. 1993; MAEDA et al. 1994). Inhibition of NO synthase has also been reported to decrease the microvascular hyperpermeability induced by histamine, bradykinin and several other vasoactive mediators (BOUGHTON-SMITH et al. 1993; FUJII et al. 1994; KUBES 1995; RAMIREZ et al. 1995; NAKANO et al. 1996), whereas it augmented the increase in permeability induced by serotonin and platelet-activating factor (PAF) (KUBES and GRANGER 1992; FILEP and FOLDES-FILEP 1993; KUBES 1995). These different results may reflect differences in the tissues, species and agonists that were tested (FUKUMURA et al. 1997). Recently, however, MUROHARA et al. demonstrated that VPF/VEGF stimulated the synthesis of NO by cultured microvascular endothelial cells and that VPF/VEGF-induced hyperpermeability in the Miles assay was significantly attenuated by nitric oxide synthase inhibitors [MUROHARA, #1407]. These results implicate NO in the signaling pathway by which VPF/VEGF induces increased microvascular permeability. Also, VPF/VEGF augments NO release from vascular endothelium (VAN DER ZEE et al. 1997) and NO synthase is reported to increase during VPF/VEGF-induced endothelial cell mitogenesis (MORBIDELLI et al. 1996) and angiogenesis (ZICHE et al. 1997). The effects are likely to be complex as NO has been shown to attenuate the expression of VPF/VEGF by inhibition of protein-kinase C (PKC)-induced binding of the transcription factor activator protein-1 (AP-1) to the VPF/VEGF promoter (TSURUMI et al. 1997). More work will be required to determine the role, that, NO has in the vascular hyperpermeability induced by VPF/VEGF.

## 8 The Structural Basis of VPF/VEGF-Induced Microvascular Hyperpermebililty

The newly formed vessels that supply tumors and other examples of pathological and physiological angiogenesis and new stroma generation are hyperpermeable to plasma proteins and other circulating macromolecules. However, before turning to a discussion of the anatomic pathways by which macromolecules extravasate from these leaky blood vessels, it is necessary to consider briefly the pathways by which plasma water and solutes of varying size egress from normal blood vessels. This problem has engaged several generations of physiologists, pathologists and anatomists and has generated a lengthy and controversial literature. For a more detailed discussion, please see review articles, such as SIMIONESCU 1983; PALADE 1988; RIPPE and HARALDSSON 1994.

### 8.1 Extravasation of Plasma and Its Solutes from Normal Blood Vessels

Capillaries are the most important vessels for fluid exchange in normal tissues and the endothelial cells that line these capillaries, not the underlying basal lamina or pericytes, provide the rate-limiting barrier to the passage of fluid and solutes (Fig. 4). Water, glucose, ions and most other small hydrophilic molecules with an effective diameter of less than ~2 nm freely cross normal capillaries that are lined by continuous endothelium. It is thought that they do so largely by passing through the junctions between adjacent endothelial cells, and their passage is driven by convection and diffusion according to Starling's law (RIPPE and HARALDSSON 1994). Small proteins, such as horseradish peroxidase (HRP, diameter ~5 nm), also have a limited capacity to pass through the intact junctions of normal microvascular endothelium (KARNOVSKY 1967).

In contrast to plasma water and low-molecular weight solutes, only a very small fraction of plasma proteins extravasate from plasma on passage through typical capillary beds. Plasma proteins are too large to pass through intact inter-endothelial junctions and such proteins that do escape are thought to do so largely by means of transcytosis (CLEMENTI and PALADE 1969; SIMIONESCU 1983; PALADE 1988). Transcytosis is a process in which uncoated plasmalemmal vesicles (caveolae) bud off from the luminal plasma membrane of capillary endothelium, taking with them as 'cargo' a certain amount of plasma, including plasma proteins. Plasma-filled vesicles are then thought to shuttle across endothelial cytoplasm to the abluminal surface where they discharge their contents into the extracellular space. Sometimes two or three vesicles may be linked together in a chain that

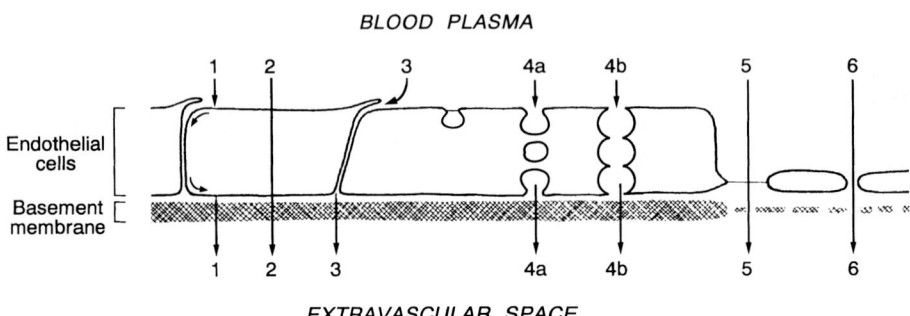

**Fig. 4.** Schematic diagram of the pathways by which solutes cross capillary endothelium. Most capillaries supplying normal tissues are lined by continuous endothelium, and solutes and water cross such endothelial cells by mechanisms *1–4b*: *1* Lipophilic molecules may dissolve in the luminal plasma membrane and diffuse laterally in that membrane to reach the abluminal surface. *2* Some molecules (e.g., amino acids) may be transported across endothelium by a receptor mediated process. *3* Water and small hydrophilic solutes (<2 nm) may pass through intact inter-endothelial cell junctions. *4a* Transcytotic pathway by which plasma proteins and other large molecules may traverse capillary endothelium by means of caveolae that bud from the luminal plasma membrane and shuttle across the endothelial cytoplasm to the ablumen where they fuse with the plasma membrane, discharging their cargo. *4b* Sometimes two or three vesicles link together to form a trans-endothelial cell pathway that does not require vesicle shuttling. *5* Some capillaries found in kidney, adrenal cortex, gastrointestinal tract, etc. are lined by fenestrated endothelium. Fenestrae are intermittent zones of radical thinning of endothelial cytoplasm such that only a thin diaphragm separates the vascular lumen from the underlying basal lamina. Endothelial fenestration has been associated with increased permeability to water and small molecules but not, it is thought, to macromolecules such as plasma proteins (CLEMENTI and PALADE 1969; SIMIONESCU 1983). *6* Discontinuous endothelium lines the sinusoids of liver and spleen

extends across capillary endothelium from lumen to ablumen, making shuttling of caveolae unnecessary.

When normal tissues are exposed to vasoactive mediators, such as histamine, serotonin, bradykinin or VPF/VEGF, large amounts of plasma proteins extravasate along with plasma water. However, this leakage takes place primarily from venules, not capillaries.

## 8.2 Extravasation of Plasma and Its Solutes From Hyperpermeable Tumor Blood Vessels

In contrast to studies of the normal vasculature, investigations into the mechanisms of tumor blood vessel hyperpermeability are at a relatively early stage. Thus far, electron microscopy has been employed to follow the passage of macromolecular tracers across the vascular endothelium in only a handful of tumors. However, these few studies have revealed important new information. Leakage is particularly prominent from vessels at the tumor-host interface (DVORAK et al. 1979, 1988; BROWN et al. 1987, 1988). Especially in larger tumors, the flow of blood is impaired within the tumor mass in large part because of increased interstitial tissue pressure (JAIN 1990), and it follows that leakage of circulating tracers from internal vessels is also less substantial than from vessels at the tumor edge (DVORAK et al. 1988).

The structure of tumor vessels does not correspond precisely to that of normal vessels. Tumor vessels often have a diameter that is similar to that of normal venules, but with a thinner wall that is more characteristic of capillaries (WARREN 1979). In a number of tumors, the endothelial cells that line leaky vessels contain extensive deposits of VPF/VEGF that are associated with their abluminal and luminal surfaces and also with internal vesicles and vacuoles (DVORAK et al. 1991; PLATE et al. 1992; BROWN et al. 1993a,b; QU-HONG et al. 1995). The VPF/VEGF that decorates tumor vascular endothelium is synthesized and secreted by tumor cells; endothelial cells themselves do not express detectable VPF/VEGF mRNA except under hypoxic conditions and, even then, do so at levels that are substantially lower than those of epithelial cells or fibroblasts (DETMAR et al. 1997).

Electron microscopic studies to elucidate the anatomical basis of tumor vessel hyperpermeability to macromolecules were undertaken in our laboratory with several different transplantable animal tumors, the guinea-pig line-1 and line-10 hepato (bile duct) carcinomas and the mouse ovarian tumor (MOT); tumors were studied at early stages of growth, 5 days after transplant and before the onset of significant tumor necrosis. Four macromolecular tracers were employed: HRP (diameter $\sim 5$ nm), ferritin (a normal plasma protein, diameter $\sim 11$ nm), 150-kDa dextran (diameter $\sim 17$ nm) and gold-bovine serum albumin (diameter $\sim 20$ nm).

An important finding to emerge from this work was the discovery of a new organelle in tumor vessel endothelium that provides a major pathway for extravasation of circulating macromolecules and to which we have given the name vesiculo-vacuolar organelle or VVO. VVOs are a bunch of grape-like clusters of membrane-lined vesicles and vacuoles in endothelial cytoplasm. Together these vesicles and vacuoles provide a system of interconnected vesicles and vacuoles that traverse endothelial cell cytoplasm from lumen to ablumen (KOHN et al. 1992; DVORAK et al. 1996). Ferritin and other circulating macromolecular tracers entered VVOs by way of vesicles that opened to the luminal plasma membrane and proceeded to pass through successive vesicles and vacuoles until they reached the abluminal surface where they were discharged into the underlying basal lamina (Fig. 5).

Junctions between adjacent endothelial cells were normally closed and gave no evidence of a widening of the inter-endothelial cleft. Of the tracers studied, only HRP was detectable while escaping from tumor venules by passing through apposed intercellular junctions. Even for HRP, however, the inter-endothelial cell pathway was slow to develop, did not exceed that of normal vessels (KARNOVSKY 1967) and accounted for much less HRP transport than the VVO pathway. Transendothelial cell gaps, as have been described in other tumor systems (ROBERTS and PALADE 1995), were not observed in these tumors which were studied during the early stages of growth.

The endothelium lining the microvessels that supply tumors may be of the continuous or the fenestrated type (KOHN et al. 1992; ROBERTS and PALADE 1995). Microvessels supplying the guinea-pig line-1 and line-10 tumors, as well as the mouse TA3/St mammary tumor, were lined with continuous endothelium, whereas MOTs elicited vessels that were, in part, fenestrated (KOHN et al. 1992). Ferritin and other tracers were sometimes found in the basal lamina immediately beneath fenestrae in MOT vessels, suggesting that they had passed through fenestrae, the diaphragms of which had opened. However, even in MOTs with fenestrated endothelium, VVOs provided the most important pathway for the extravasation of ferritin and of the other tracers studied.

## 8.3 Vesiculo-vacuolar Organelles (VVOs)

Vesiculo-vacuolar organelles (VVOs) with anatomical properties identical to those of tumor vessels have now been found (and with equivalent frequency) in the venular endothelium of several normal tissues, including skin, skeletal muscle and subcutis (KOHN et al. 1992; DVORAK et al. 1996) (Fig. 6). In both tumor and normal venular endothelium, individual vesicles and vacuoles comprising VVOs are

**Fig. 5.** Endothelium of a venule supplying a mouse ovarian tumor (MOT) grown in the subcutaneous space of a syngeneic mouse and harvested 30 min after intravenous injection of an anionic ferritin tracer. Ferritin molecules (*small black dots*) have entered cytoplasmic vesicles and vacuoles of a vesiculo-vacuolar organelle (VVO) and have passed through this structure to the endothelial cell ablumen, where tracer has been discharged into the basal lamina (*BL*) and the underlying edematous extracellular matrix. Individual vesicles and vacuoles are bridged, in part, by fenestral diaphragms (*open arrowhead*). The diaphragm closing the fenestra of one vesicle contains a central density (*open arrow*); another vesicle (*closed arrow*) contains eight individual ferritin particles; still another vesicle (*closed arrowhead*) appears empty. *L* vascular lumen. Reprinted from KOHN et al. 1992 with permission

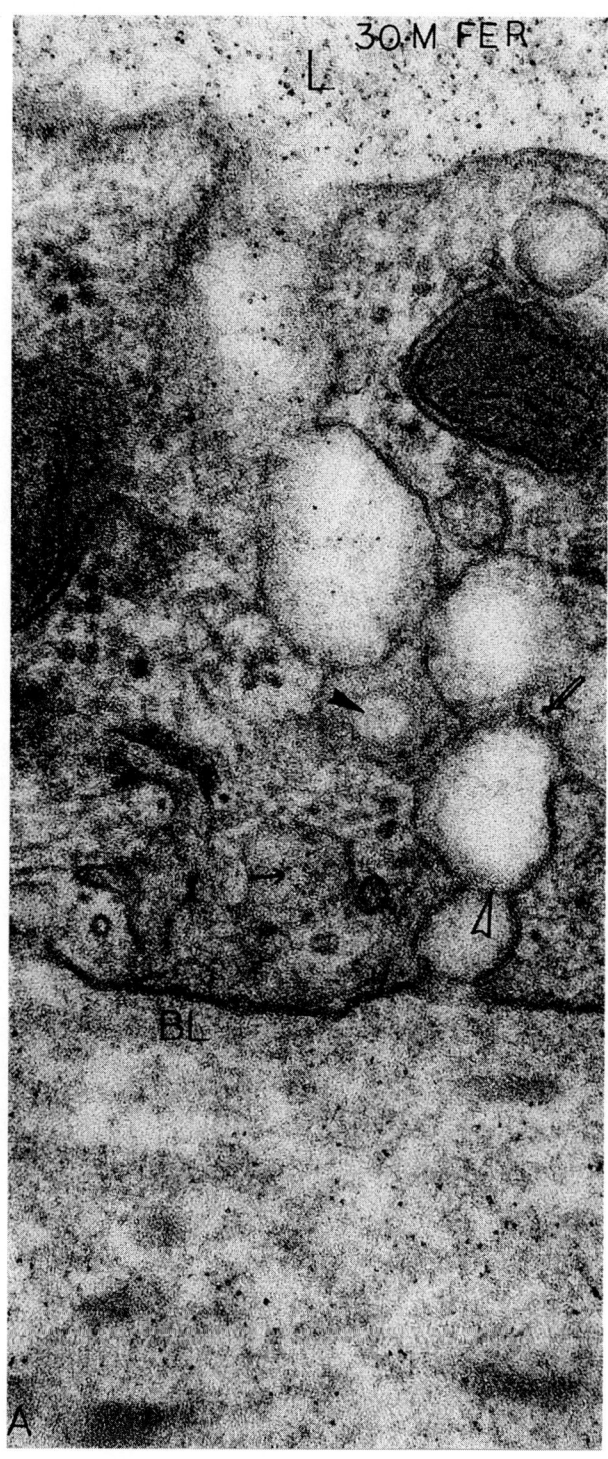

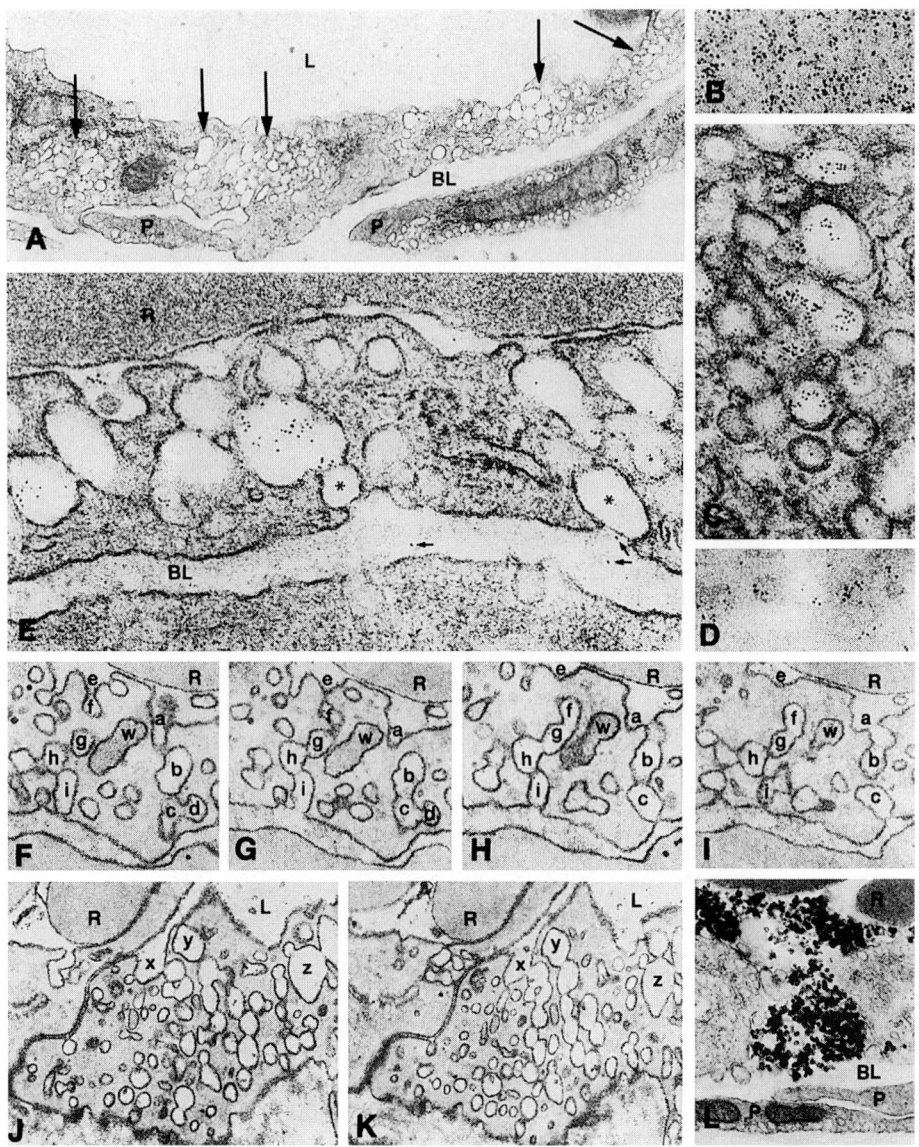

bounded by trilaminar unit membranes; VVO vesicles and vacuoles interconnect with each other and with the endothelial cell plasma membrane by means of fenestrae (also referred to as stomata) that are generally closed by thin diaphragms. [2]

---

[2] The terminology is confusing. The term fenestra (or stomata) is used interchangeably to describe portions of vascular endothelium that are radically thinned to the dimensions of 7 nm-thick diaphragms (Fig. 4) as well as the similar but significantly smaller diameter openings that interconnect caveolae (and larger cytoplasmic vacuoles) with each other and with the endothelial plasma membrane as are found in VVOs. Similarly, the term diaphragm is used to describe the thin structure that closes both types of fenestrae, even though they have significantly different diameters and bind cations with different affinities (SIMIONESCU 1983).

**Fig. 6A–L.** Electron micrographs of venules from mouse or guinea-pig skin (**A-K**) or from rat cremaster (**L**). As indicated, ferritin or colloidal carbon was injected intravenously., followed immediately by local injection of 12.5–100 ng vascular permeability factor (VPF). Tissues were harvested 5 min later. **A** Venule in normal mouse skin illustrating five distinct vesiculo-vacuolar organelles (VVOs) (*arrows*) in the cytoplasm of a single endothelial cell. **B–E** VPF injection sites in guinea-pig skin, illustrating progressive extravasation of circulating ferritin from plasma across venular endothelium via VVOs. **B** Ferritin particles (*black dots*) in plasma. **C** A succession of ferritin-containing VVOs and vacuoles at mid-level of endothelial cell cytoplasm. **D** Ferritin particles in subjacent extracellular matrix. **E** VVO with ferritin-containing vesicles and vacuoles. Ferritin is present in several vesicles that open to the vascular lumen or are separated from it by thin diaphragms. Two vesicles (*) in continuity with the abluminal plasma membrane do not contain ferritin, but exhibit stomata closed by diaphragms. One starred vesicle (*left* *) is also separated by a diaphragm from a more proximal (more luminal) vacuole that contains numerous ferritin particles; micrographs such as this suggest that stomatal diaphragms represent barriers that restrict the trans-cellular passage of macromolecules. Scattered ferritin particles are also present in the underlying basal lamina, several of which are indicated by *arrows*. **F–I** Four (of a set of 36) consecutive serial ultrathin (14 nm) sections illustrate two sequences of interconnecting VVO vesicles-vacuoles (*a–d*; *e–i*) that traverse venule endothelial cell cytoplasm from vascular lumen to ablumen. **J, K** Two consecutive serial ultrathin sections of a single VVO. Three sequences of interconnecting vesicles-vacuoles are illustrated, the most luminal of which are designated *x*, *y* and *z*; each sequence nearly forms a transendothelial channel in just two consecutive sections. Note closed inter-endothelial junction to the left of figure. **L** VPF-induced endothelial gap allows carbon extravasation from venule in rat cremaster muscle. *L* lumen, *R* red blood cell, *BL* basal lamina, *p* pericyte, *W* Weibel-Palade body. Magnifications: **A** ×25,000; **B–D** ×145,000; **E** ×105,000; **F-I** ×60,000; **J, K** ×50,000; **L** ×26,000. Reprinted from FENG et al. 1996 with permission

The major difference, thus far, recognized between VVOs in tumor and normal venules is one of functional activity, i.e., the capacity to allow trans-endothelial passage of macromolecules. In tumor vessels, circulating tracers such as ferritin readily entered VVOs, and their passage across the endothelial cell cytoplasm can be followed sequentially from the vascular lumen to the ablumen and beyond into the underlying basal lamina. In contrast, the VVOs of normal venules permit only minimal entry and passage of macromolecular tracers (KOHN et al. 1992; DVORAK et al. 1996).

We postulated that the relatively free passage of macromolecular tracers through tumor vessel VVOs was attributable to tumor cell-secreted VPF/VEGF. If this were the case, then the VVOs of normal venules would also be expected to become hyperpermeable upon exposure to VPF/VEGF. In fact, this hypothesis proved to be correct. Injection of VPF/VEGF into normal flank or scrotal skin led to a prompt increase in venule permeability, and electron microscopy revealed that circulating tracer ferritin exited venules primarily by way of VVOs, just as in tumor microvessels (FENG et al. 1996). Moreover, the capacity to activate VVO function proved not to be unique to VPF/VEGF. Two other vasoactive mediators, histamine and serotonin, also opened VVOs to the passage of macromolecular tracers.

## 8.4 Structural Properties of VVOs

VVOs are three-dimensional organelles and their structure cannot be fully appreciated from standard electron micrographs (Fig. 6). Therefore, we prepared serial

14 nm-thick sections of VPF-injected and control mouse skin (FENG et al. 1996). Study of these sets of serial ultrathin sections, along with computer-assisted three-dimensional reconstructions, revealed a network of interconnecting vesicles and vacuoles and established that VVOs provided a continuous if serpentine pathway across venular endothelium, opening to both the lumen and ablumen at multiple sites (Fig. 6). Subsequent studies have revealed that the VVOs of adjacent endothelial cells may sometimes interconnect across the intercellular cleft, thereby raising the possibility that macromolecular tracers might extravasate by a pathway that extends across two overlapping endothelial cells (FENG et al. 1997).

Morphometric measurements revealed that VVOs occupied a substantial portion (16–18%) of venular endothelial cell cytoplasm in both normal and VPF-injected mouse skin (FENG et al. 1996). Individual VVOs commonly extended through endothelial cell cytoplasm for distances of 1–2 µm and were typically comprised of more than 100 individual vesicles and vacuoles. The smallest VVO vesicles resembled the caveolae found in capillaries (SIMIONESCU 1983; PALADE 1988; KOHN et al. 1992; DVORAK et al. 1996). However, the vesicles and vacuoles comprising VVOs had a much greater range of sizes and were significantly larger on average and more heterogeneous than capillary caveolae (FENG et al. 1996). Thus, the vesicles and vacuoles comprising VVOs differ from the caveolae of capillary endothelium in several important respects: significantly larger average size, greater size heterogeneity and organization into an organelle (the VVO) that occupies nearly one-fifth of venular endothelial cell cytoplasm and that links the vascular lumen with the ablumen and the inter-endothelial interface.

The large size and complexity of VVOs suggest that they are sessile structures and that their component vesicles and vacuoles are unlikely to shuttle back and forth across endothelial cytoplasm as has been proposed for capillary caveolae (SIMION-ESCU 1983; PALADE 1988). VVOs more closely resemble the vesicular-vacuolar networks that have been described in amphibian capillary cytoplasm (BUNDGAARD 1980; FROKJAER-JENSEN 1991), but differ in several respects: they are present in venules, not capillaries; they traverse the entire thickness of endothelial cell cytoplasm, opening to lumen, ablumen and, at least in some instances, to the inter-endothelial cleft, permitting interconnections between VVOs in adjacent endothelial cells; and, when activated by VPF/VEGF or other vasoactive mediator, they play a prominent role in the extravasation of macromolecules from the circulation.

## 8.5 Trans-Endothelial Openings. Are They Inter-Endothelial Cell Gaps or Trans-Endothelial Cell Pores?

VVOs do not provide a complete explanation for the extravasation of macromolecules in response to vasoactive mediators. Many authors have described the escape of very large particulates, such as colloidal carbon (diameter ~50 nm), liposomes (diameter ~600 nm) and even erythrocytes (diameter ~7 µm!!), from tumor vessels and normal venules in response to histamine, serotonin, VPF/VEGF and other vasoactive mediators (MAJNO et al. 1961, 1967, 1969; SENGER et al. 1983; JORIS et al.

1987; YUAN et al. 1995; BALUK et al. 1997; FENG et al. 1997). The stomata interconnecting VVO vesicles and vacuoles with each other and with the endothelial plasma membrane are too small (diameter < 50 nm) to admit particles the size of even the smallest of these particulates, colloidal carbon (DVORAK et al. 1988, 1996; FENG et al. 1996, 1997). Over the years, many investigators, beginning with Majno and his associates, have shown that in response to vasoactive mediators, such as histamine, colloidal carbon and other particulates escape from venules by passing through variably sized trans-endothelial openings (Fig. 6L). Similar types of trans-endothelial openings have been described in tumor vessel endothelium (ROBERTS and PALADE 1995).

All workers agree that trans-endothelial openings develop in venular endothelium in response to vasoactive mediators and that these openings occur close to the interface between adjacent endothelial cells. However, the exact site of their formation, whether intercellular or trans-cellular, is a subject of dispute. Early investigators (ALKSNE 1959; HASHIMOTO et al. 1974) had suggested that the openings were trans-cellular, i.e., passed through endothelial cell cytoplasm. In support of this view, specialized junctions are found to be intact immediately adjacent to trans-endothelial openings and inter-endothelial cell clefts are not noticeably widened. These data favored the possibility that carbon extravasated by passing through endothelial cells, not through open inter-endothelial cell junctions. Nonetheless, most workers hold the view that the trans-endothelial openings elicited by vasoactive mediators represent inter-endothelial cell gaps that resulted from the contraction and pulling apart of adjacent endothelial cells (reviewed in MAJNO et al. 1969; JORIS et al. 1987; MICHEL 1988; RIPPE and HARALDSSON 1994).

Few attempts have been made over the years to address this question in depth. Given the complex membrane interdigitations between adjacent venular endothelial cells, it was recognized that precise localization of trans-endothelial openings might best be achieved by evaluation of serial electron microscopic sections and three-dimensional reconstructions. However, because of the high degree of technical difficulty and the extensive labor required for this approach, few such studies have been performed and they have not been conclusive.

Recently, NEAL and MICHEL (1995) performed experiments in which individual frog mesenteric capillaries and rat mesenteric venules were cannulated and perfused with the calcium ionophore A21387. Microvessels, so treated, developed increased permeability and formed endothelial gaps. Serial electron microscopic sections revealed that 39 of 40 gaps in frog capillaries and 15 of 16 in rat venules were trans-endothelial pores that were clearly separate from intercellular junctions. Subsequently, these authors identified the sites of endothelial disjunction in frog mesenteric venules, following the application of supraphysiological intraluminal pressure. On the basis of serial electron microscopic sections, they found that in a majority of instances increased intraluminal pressure led to vessel rupture through trans-endothelial pores rather than through intercellular junctions (NEAL and MICHEL 1996).

In contrast, McDonald and his colleagues, making use of light and scanning electron microscopy, concluded that the great majority of endothelial openings form between endothelial cells (HIRATA et al. 1995; BALUK et al. 1997). In a collaborative study with Neal and Michel, BALUK et al. (1997) reported that the great majority of endothelial openings that developed in rat tracheal venules were between endothelial cells. An important feature of these studies was the administration of a vasoactive mediator (substance P) by the intravenous route; under these circumstances, the permeability increase was transient, peaking at 1 min, with return to near normal permeability levels by 3 min.

We also recently investigated the nature of the openings that form in response to vasoactive mediators (VPF/VEGF, histamine and serotonin) in guinea-pig, rat and mouse skin and in rat cremaster muscle (FENG et al. 1997). Serial electron microscopic sections defined the properties and location of 92 such openings. Maximum diameter varied from 0.08 µm to 2.0 µm; all openings were filled with particulates (84 with colloidal carbon, 6 with extravasating erythrocytes and 2 with platelets).

Endothelial openings were nearly always in close proximity to the interface between adjacent endothelial cells. Sometimes these openings followed a simple course, passing through the peripheral cytoplasm of a single endothelial cell. Often, however, the pathway across endothelium was tortuous, involving passage through endothelial cell cytoplasm and portions of the intercellular cleft (Fig. 7). Forty of the 92 openings followed in serial sections passed through a single endothelial cell and, in every case, they were distinctly separate from specialized inter-endothelial cell junctions at all levels of sectioning. Computer-assisted three-dimensional reconstructions confirmed these findings (Fig. 8). An additional 49 openings (53%) were also trans-endothelial but, like some VVOs, passed through overlapping portions of two adjacent endothelial cells. Such pores opened to the vascular lumen of one endothelial cell, but did not extend through that cell to the underlying basal lamina. Instead, they crossed through the loosely apposed intercellular cleft portion of the inter-endothelial cell interface, passing above or below specialized (occludens-type) junctions that were characterized by close membrane apposition; pores then extended across the second endothelial cell, exiting to its vascular ablumen and the underlying basal lamina. Together, these 49 openings were also trans-cellular through most of their course, extending from the vascular lumen to the ablumen by passing through portions of two adjacent, overlapping endothelial cells and involving portions of the intercellular cleft.

Thus, 89 of the 92 pores completely encompassed in serial sections passed through endothelial cells over all or much of their course. Of these, 26 (28%) passed through a single endothelial cell, entirely apart from an endothelial cell interface. Fourteen (16%) passed through a single endothelial cell, but also made use of the intercellular cleft above or below specialized inter-endothelial junctional areas for a portion of their course. Forty-nine (53%) passed through two overlapping endothelial cells and the intercellular cleft separating these cells. In three pores of unusual complexity, we could not determine whether the pathway taken was transcellular or entirely paracellular.

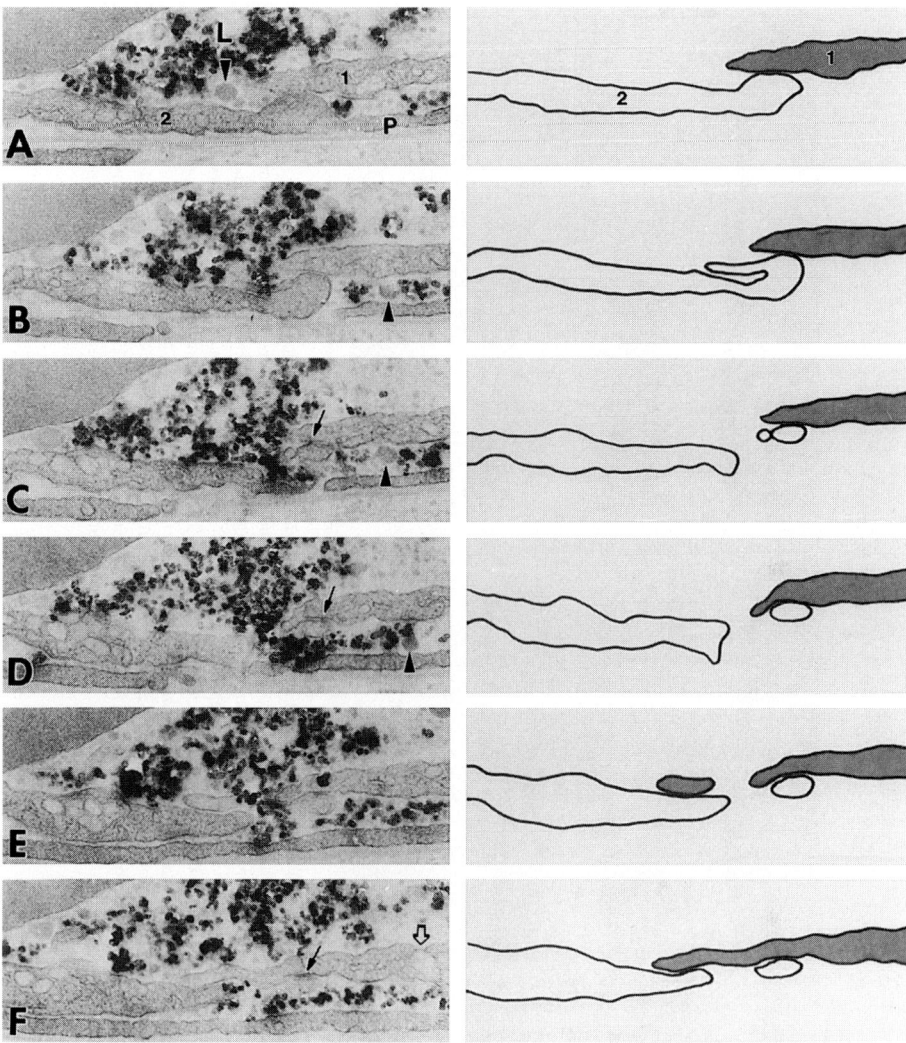

**Fig. 7A–F.** Colloidal carbon-filled trans-endothelial cell pore in a venule of rat cremaster muscle. Colloidal carbon was injected intravenously and tissue was harvested 1 min. after local injection of 50 ng vascular permeability factor/vascular endothelial growth factor (VPF/VEGF). *Left column* illustrates six consecutive 100 nm-thick serial sections taken from of a series of 41 consecutive serial sections that, together, encompassed the pore in its entirety. *Right column* presents tracings outlining endothelial cells illustrated in each corresponding left panel. The pore passes through endothelial cell no. 2 in *panels C–F* and beyond (not shown). In *panel E* only, the pore also passes through endothelial cell no. 1. Chylomicra also extravasated and are indicated by *arrowheads*; occludens-type junctions (all normally closed) are indicated by *arrows*. Note large vacuole (*open arrowhead, panel F*) that spans the entire thickness of endothelial cell no. 1. This vacuole is part of a vesiculo-vacuolar organelle (VVO) that extends to the right beyond the sections illustrated; its interfaces with the luminal and abluminal surfaces are closed by diaphragms, though carbon particles are found immediately beneath the abluminal interface, suggesting that it had been open at one time. *p* pericyte. Magnification ×27,000

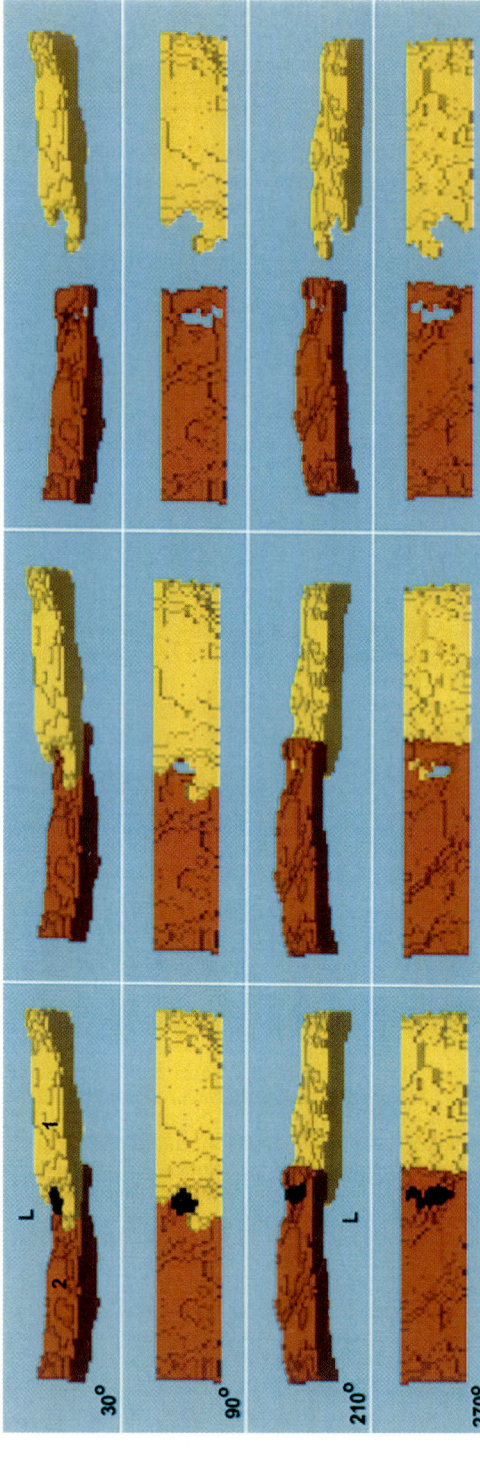

**Fig. 8.** Computer-generated three dimensional reconstruction of the pore illustrated in Fig. 7, but including additional consecutive serial sections. The panels portray successive rotations toward the viewer around a horizontal axis at angles of 30°, 90°, 210° and 270° as indicated. 0° (not shown) represents a vascular cross section at right angles to the direction of blood flow; 90° a view looking down on the luminal surface; and 270° a view facing the abluminal surface. Endothelial cells labeled no. 1 (*yellow*) and no. 2 (*orange-brown*) represent corresponding cells in Fig. 7. Carbon is represented in *black* in the *left column*. In *middle column* carbon was subtracted from the reconstruction to facilitate visualization into the interior of the pore, which is seen to pass cleanly through the cytoplasm of cell no. 2 (*orange-brown*), revealing underlying turquoise background. At its lowermost point in the view at 90° (*middle panel*), the pore passes through a peripheral portion of cell no. 1 (as illustrated also in Fig. 7E). The 270° view (*middle column*) demonstrates that while most of the trans-cellular pore through cell no. 2 provides continuity between lumen and ablumen (*turquoise background*), a small portion is occluded by overlapping cell no. 1 (*yellow background*); this corresponds to Figs. 7E,F and subsequent panels beyond those included in Fig. 7. *Right column* depicts endothelial cells no. 1 and no. 2 separately, in panels corresponding to those in the *left and middle columns*. *L* lumen

## 8.6 The Controversy Regarding Trans-Endothelial Openings: Gaps, Pores and VVOs

It is obvious from the foregoing discussion that different authors have come to very different conclusions regarding the anatomical pathways by which macromolecular tracers cross tumor and normal venular endothelium in response to vasoactive mediators. How are these differences to be interpreted and resolved? To some extent the apparent contradictions may be attributed to differences in study design and technique. Important differences include the study of different species (mice, rats, guinea pigs); different mediators (substance P, calcium ionophore A21387, VPF/VEGF, histamine, serotonin); different routes of mediator administration (vagal nerve stimulation, intravenous substance P or ionophore injection, intradermal injection of VPF/VEGF and other mediators); different fixation protocols; study at different time points after mediator injection, and different morphological approaches (light and scanning electron microscopy versus serial-section transmission electron microscopy with three-dimensional reconstructions); different vascular beds (trachea, skin, cremaster muscle); and the duration of vascular hyperpermeability (tracheal vessel permeability to a particulate was complete by ~3 min after intravenous administration of substance P, whereas skin and cremaster venules remained hyperpermeable to plasma proteins for up to 20 min after intradermal injection of VPF/VEGF, histamine or serotonin).

Resolution of differences may, therefore, be possible when each of these variables is isolated and investigated separately. For example, BALUK et al. (1997) did find trans-endothelial cell pores 3 min after intravenous administration of substance P, though with much lower frequency than NEAL and MICHEL (1995, 1996) and we (FENG et al. 1997) have reported. Also, BALUK et al. (1997) reported that tracheal venule permeability to a particulate, Monastral Blue, was complete by ~3 min after intravenous administration of substance P; however, skin and cremaster venules remain hyperpermeable to plasma proteins for up to 20 min after intradermal injection of VPF/VEGF, histamine or serotonin. Some of these disparate results could be harmonized if, as is likely, the trans-endothelial openings induced by intravenous substance P closed with great rapidity, perhaps within a time frame of 1–3 min, whereas the more prolonged (up to ~20 min) hyperpermeability of venular endothelium following intradermal injection of vasoactive mediators involved passage of tracers through VVOs.

## 8.7 Possible Relationships Between VVOs and Trans-Endothelial Cell Pores

There is some evidence to suggest that VVOs and trans-endothelial cell pores may be related structures (FENG et al. 1996, 1997). They share a common anatomical distribution in the microvasculature. Both are most numerous in venular endothelium, both concentrate in parajunctional zones of individual endothelial cells and both may open to the intercellular clefts between lateral surfaces of adjacent endothelial cells as well as to the luminal and abluminal surfaces. In addition, both VVOs and pores form pathways for extravasation of macromolecules that can extend across adjacent endothelial cells (FENG et al. 1997). Taken together, these findings raise the possibility that pores may develop from VVOs in response to vasoactive mediators (Fig. 9). One can imagine that opening of the diaphragms that separate individual VVO vesicles and vacuoles from each other would lead to the formation of larger membrane-lined vacuolar structures. If the diaphragms linking individual vesicles or vacuoles to the endothelial cell luminal, lateral and abluminal plasma membranes were also opened or deleted, trans-endothelial channels would form and, if allowed to expand, might readily accommodate the passage of particulate tracers, such as colloidal carbon, and endogenous tracers, such as erythrocytes.

## 9 Summary

This Chapter has reviewed the literature concerning VPF/VEGF as a potent vascular permeabilizing cytokine. In accord with this important role, microvessels have been found to be hyperpermeable to plasma proteins and other circulating macromolecules at sites where VPF/VEGF and its receptors are overexpressed, i.e., in tumors, healing wounds, retinopathies, many important inflammatory conditions and in certain physiological processes, such as ovulation and corpus luteum formation. Moreover, microvascular hyperpermeability to plasma proteins was shown to have an important consequence: the laying down of a fibrin-rich extracellular matrix. This provisional matrix, in turn, favors and supports the ingrowth of fibroblasts and

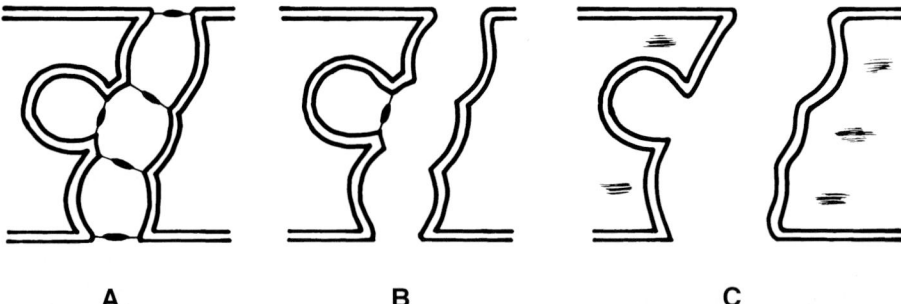

**Fig. 9A–C.** Schematic diagram indicating a possible mechanism by which vesiculo-vacuolar organelles (VVOs) could be transformed into trans-endothelial cell pores. **A** Portion of a VVO comprised of vesicles that are interconnected to each other and to the luminal and abluminal plasma membrane of venular endothelium by fenestrae (stomata) that are closed by thin diaphragms with characteristic central thickenings termed "knobs". **B** Diaphragms interconnecting vesicles with each other and with the plasma membrane have opened to form a trans-endothelial cell pore. **C** Pore has enlarged as the result of contraction of VVO-associated actin-myosin filaments

endothelial cells which, together, transform the provisional matrix into the mature stroma characteristic of tumors and healed wounds. Finally, we have considered the pathways by which these and other circulating macromolecules cross the endothelium of normal and VPF/VEGF-permeabilized microvessels. These pathways include VVOs and trans-endothelial openings that have been variously interpreted as inter-endothelial cell gaps or trans-endothelial cell pores. At least some trans-endothelial cell pores may arise from VVOs. In conclusion, these data provide new insights into the mechanisms of angiogenesis and stroma formation, insights which are potentially applicable to a wide variety of disease states and which may lead to identification of new targets for therapeutic intervention.

# References

Abu-Jawdeh GM, Faix JD, Niloff J, Tognazzi K, Manseau E, Dvorak HF, Brown LF (1996) Strong expression of vascular permeability factor (vascular endothelial growth factor) and its receptors in ovarian borderline and malignant neoplasms. Lab Invest 74:1105–1115

Adamis A, Shima D, Yeo K-T, Yeo T-K, Brown L, Berse B, D'Amore P, Folkman J (1993) Synthesis and secretion of vascular permeability factor/vascular endothelial growth factor by human retinal pigment epithelial cells. Biochem Biophys Res Commun 193:631–638

Adamis AP, Miller JW, Bernal MT, D'Amico DJ, Folkman J, Yeo TK, Yeo KT (1994) Increased vascular endothelial growth factor levels in the vitreous of eyes with proliferative diabetic retinopathy. Am J Ophthalmol 118:445–450

Aiello LP, Avery RL, Arrigg PG, Keyt BA, Jampel HD, Shah ST, Pasquale LR, Thieme H, Iwamoto MA, Park JE, Nguyen MS, Aiello LM, Ferrara N, King GL (1994) Vascular endothelial growth factor in ocular fluid of patients with diabetic retinopathy and other retinal disorders. N Engl J Med 331:1480–1487

Alksne J (1959) The passage of colloidal particles across the dermal capillary wall under the influence of histamine. Q J Exp Physiol 44:51–66

Alon T, Hemo I, Itin A, Pe'er J, Stone J, Keshet E (1995) Vascular endothelial growth factor acts as a survival factor for newly formed retinal vessels and has implications for retinopathy of prematurity. Nat Med 1:1024–1028

Baluk P, Hirata A, Thurston G, Fujiwara T, Neal CR, Michel CC, McDonald DM (1997) Endothelial gaps: time course of formation and closure in inflamed venules of rats. Am J Physiol 272:L155–L170

Banai S, Shweiki D, Pinson A, Chandra M, Lazarovici G, Keshet E (1994) Upregulation of vascular endothelial growth factor expression induced by myocardial ischaemia: implications for coronary angiogenesis. Cardiovasc Res 28:1176–1179

Bates DO, Curry FE (1996) Vascular endothelial growth factor increases hydraulic conductivity of isolated perfused microvessels. Am J Physiol 271:H2520–H2528

Berse B, Brown LF, Van De Water L, Dvorak HF, Senger DR (1992) Vascular permeability factor (vascular endothelial growth factor) gene is expressed differentially in normal tissues, macrophages, and tumors. Mol Biol Cell 3:211 220

Boughton-Smith NK, Evans SM, Laszlo F, Whittle BJ, Moncada S (1993) The induction of nitric oxide synthase and intestinal vascular permeability by endotoxin in the rat. Br J Pharmacol 110:1189–1195

Breier G, Albrecht U, Sterrer S, Risau W (1992) Expression of vascular endothelial growth factor during embryonic angiogenesis and endothelial cell differentiation. Development 114:521–532

Breuing K, Eriksson E, Liu P, Miller DR (1992) Healing of partial thickness porcine skin wounds in a liquid environment. J Surg Res 52:50–58

Brock TA, Dvorak HF, Senger DR (1991) Tumor-secreted vascular permeability factor increases cytosolic $Ca^{2+}$ and von Willebrand factor release in human endothelial cells. Am J Pathol 138:213–221

Brown LF, Chester JF, Malt RA, Dvorak HF (1987) Fibrin deposition in autochthonous Syrian hamster pancreatic adenocarcinomas induced by the chemical carcinogen N-nitroso-bis(2-oxopropyl)amine. J Natl Cancer Inst 78:979–986

Brown LF, Asch B, Harvey VS, Buchinski B, Dvorak HF (1988a) Fibrinogen influx and accumulation of cross-linked fibrin in mouse carcinomas. Cancer Res 48:1920–1925

Brown LF, Van De Water L, Harvey VS, Dvorak HF (1988b) Fibrinogen influx and accumulation of cross-linked fibrin in healing wounds and in tumor stroma. Am J Pathol 130:455–465

Brown LF, Yeo K-T, Berse B, Yeo T-K, Senger DR, Dvorak HF, Van De Water L (1992) Expression of vascular permeability factor (vascular endothelial growth factor) by epidermal keratinocytes during wound healing. J Exp Med 176:1375–1379

Brown LF, Berse B, Jackman RW, Tognazzi K, Manseau EJ, Dvorak HF, Senger DR (1993a) Increased expression of vascular permeability factor (vascular endothelial growth factor) and its receptors in kidney and bladder carcinomas. Am J Pathol 143:1255–1262

Brown LF, Berse B, Jackman RW, Tognazzi K, Manseau EJ, Senger DR, Dvorak HF (1993b) Expression of vascular permeability factor (vascular endothelial growth factor) and its receptors in adenocarcinomas of the gastrointestinal tract. Cancer Res 53:4727–4735

Brown LF, Yeo K-T, Berse B, Morgentaler A, Dvorak H, Rosen S (1995a) Vascular permeability factor (vascular endothelial growth factor) is strongly expressed in the normal male genital tract and is present in substantial quantities in semen. J Urol 154:576–579

Brown LF, Berse B, Jackman RW, Tognazzi K, Guidi AJ, Dvorak HF, Senger DR, Connolly JL, Schnitt SJ (1995b) Expression of vascular permeability factor (vascular endothelial growth factor) and its receptors in breast cancer. Hum Pathol 26:86–91

Brown LF, Harrist TJ, Yeo KT, Stahle BM, Jackman RW, Berse B, Tognazzi K, Dvorak HF, Detmar M (1995c) Increased expression of vascular permeability factor (vascular endothelial growth factor) in bullous pemphigoid, dermatitis herpetiformis, and erythema multiforme. J Invest Dermatol 104:744–749

Brown LF, Olbricht SM, Berse B, Jackman RW, Matsueda G, Tognazzi KA, Manseau EJ, Dvorak HF, Van De Water L (1995d) Overexpression of vascular permeability factor (VPF/VEGF) and its endothelial cell receptors in delayed hypersensitivity skin reactions. J Immunol 154:2801–2807

Brown LF, Tognazzi K, Dvorak H, Harrist T (1996) Strong expression of KDR, a vascular permeability factor/vascular endothelial growth factor receptor in AIDS-associated Kaposi's sarcoma and cutaneous angiosarcoma. Am J Pathol 148:1065–1074

Brown LF, Detmar M, Claffey K, Nagy J, Feng D, Dvorak A, Dvorak H (1997a) Vascular permeability factor/vascular endothelial growth factor: a multifunctional angiogenic cytokine. In: Goldberg I, Rosen E (eds) Regulation of angioigenesis. Birkhauser, Basel

Brown LF, Detmar M, Tognazzi K, Abu-Jawdeh G, Iruela-Arispe ML (1997b) Uterine smooth muscle cells express functional receptors (flt-1 and KDR) for vascular permeability factor/vascular endothelial growth factor. Lab Invest 76:245–255

Bundgaard M (1980) Transport pathways in capillaries – in search of pores. Annu Rev Physiol 42:325–336

Chow NH, Hsu PI, Lin XZ, Yang HB, Chan SH, Cheng KS, Huang SM, Su IJ (1997) Expression of vascular endothelial growth factor in normal liver and hepatocellular carcinoma:An immunohistochemical study. Hum Pathol 28:698–703

Claffey KP, Brown LF, del Aguila LF, Tognazzi K, Yeo KT, Manseau EJ, Dvorak HF (1996) Expression of vascular permeability factor/vascular endothelial growth factor by melanoma cells increases tumor growth, angiogenesis, and experimental metastasis. Cancer Res 56:172–181

Clauss M, Gerlach M, Gerlach H, Brett J, Wang F, Familletti PC, Pan Y-CE, Olander JV, Connolly DT, Stern D (1990) Vascular permeability factor:A tumor-derived polypeptide that induces endothelial cell and monocyte procoagulant activity, and promotes monocyte migration. J Exp Med 172:1535–1545

Clementi F, Palade GE (1969) Intestinal capillaries. I. Permeability to peroxidase and ferritin. J Cell Biol 41:33–58

Colvin R, Dvorak H (1975) Role of the clotting system in cell-mediated hypersensitivity. II. Kinetics of fibrinogen/fibrin accumulation and vascular permeability changes in tuberculin and cutaneous basophil hypersensitivity reactions. J Immunol 114:377–387

Conn G, Bayne ML, Soderman DD, Kwok PW, Sullivan KA, Palisi TM, Hope DA, Thomas KA (1990a) Amino acid and cDNA sequences of a vascular endothelial cell mitogen that is homologous to platelet-derived growth factor. Proc Natl Acad Sci USA 87:2628–2632

Conn G, Soderman DD, Schaeffer M-T, Wile M, Hatcher VB, Thomas KA (1990b) Purification of a glycoprotein vascular endothelial cell mitogen from a rat glioma-derived cell line. Proc Natl Acad Sci USA 87:1323–1327

Connolly DT, Heuvelman DM, Nelson R, Olander JV, Eppley BL, Delfino JJ, Siegel NR, Leimgruber RM, Feder J (1989) Tumor vascular permeability factor stimulates endothelial cell growth and angiogenesis. J Clin Invest 84:1470–1478

Cullinan-Bove K, Koos RD (1993) Vascular endothelial growth factor/vascular permeability factor expression in the rat uterus:rapid stimulation by estrogen correlates with estrogen-induced increases in uterine capillary permeability and growth. Endocrinology 133:829–837

de Vries C, Escobedo JA, Ueno H, Houck K, Ferrara N, Williams LT (1992) The fms-like tyrosine kinase, a receptor for vascular endothelial growth factor. Science 255:989–991

Detmar M, Brown LF, Claffey KP, Yeo KT, Kocher O, Jackman RW, Berse B, Dvorak HF (1994) Overexpression of vascular permeability factor/vascular endothelial growth factor and its receptors in psoriasis. J Exp Med 180:1141–1146

Detmar M, Brown LF, Berse B, Jackman RW, Elicker BM, Dvorak HF, Claffey KP (1997) Hypoxia regulates the expression of vascular permeability factor/vascular endothelial growth factor (VPF/VEGF) and its receptors in human skin. J Invest Dermatol 108:263–268

Dewey WC (1959) Vascular-extravascular exchange of $^{131}$I plasma proteins in the rat. Am J Physiol 197:423–431

Dougher-Vermazen M, Hulmes JD, Bohlen P, Terman BI (1994) Biological activity and phosphorylation sites of the bacterially expressed cytosolic domain of the KDR VEGF-receptor. Biochem Biophys Res Commun 205:728–738

Dvorak AM, Kohn S, Morgan ES, Fox P, Nagy JA, Dvorak HF (1996) The vesiculo-vacuolar organelle (VVO): a distinct endothelial cell structure that provides a transcellular pathway for macromolecular extravasation. J Leukoc Biol 59:100–115

Dvorak HF (1986) Tumors: wounds that do not heal. Similarities between tumor stroma generation and wound healing. N Engl J Med 315:1650–1659

Dvorak HF (1990) Leaky tumor vessels:consequences for tumor stroma generation and for solid tumor therapy. Prog Clin Biol Res 317–330

Dvorak HF, Dvorak AM, Manseau EJ, Wiberg L, Churchill WH (1979a) Fibrin gel investment associated with line 1 and line 10 solid tumor growth, angiogenesis, and fibroplasia in guinea pigs. Role of cellular immunity, myofibroblasts, microvascular damage, and infarction in line 1 tumor regression. J Natl Cancer Inst 62:1459–1472

Dvorak HF, Orenstein NS, Carvalho AC, Churchill WH, Dvorak AM, Galli SJ, Feder J, Bitzer AM, Rypysc J, Giovinco P (1979b) Induction of a fibrin-gel investment: an early event in line 10 hepatocarcinoma growth mediated by tumor-secreted products. J Immunol 122:166–174

Dvorak HF, Senger DR, Dvorak AM (1983) Fibrin as a component of the tumor stroma:Origins and biological significance. Cancer Metastasis Rev 2:41–73

Dvorak HF, Harvey VS, McDonagh J (1984) Quantitation of fibrinogen influx and fibrin deposition and turnover in line 1 and line 10 guinea pig carcinomas. Cancer Res 44:3348–3354

Dvorak HF, Senger DS, Dvorak AM, Harvey VS, McDonagh J (1985) Regulation of extravascular coagulation by microvascular permeability. Science 227:1059–1061

Dvorak HF, Nagy JA, Dvorak JT, Dvorak AM (1988) Identification and characterization of the blood vessels of solid tumors that are leaky to circulating macromolecules. Am J Pathol 133:95–109

Dvorak HF, Sioussat TM, Brown LF, Nagy JA, Sotrel A, Manseau E, Van De Water L, Senger DR (1991a) Distribution of vascular permeability factor (vascular endothelial growth factor) in tumors:concentration in tumor blood vessels. J Exp Med 174:1275–1278

Dvorak HF, Nagy JA, Dvorak AM (1991b) Structure of solid tumors and their vasculature: implications for therapy with monoclonal antibodies. Cancer Cells 3:77–85

Dvorak HF, Nagy JA, Berse B, Brown LF, Yeo KT, Yeo TK, Dvorak AM, van de Water L, Sioussat TM, Senger DR (1992) Vascular permeability factor, fibrin, and the pathogenesis of tumor stroma formation. Ann N Y Acad Sci 667:101–111

Dvorak HF, Brown LF, Detmar M, Dvorak AM (1995) Vascular permeability factor/vascular endothelial growth factor, microvascular hyperpermeability, and angiogenesis. Am J Pathol 146:1029–1039

Fava RA, Olsen NJ, Spencer GG, Yeo KT, Yeo TK, Berse B, Jackman RW, Senger DR, Dvorak HF, Brown LF (1994) Vascular permeability factor/endothelial growth factor (VPF/VEGF): accumulation and expression in human synovial fluids and rheumatoid synovial tissue. J Exp Med 180:341–346

Feng D, Nagy J, Hipp J, Dvorak H, Dvorak A (1996) Vesiculo-vacuolar organelles and the regulation of venule permeability to macromolecules by vascular permeability factor, histamine, and serotonin. J Exp Med 183:1981–1986

Feng D, Nagy J, Hipp J, Pyne K, Dvorak H, Dvorak A (1997) Reinterpretation of endothelial cell gaps induced by vasoactive mediators in guinea-pig, mouse and rat: many are transcellular pores. J Physiol (Lond) 504:747–761

Ferrara N, Henzel WJ (1989) Pituitary follicular cells secrete a novel heparin-binding growth factor specific for vascular endothelial cells. BBRC 161:851–858

Ferrara N, Houck K, Jakeman L, Leung DW (1992) Molecular and biological properties of the vascular endothelial growth factor family of proteins. Endocr Rev 13:18–32

Filep JG, Foldes-Filep E (1993) Modulation by nitric oxide of platelet-activating factor-induced albumin extravasation in the conscious rat. Br J Pharmacol 110:1347–1352

Frokjaer-Jensen J (1991) The endothelial vesicle system in cryofixed frog mesenteric capillaries analysed by ultrathin serial sectioning. J Electron Microsc Tech 19:291–304

Fujii E, Irie K, Uchida Y, Tsukahara F, Muraki T (1994) Possible role of nitric oxide in 5-hydroxytryptamine-induced increase in vascular permeability in mouse skin. Naunyn Schmiedebergs Arch Pharmacol 350:361–364

Fukumura D, Yuan F, Endo M, Jain RK (1997) Role of nitric oxide in tumor microcirculation. Blood flow, vascular permeability, and leukocyte-endothelial interactions. Am J Pathol 150:713–725

Gerlowski LE, Jain RK (1986) Microvascular permeability of normal and neoplastic tissues. Microvasc Res 31:288–305

Goldberg M, Schneider T (1994) Similarities between the oxygen sensing mechanisms regulating the expression of vascular endothelial growth factor and erythropoietin. J Biol Chem 269:4355–4359

Gospodarowicz D, Abraham JA, Schilling J (1989) Isolation and characterization of a vascular endothelial cell mitogen produced by pituitary-derived folliculo stellate cells. Proc Natl Acad Sci USA 86:7311–7315

Greenbaum LM, Semente G, Prakash A, Roffman S (1978) Leukokinins; their pharmacological properties and role in the pathology of fluid (ascites) accumulation. Agents Actions 8:80–84

Guidi A, Abu-Jawdeh G, Berse B, Jackman R, Tognazzi K, Dvorak H, Brown L (1995) Vascular permeability factor (vascular endothelial growth factor) expression and angiogenesis in cervical neoplasia. J Natl Cancer Inst 87:1237–1245

Guo D, Jia Q, Song HY, Warren RS, Donner DB (1995) Vascular endothelial cell growth factor promotes tyrosine phosphorylation of mediators of signal transduction that contain SH2 domains. Association with endothelial cell proliferation. J Biol Chem 270:6729–6733

Hashimoto E, Ogita T, Nakaoka T, Matsuoka R, Takao A, Kira Y (1994) Rapid induction of vascular endothelial growth factor expression by transient ischemia in rat heart. Am J Physiol 267:H1948–1954

Hashimoto PH, Takaesu S, Chazono M, Amano T (1974) Vascular leakage through intraendothelial channels induced by cholera toxin in the skin of guinea pigs. Am J Pathol 75:171–180

Hata Y, Nakagawa K, Ishibashi T, Inomata H, Ueno H, Sueishi K (1995) Hypoxia-induced expression of vascular endothelial growth factor by retinal glial cells promotes in vitro angiogenesis. Virchows Arch 426:479–486

Hatva E, Kaipainen A, Mentula P, Jaaskelainen J, Paetau A, Haltia M, Alitalo K (1995) Expression of endothelial cell-specific receptor tyrosine kinases and growth factors in human brain tumors. Am J Pathol 146:368–378

Hirata A, Baluk P, Fujiwara T, McDonald DM (1995) Location of focal silver staining at endothelial gaps in inflamed venules examined by scanning electron microscopy. Am J Physiol 269:L403–418

Houck KA, Ferrara N, Winer J, Cachianes G, Li B, Leung DW (1991) The vascular endothelial growth factor family:Identification of a fourth molecular species and characterization of alternative splicing of RNA. Mol Endocrinol 5:1806–1814

Jain RK (1990) Vascular and interstitial barriers to delivery of therapeutic agents in tumors. Cancer Metastasis Rev 9:253–266

Jain RK (1996) 1995 Whitaker Lecture: delivery of molecules, particles, and cells to solid tumors. Ann Biomed Eng 24:457–473

Jeltsch M, Kaipainen A, Joukov V, Meng X, Lakso M, Rauvala H, Swartz M, Fukumura D, Jain RK, Alitalo K (1997) Hyperplasia of lymphatic vessels in VEGF-C transgenic mice. Science 276:1423–1425

Joris I, Majno G, Lorey EJ, Lewis RA (1987) The mechanism of vascular leakage induced by leukotriene $E_4$ endothelial contraction. Am J Pathol 126:19–24

Joukov V, Pajusola K, Kaipainen A, Chilov D, Lahtinen I, Kukk E, Saksela O, Kalkkinen N, Alitalo K (1996) A novel vascular endothelial growth factor, VEGF-C, is a ligand for the Flt4 (VEGFR-3) and KDR (VEGFR-2) receptor tyrosine kinases. EMBO J 15:290–298

Kamat BR, Brown LF, Manseau EJ, Senger DR, Dvorak HF (1995) Expression of vascular permeability factor/vascular endothelial growth factor by human granulosa and theca lutein cells. Role in corpus luteum development. Am J Pathol 146:157–165

Karnovsky MJ (1967) The ultrastructural basis of capillary permeability studied with peroxidase as a tracer. J Cell Biol 35:213–236

Keck PJ, Hauser SD, Krivi G, Sanzo K, Warren T, Feder J, Connolly DT (1989) Vascular permeability factor, an endothelial cell mitogen related to PDGF. Science 246:1309–1312

Kim J, Li B, Weiner J, Houck K, Ferrara N (1992) The vascular endothelial cell growth factor family. Identification of biologically relevant regions by neutralizing monoclonal antibodies. Growth Factors 7:53–64

Knighton DR, Hunt TK, Scheuenstuhl H, Halliday BJ, Werb Z, Banda MJ (1983) Oxygen tension regulates the expression of angiogenesis factor by macrophages. Science 221:1283–1285

Koch AE, Harlow LA, Haines GK, Amento EP, Unemori EN, Wong WL, Pope RM, Ferrara N (1994) Vascular endothelial growth factor. A cytokine modulating endothelial function in rheumatoid arthritis. J Immunol 152:4149–4156

Kohn S, Nagy JA, Dvorak HF, Dvorak AM (1992) Pathways of macromolecular tracer transport across venules and small veins. Structural basis for the hyperpermeability of tumor blood vessels. Lab Invest 67:596–607

Koos RD, Olson CE (1991) Hypoxia stimulates expression of the gene for vascular endothelial growth factor (VEGF), a putative angiogenic factor, by granulosa cells of the ovarian follicle, a site of angiogenesis (abstract). J Cell Biol 115:421a

Ku DD, Zaleski JK, Liu S, Brock TA (1993) Vascular endothelial growth factor induces EDRF-dependent relaxation in coronary arteries. Am J Physiol 265:H586–592

Kubes P (1995) Nitric oxide affects microvascular permeability in the intact and inflamed vasculature. Microcirculation 2:235–244

Kubes P, Granger DN (1992) Nitric oxide modulates microvascular permeability. Am J Physiol 262:H611–615

Ladoux A, Frelin C (1993) Hypoxia is a strong inducer of vascular endothelial growth factor mRNA expression in the heart. BBRC 195:1005–1010

Leung DW, Cachianes G, Kuang W-J, Goeddel DV, Ferrara N (1989) Vascular endothelial growth factor is a secreted angiogenic mitogen. Science 246:1306–1309

Li J, Perrella MA, Tsai JC, Yet SF, Hsieh CM, Yoshizumi M, Patterson C, Endege WO, Zhou F, Lee ME (1995) Induction of vascular endothelial growth factor gene expression by interleukin-1 beta in rat aortic smooth muscle cells. J Biol Chem 270:308–312

Li J, Brown L, Hibbeerd M, Grossman J, Morgan J, Simons M (1996) VEGF, flk-1, and flt-1 expression in a rat myocardial infarction model of angiogenesis. Am J Physiol 270:H1803–1811

Lin K, Nagy JA, Xu H, Shockley TR, Yarmush ML, Dvorak HF (1994) Compartmental distribution of tumor-specific monoclonal antibodies in human melanoma xenografts. Cancer Res 54:2269–2277

Maeda H, Noguchi Y, Sato K, Akaike T (1994) Enhanced vascular permeability in solid tumor is mediated by nitric oxide and inhibited by both new nitric oxide scavenger and nitric oxide synthase inhibitor. Jpn J Cancer Res 85:331–334

Maillard JL, Pick E, Turk JL (1972) Interaction between 'sensitized lymphocytes' and antigen in vitro. V. Vascular permeability induced by skin-reactive factor. Int Arch Allergy 42:50–68

Majno G, Palade G, Schoefl G (1961) Studies on inflammation. II. The site of action of histamine and serotonin along the vascular tree: a topographic study. J Biophys Biochem Cytol 11:607–626

Majno G, Gilmore V, Leventhal M (1967) On the mechanism of vascular leakage caused by histamine type mediators. A microscopic study in vivo. Circ Res 21:833–847

Majno G, Shea SM, Leventhal M (1969) Endothelial contraction induced by histamine-type mediators: an electron microscopic study. J Cell Biol 42:647–672

Matsumura Y, Kimura M, Yamamoto T, Maeda H (1988) Involvement of the kinin-generating cascade in enhanced vascular permeability in tumor tissue. Jpn J Cancer Res 79:1327–1334

Matthews W, Jordan C, Gavin M, Jenkins N, Copeland N, Lemischka I (1991) A receptor tyrosine kinase cDNA isolated from a population of enriched primitive hematopoietic cells and exhibiting close genetic linkage to c-kit. Proc Natl Acad Sci USA 88:9026–9030

Michel CC (1988) Capillary permeability and how it may change. J Physiol (Lond) 404:1–29

Miles AA, Miles EM (1952) Vascular reaction to histamine, histamine-liberator, and leukotaxine in the skin of guinea pigs. J Physiol (Lond) 118:228–257

Millauer B, Wizigmann-Voos S, Schnurch H, Martinez R, Moller N, Risau W, Ullrich A (1993) High affinity VEGF binding and developmental expression suggest Flk-1 as a major regulator of vasculogenesis and angiogenesis. Cell 72:835–846

Miller JW, Adamis AP, Shima DT, D'Amore PA, Moulton RS, O'Reilly MS, Folkman J, Dvorak HF, Brown LF, Berse B, Yeo T-K, Yeo K-T (1994) Vascular endothelial growth factor/vascular permeability factor is temporally and spatially correlated with ocular angiogenesis in a primate model. Am J Pathol 145:574–584

Minchenko A, Bauer T, Salceda S, Caro J (1994) Hypoxic stimulation of vascular endothelial growth factor expression in vitro and in vivo. Lab Invest 71:374–379

Monacci WT, Merrill MJ, Oldfield EH (1993) Expression of vascular permeability factor/vascular endothelial growth factor in normal rat tissues. Am J Physiol 264:C995–1002

Moor RM, Seamark RF (1986) Cell signaling, permeability, and microvasculatory changes during antral follicle development in mammals. J Dairy Sci 69:927–943

Morbidelli L, Chang CH, Douglas JG, Granger HJ, Ledda F, Ziche M (1996) Nitric oxide mediates mitogenic effect of VEGF on coronary venular endothelium. Am J Physiol 270:H411–415

Morris B, Sass M (1966) The formation of lymph in the ovary. Proc R Soc Lond Biol 164:577–591

Murata T, Ishibashi T, Khalil A, Hata Y, Yoshikawa H, Inomata H (1995) Vascular endothelial growth factor plays a role in hyperpermeability of diabetic retinal vessels. Ophthalmic Res 27:48–52

Murohara T, Horowitz JR, Silver M, Tsurumi Y, Chen D, Sullivan A, Isner JM Vascular endothelial growth factor/vascular permeability factor enhances vascular permeability via nitric oxide and prostacyclin. Circulation 97:99–107

Nagy JA, Brown LF, Senger DR, Lanir N, Van De Water L, Dvorak AM, Dvorak HF (1988) Pathogenesis of tumor stroma generation:A critical role for leaky blood vessels and fibrin deposition. Biochim Biophys Acta 948:305–326

Nagy JA, Herzberg KT, Dvorak JM, Dvorak HF (1993) Pathogenesis of malignant ascites formation: initiating events that lead to fluid accumulation. Cancer Res 53:2631–2643

Nagy JA, Morgan E, Herzberg K, Manseau E, Dvorak A, Dvorak H (1995a) Pathogenesis of ascites tumor growth. Angiogenesis, vascular remodeling and stroma formation in the peritoneal lining. Cancer Res 55:376–385

Nagy JA, Masse EM, Herzberg KT, Meyers MS, Yeo K-T, Yeo T-K, Sioussat TM, Dvorak HF (1995b) Pathogenesis of ascites tumor growth. Vascular permeability factor, vascular hyperpermeability and ascites fluid accumulation. Cancer Res 55:360–368

Nagy JA, Meyers MS, Masse EM, Herzberg KT, Dvorak HF (1995c) Pathogenesis of ascites tumor growth. Fibrinogen influx and fibrin accumulation in tissues lining the peritoneal cavity. Cancer Res 55:369–375

Nakano S, Matsukado K, Black KL (1996) Increased brain tumor microvessel permeability after intracarotid bradykinin infusion is mediated by nitric oxide. Cancer Res 56:4027–4031

Neal CR, Michel CC (1995) Transcellular gaps in microvascular walls of frog and rat when permeability is increased by perfusion with the ionophore A23187. J Physiol (Lond) 488:427–437

Neal C, Michel C (1996) Openings in frog microvascular endothelium induced by high intravascular pressure. J Physiol (Lond) 492:39–52

O'Connor SW, Bale WF (1984) Accessibility of circulating immunoglobulin G to the extravascular compartment of solid rat tumors. Cancer Res 44:3719–3723

Okuda Y, Okamura H, Kanzaki H, Takenaka A, Morimoto K, Nishimura T (1980) An ultrastructural study of capillary permeability of rabbit ovarian follicles during ovulation using carbon tracer. Acta Obstet Gynecol Jpn 32:859–867

Olofsson B, Pajusola K, Kaipainen A, Voneuler G, Joukov V, Saksela O, Orpana A, Petersson RF, Alitalo K, Eriksson U (1996) Vascular endothelial growth factor B, a novel growth factor for endothelial cells. Proc Natl Acad Sci USA 93:2576–2581

Palade GE (1988) The microvascular endothelium revisited. In: Simionescu N, Simionescu M (eds) Endothelial cell biology in health and disease. Plenum, New York

Park J, Keller G-A, Ferrara N (1993) The vascular endothelial growth factor (VEGF) isoforms:differential deposition into the subepithelial extracellular matrix and bioactivity of extracellular matrix-bound VEGF. Mol Biol Cell 4:1317–1326

Payer A (1975) Permeability of ovarian follicles and capillaries in mice. Am J Anat 142:295–301

Pe'er J, Shweiki D, Itin A, Hemo I, Gnessin H, Keshet E (1995) Hypoxia-induced expression of vascular endothelial growth factor by retinal cells is a common factor in neovascularizing ocular diseases. Lab Invest 72:638–645

Peters K, DeVries C, Williams L (1993) Vascular endothelial growth factor receptor expression during embryogenesis and tissue repair suggests a role in endothelial differentiation and blood vessel growth. Proc Natl Acad Sci USA 90:8915–8919

Phillips HS, Hains J, Leung DW, Ferrara N (1990) Vascular endothelial growth factor is expressed in rat corpus luteum. Endocrinology 127:965–967

Pierce EA, Avery RL, Foley ED, Aiello LP, Smith LE (1995) Vascular endothelial growth factor/vascular permeability factor expression in a mouse model of retinal neovascularization. Proc Natl Acad Sci USA 92:905–909

Plate KH, Breier G, Weich HA, Risau W (1992) Vascular endothelial growth factor is a potential tumour angiogenesis factor in human gliomas in vivo. Nature 359:845–848

Qu-Hong, Nagy JA, Senger DR, Dvorak HF, Dvorak AM (1995) Ultrastructural localization of vascular permeability factor/vascular endothelial growth factor (VPF/VEGF) to the abluminal plasma membrane and vesiculo-vacuolar organelles of tumor microvascular endothelium. J Histochem Cytochem 43:381–389

Quinn T, Peters K, DeVries C, Ferrara N, Williams L (1993) Fetal liver kinase 1 is a receptor for vascular endothelial growth factor and is selectively expressed in vascular endothelium. Proc Natl Acad Sci USA 90:7533–7537

Ramirez MM, Quardt SM, Kim D, Oshiro H, Minnicozzi M, Duran WN (1995) Platelet activating factor modulates microvascular permeability through nitric oxide synthesis. Microvasc Res 50:223–234

Ravindranath N, Little-Ihrig L, Phillips HS, Ferrara N, Zeleznik AJ (1992) Vascular endothelial growth factor messenger ribonucleic acid expression in the primate ovary. Endocrinology 131:254–260

Rippe B, Haraldsson B (1994) Transport of macromolecules across microvascular walls:the two-pore theory. Physiol Rev 74:163–219

Roberts WG, Hasan T (1993) Tumor-secreted vascular permeability factor/vascular endothelial growth factor influences photosensitizer uptake. Cancer Res 53:153–157

Roberts WG, Palade GE (1995) Increased microvascular permeability and endothelial fenestration induced by vascular endothelial growth factor. J Cell Sci 108:2369–2379

Seetharam L, Gotoh N, Maru Y, Neufeld G, Yamaguchi S, Shibuya M (1995) A unique signal transduction from FLT tyrosine kinase, a receptor for vascular endothelial growth factor VEGF. Oncogene 10:135–147

Senger DR, Galli SJ, Dvorak AM, Perruzzi CA, Harvey VS, Dvorak HF (1983) Tumor cells secrete a vascular permeability factor that promotes accumulation of ascites fluid. Science 219:983–985

Senger DR, Perruzzi CA, Feder J, Dvorak HF (1986) A highly conserved vascular permeability factor secreted by a variety of human and rodent tumor cell lines. Cancer Res 46:5629–5632

Senger DR, Connolly D, Perruzzi CA, Alsup D, Nelson R, Leimgruber R, Feder J, Dvorak HF (1987) Purification of a vascular permeability factor (VPF) from tumor cell conditioned medium. Fed Proc 46:2102

Senger DR, Connolly DT, Van De Water L, Feder J, Dvorak HF (1990) Purification and $NH_2$-terminal amino acid sequence of guinea pig tumor-secreted vascular permeability factor. Cancer Res 50:1774–1778

Senger DR, Van De Water L, Brown L, Nagy J, Yeo K-T, Yeo T-K, Berse B, Jackman R, Dvorak A, Dvorak H (1993) Vascular permeability factor (VPF, VEGF) in tumor biology. Cancer Metastasis Rev 12:303–324

Shweiki D, Itin A, Soffer D, Keshet E (1992) Vascular endothelial growth factor induced by hypoxia may mediate hypoxia-initiated angiogenesis. Nature 359:843–845

Shweiki D, Itin A, Neufeld G, Gitay GH, Keshet E (1993) Patterns of expression of vascular endothelial growth factor (VEGF) and VEGF receptors in mice suggest a role in hormonally regulated angiogenesis. J Clin Invest 91:2235–2243

Simionescu N (1983) Cellular aspects of transcapillary exchange. Physiol Rev 63:1536–1579

Simon M, Grone HJ, Johren O, Kullmer J, Plate KH, Risau W, Fuchs E (1995) Expression of vascular endothelial growth factor and its receptors in human renal ontogenesis and in adult kidney. Am J Physiol 268:F240–250

Sioussat TM, Dvorak HF, Brock TA, Senger DR (1993) Inhibition of vascular permeability factor (vascular endothelial growth factor) with anti-peptide antibodies. Arch Biochem Biophys 301:15–20

Sobel A, LaGrue G (1980) Role of a vascular permeability-increasing factor released by lympyhocytes in renal pathology. Lymphokine Rep 1:211–230

Song CW, Levitt SH (1971) Quantitative study of vascularity in Walker carcinoma 256. Cancer Res 31:587–589

Spar IL, Bale WF, Marrack D, Dewey WC, McCardle RJ, Harper PV (1967) [131]I-labeled antibodies to human fibrinogen: diagnostic studies and therapeutic trials. Cancer 20:865–870

Terman BI, Carrion ME, Kovacs E, Rasmussen BA, Eddy RL, Shows TB (1991) Identification of a new endothelial cell growth factor receptor tyrosine kinase. Oncogene 6:1677–1683

Terman BI, Dougher-Vermazen M, Carrion ME, Dimitrov D, Armellino DC, Gospodarowicz D, Böhlen P (1992) Identification of the KDR tyrosine kinase as a receptor for vascular endothelial cell growth factor. BBRC 187:1579–1586

Tsurumi Y, Murohara T, Krasinski K, Chen D, Witzenbichler B, Kearney M, Couffinhal T, Isner JM (1997) Reciprocal relation between VEGF and NO in the regulation of endothelial integrity. Nat Med 3:879–886

van der Zee R, Murohara T, Luo Z, Zollmann F, Passeri J, Lekutat C, Isner JM (1997) Vascular endothelial growth factor/vascular permeability factor augments nitric oxide release from quiescent rabbit and human vascular endothelium. Circulation 95:1030–1037

Waltenberger J, Claesson WL, Siegbahn A, Shibuya M, Heldin CH (1994) Different signal transduction properties of KDR and Flt1, two receptors for vascular endothelial growth factor. J Biol Chem 269:26988–26995

Warren BA (1979) The vascular morphology of tumors. In: Peterson H-I (ed) Tumor blood circulation:angiogenesis, vascular morphology and blood flow of experimental and human tumors. CRC Press, Boca Raton

Watanabe Y, Dvorak HF (1997) Vascular permeability factor/vascular endothelial growth factor inhibits anchorage-disruption-induced apoptosis in microvessel endothelial cells by inducing scaffold formation. Exp Cell Res 233:340–349

Watanabe Y, Lee SW, Detmar M, Ajioka I, Dvorak HF (1997) Vascular permeability factor/vascular endothelial growth factor (VPF/VEGF) delays and induces escape from senescence in human dermal microvascular endothelial cells. Oncogene 14:2025–2032

Weindel K, Moringlane JR, Marme D, Weich HA (1994) Detection and quantification of vascular endothelial growth factor/vascular permeability factor in brain tumor tissue and cyst fluid:the key to angiogenesis? Neurosurgery 35:439–448

Yeo T-K, Dvorak HF (1995) Tumor stroma. In: Colvin R, Bhan A, McCluskey R (eds) Diagnostic immunopathology. Raven, New York

Yeo K-T, Wang HH, Nagy JA, Sioussat TM, Ledbetter SR, Hoogewerf AJ, Zhou Y, Masse EM, Senger DR, Dvorak HF, Yeo T-K (1993) Vascular permeability factor (vascular endothelial growth factor) in guinea pig and human tumor and inflammatory effusions. Cancer Res 53:2912–2918

Yoshiji H, Harris SR, Thorgeirsson UP (1997) Vascular endothelial growth factor is essential for initial but not continued in vivo growth of human breast carcinoma cells. Cancer Res 57:3924–3928

Yuan Y, Granger HJ, Zawieja DC, DeFily DV, Chilian WM (1993) Histamine increases venular permeability via a phospholipase C-NO synthase-guanylate cyclase cascade. Am J Physiol 264:H1734–1739

Yuan F, Dellian M, Fukumura D, Leunig M, Berk DA, Torchilin VP, Jain RK (1995) Vascular permeability in a human tumor xenograft: molecular size dependence and cutoff size. Cancer Res 55:3752–3756

Yuan F, Chen Y, Dellian M, Safabakhsh N, Ferrara N, Jain RK (1996) Time-dependent vascular regression and permeability changes in established human tumor xenografts induced by an anti-vascular endothelial growth factor/vascular permeability factor antibody. Proc Natl Acad Sci USA 93:14765–14770

Ziche M, Morbidelli L, Choudhuri R, Zhang HT, Donnini S, Granger HJ, Bicknell R (1997) Nitric oxide synthase lies downstream from vascular endothelial growth factor-induced but not basic fibroblast growth factor-induced angiogenesis. J Clin Invest 99:2625–2634

# Role of Vascular Endothelial Growth Factor and Vascular Endothelial Growth Factor Receptors in Vascular Development

P. Carmeliet and D. Collen

| | | |
|---|---|---|
| 1 | Assembly of Endothelial Cells in Embryonic Blood Vessels | 133 |
| 2 | The VEGF Family | 140 |
| 2.1 | VEGF-A: Structure and Gene Organization | 140 |
| 2.2 | VEGF-A: Biological Role | 141 |
| 2.3 | VEGF-A: Expression | 143 |
| 2.4 | VEGF-A: Transgenesis | 144 |
| 2.5 | VEGF-A: Regulation by Hypoxia | 145 |
| 2.6 | VEGF-A (Gene) Therapy | 147 |
| 2.7 | Placental Growth Factor | 147 |
| 2.8 | VEGF-B | 148 |
| 2.9 | VEGF-C | 149 |
| 2.10 | VEGF-D | 149 |
| 2.11 | VEGF Receptors | 150 |
| 3 | Conclusion | 151 |
| References | | 152 |

## 1 Assembly of Endothelial Cells in Embryonic Blood Vessels

Distinct cellular processes mediate blood-vessel formation during embryogenesis (Beck and D'Amore 1997; Carmeliet and Collen 1998a; Folkman and D'Amore 1996; Noden 1989; Risau 1997; Wilting and Christ 1996) (Fig. 1). Initially, mesodermal cells differentiate in situ into early haemangioblasts and form cellular aggregates (blood islands), in which the inner-cell population develops into haematopoietic precursors and the outer-cell population gives rise to the primitive endothelial cells. In vitro findings suggest that basic fibroblast growth factor (bFGF or FGF2) may participate in angioblast differentiation via induction of a cellular receptor for vascular endothelial growth factor (VEGF-A) (Flamme and Risau 1992).

The second phase involves "vasculogenesis", during which endothelial cells fuse and form a primordial vascular network (Fig. 1); the larger vessels of the

---

Center for Transgene Technology and Gene Therapy, Campus Gasthuisberg, Herestraat 49, Flanders Interuniversity Institute for Biotechnology, University of Leuven, KU Leuven, Leuven, 3000, Belgium

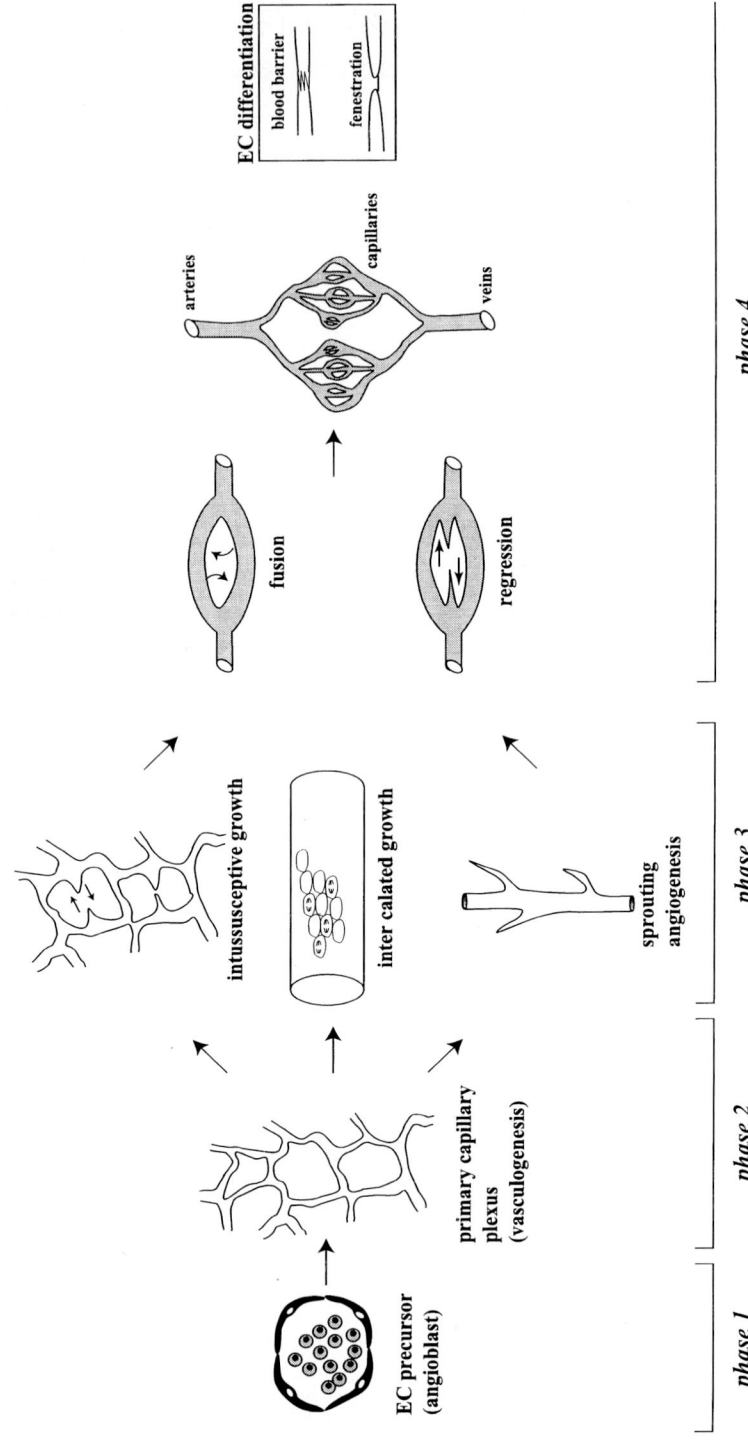

**Fig. 1.** Different phases of embryonic vascular development. Endothelial precursors (angioblasts) differentiate to early endothelial cells (*phase 1*), which become assembled into a primitive capillary plexus (vasculogenesis) (*phase 2*). This emerging network expands via intussusceptive growth, intercalated growth and sprouting angiogenesis (*phase 3*), after which it becomes remodelled via pruning, fusion and regression of pre-existing vessels into a tree of arteries, capillaries and veins (*phase 4*). Endothelial cells further differentiate and acquire specific properties such as the formation of a tight barrier in the brain, or the formation of fenestrations in exocrine glands

embryo and the primary vascular plexus in the lung, the pancreas, the spleen, the heart and the yolk sac arise by this process (WILTING and CHRIST 1996). Lumen formation of the primitive capillaries, an essential aspect of vasculogenesis, may result from endothelial vacuolization (intracellular lumen formation), or from a continuation of a pre-existing lumen through joining of distal endothelial cells (intercellular lumen formation). Gene inactivation studies in mice indicate that these processes are controlled by VEGF-A (CARMELIET et al. 1996a; FERRARA et al. 1996), fibronectin (GEORGE et al. 1993), the $\alpha_5$ integrin receptor (YANG et al. 1993), VE-cadherin (VITTET et al. 1997), transforming growth factor (TGF)-$\beta$1 (DICKSON et al. 1995) and, possibly, FGF2. Whereas the VEGF-A receptor VEGF receptor-2 (VEGFR-2; also denoted FLK1) may be involved in positive regulation (SHALABY et al. 1995), another VEGF-A receptor, VEGFR-1 (also denoted FLT1) may antagonistically inhibit this process (FONG et al. 1995). In addition, immunoneutralization studies in the avian embryo indicate that the $a_v\beta_3$ integrin is involved in endothelial spreading and formation of endothelial protrusions required for lumen formation and vessel patterning (DRAKE and LITTLE 1995).

During the third phase, a primitive vascular plexus develops into a complex organized and interconnecting network (Fig. 1). One mechanism involves intussusceptive microvascular growth, whereby a pre-existing vessel is split into two daughter vessels by formation of trans-capillary pillars and by invagination by surrounding pericytes and extracellular matrix (PATAN et al. 1992, 1996, 1997) (Fig. 2). This process, whereby sinusoidal capillaries generate loops that remain constantly perfused, seems to be of special importance in the lung, but is probably more widespread than originally considered. It is also the predominant mechanism for VEGF-induced vascular growth in the chicken allantoic membrane (PATAN et al. 1996; WILTING et al. 1996). Although its molecular mechanisms remain largely unknown, VEGF-A (CARMELIET et al. 1996; FERRARA et al. 1996), the angiopoietins (ANG) and the TIE receptors (MAISONPIERRE et al. 1997; PATAN 1998; PURI et al. 1995; SATO et al. 1995; SURI et al. 1996), and several extracellular matrix components (GEORGE et al. 1993) may be implicated.

A second mechanism of network expansion involves sprouting of new blood vessels from a pre-existing blood vessel ("angiogenesis"), such as occurs in the brain, the kidney and the intersomitic vessels (Fig. 3). It is controversial whether endothelial cells migrate first and are followed by sprouting pericytes, or rather, whether pericytes pave the way by hollowing out the extracellular matrix, thereby allowing the sprouting endothelial cells to fill in the pericyte tube (NEHLS and DRENCKHAHN 1993). Whereas the latter mechanism may occur during adult neovascularization, the former mechanism appears to be more prevalent during embryonic vascular development.

In order for endothelial (or pericyte) cells to emigrate from their pre-existing site, they first need to proteolytically degrade the surrounding basement membrane (plasminogen activators and matrix metalloproteinases; CARMELIET and COLLEN 1996), to loosen their interendothelial cell contacts (VE-cadherin; VITTET et al. 1997) and to relieve the periendothelial cell support (ANG2; MAISONPIERRE et al. 1997). The endothelial cells need to proliferate to change their shape, to protrude

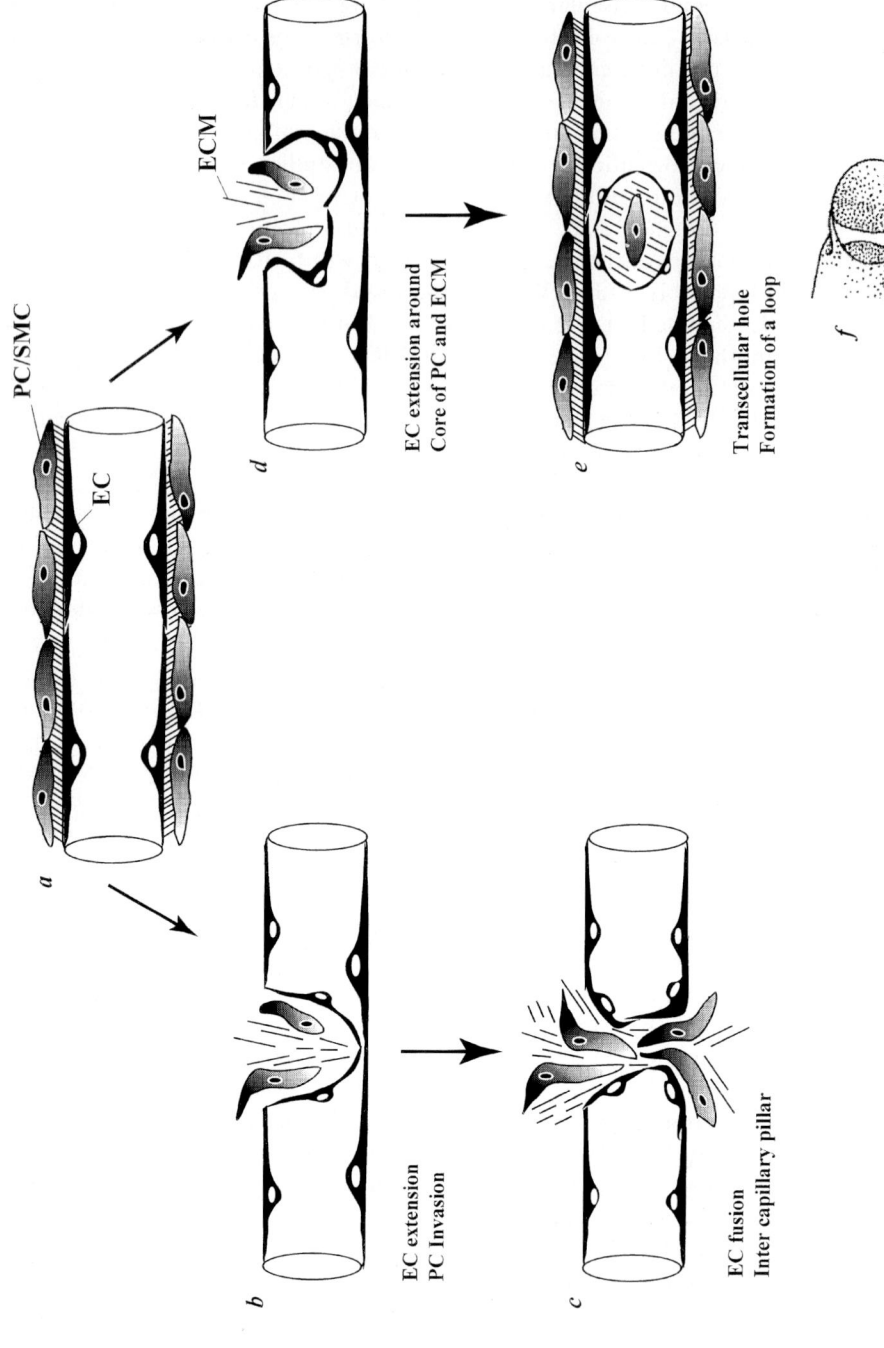

**Fig. 2a-f.** Intussusceptive vascular growth. Quiescent vessels (*panel a*) may grow by intussusception. Endothelial cell (*EC*) folds are formed by stretching and elongation of endothelial cells to the opposite wall (*panel b*) and, subsequently, fuse in the area of connection (*panel c*). Invasion of pericytes (*PC*) and subsequent stabilisation by deposition of extracellular matrix mediates the formation of an intercapillary pillar that separates two vessel segments (*panel d*). Alternatively, endothelial cells accompanied by peri-endothelial cells may extend into the centre of the vessel lumen (*panel e*) and, subsequently, become entrapped to form a trans-cellular hole, establishing a loop (*panel f*). Branching of vessels and formation of sinusoidal loops may occur via this mechanism

extensions and to migrate towards the angiogenic stimulus (VEGF-A; CARMELIET et al. 1996; FERRARA et al. 1996; the $a_v\beta_3$ integrin; DRAKE and LITTLE 1995) (initiation step). Subsequently, the endothelial cells are assembled in cords which form a lumen (VEGF-A, CARMELIET et al. 1996; FERRARA et al. 1996; the $a_v\beta_3$ integrin, DRAKE et al. 1995), and fuse with other vessels (fibronectin, GEORGE et al. 1993; VCAM-1, GURTNER et al. 1995; KWEE et al. 1995; $a_4$ integrin, YANG et al. 1995). As the endothelial tubes mature, they become surrounded by periendothelial mural cells (ANG1 and TIE2, SURI et al. 1996; tissue factor, CARMELIET et al. 1996; PDGF-B, LEVÉEN et al. 1994; LINDAHL et al. 1997; TGF-β1, DICKSON et al. 1995) (maintenance and termination steps). Somewhat surprisingly, certain extracellular matrix components (vitronectin, ZHENG et al. 1995) and proteinases (CARMELIET and COLLEN 1996) did not affect this process. This may suggest that several mechanisms of embryonic vascular development are redundant or that they are compensated for. It is conceivable that angiogenesis inhibitors are involved in the termination of sprouting, but insights into their biological role awaits future study.

A third mechanism involves the intercalated growth of blood vessels ("non-sprouting angiogenesis"), whereby pre-existing capillaries merge, or additional endothelial cells fuse into existing vessels to increase their diameter and length. This process is important for vessel growth in the heart and during healing of endothelial wounds (PATAN et al. 1996; WILTING et al. 1996), but its molecular mechanisms remain enigmatic.

Hypoxia and metabolic stress are somehow important for these early steps of embryonic vascularization, as evidenced by the vascular defects in embryos lacking the arylhydrocarbon-receptor nuclear translocator (ARNT) (KOZAK et al. 1997; MALTEPE et al. 1997), the hypoxia-inducible factor (HIF)-1α (CARMELEIT et al. 1998b, Iyer et al. 1998, Ryan et al. 1998), or the von Hippel-Lindau (VHL) gene product (GNARRA et al. 1997). Although distinct tissues have been proposed to be vascularized by either vasculogenic or angiogenic processes, more recent evidence challenges such a dogmatic classification, as primary angioblasts with resultant vasculogenesis have also been observed in the neural tube, the kidney and the somites (ROBERT et al. 1996; WILTING and CHRIST 1996).

After the onset of circulation (e.g. during the organogenetic period; fourth phase), this emerging vascular plexus becomes remodelled and pruned into a tree of veins, capillaries and arteries, the expansion of which matches the metabolic demands of the growing embryo (Fig. 1). It involves fusion and regression of blood vessels, changes in lumen diameter and vessel wall thickness, and the deposition of specialised extracellular matrix components (elastin, fibrillins), which provide the

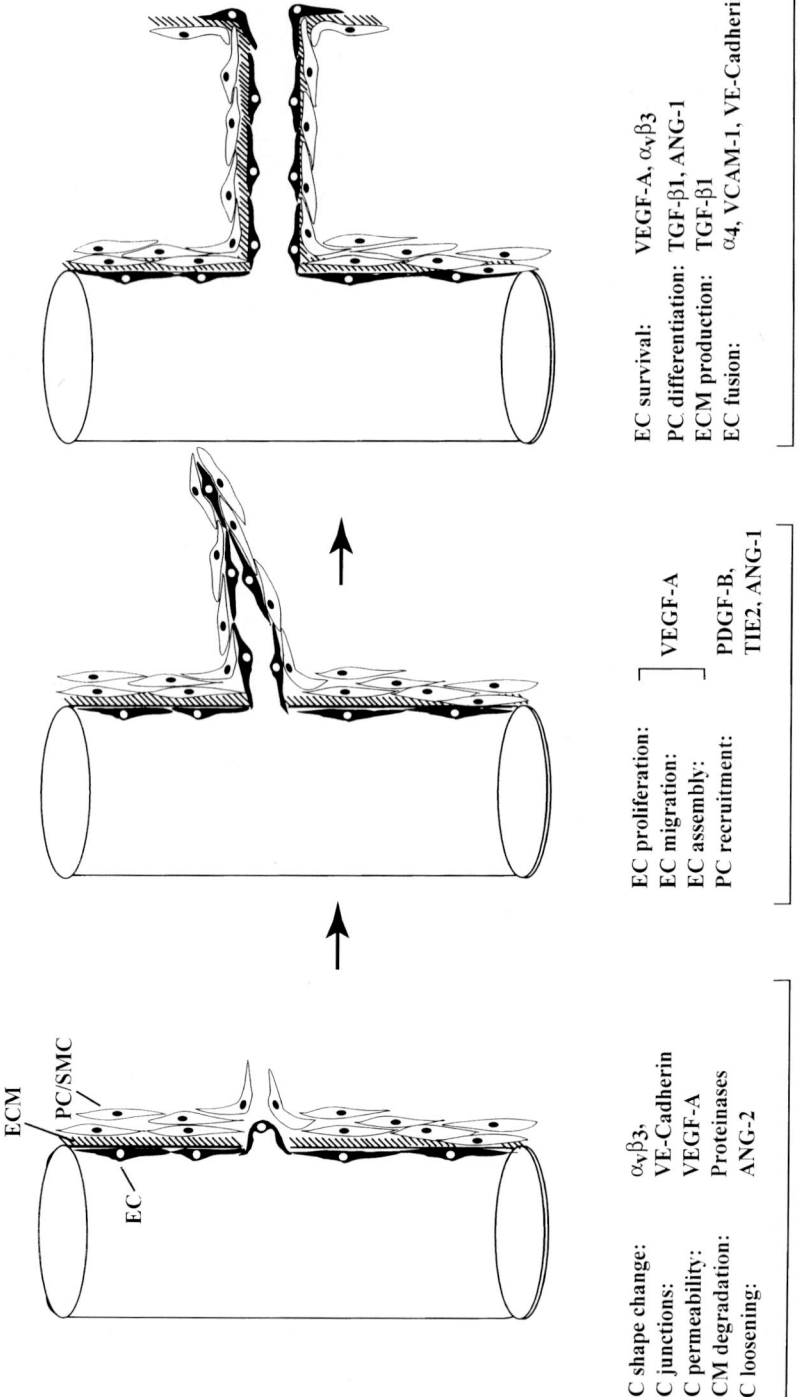

**Fig. 3.** Molecular mechanisms of angiogenic sprouting as deduced from gene-inactivation studies. Vessels may grow by sprouting from pre-existing vessels. During the initiation phase of angiogenic sprouting, quiescent endothelial cells (*EC*) change their shape (mediated via the αvβ3-integrin), loosen their intercellular junctions [dependent on vascular endothelial (VE)-cadherin] and become more permeable [in part due to the action of vascular endothelial growth factor-A (VEGF-A)]. As quiescent endothelial cells are encaged by an extracellular matrix (*ECM*) and surrounded by pericytes (*PC*) or smooth muscle cells (*SMC*), the ECM has to be degraded (proteinases) and the support by pericytes has to be relieved (possibly mediated by angiopoietin-2; ANG-2) in order to allow them to migrate to distant sites. During progression of angiogenic sprouting, VEGF-A mediates proliferation, migration and assembly of endothelial cells into lumenized tubes. Sprouting of pericytes and smooth muscle cells appears to be mediated by platelet-derived growth factor (PDGF)-B, TIE2 and angiopoietin-1 (ANG-1). Sprouting endothelial cells fuse with distant endothelial cells (meditated by α4-integrin, VCAM-1 and VE-cadherin) and require survival signals (such as VEGF and the αvβ3-integrin). Recruited pericytes and smooth muscle cells differentiate and provide structural strength to the immature sprout by deposition of the extracellular matrix [a mechanism dependent on transforming growth factor beta-1 (TGF-β1), angiopoietin-1 and ANG-1]. The role of PDGF-B, TIE2, ANG-1, ANG-2, TGF-β1 and the integrins is described in more detail elsewhere (Carmeliet and Collen 1997, 1998)

vessel novel properties (viscoelasticity). Metabolic, hydrodynamic, hypoxic or rheologic factors, and the interaction between endothelial and mural cells may determine these processes. Vessel regression may result from endothelial cell emigration, trans-differentiation or death (Wilting and Christ 1996). Both the loss of endothelial survival signals (Alon et al. 1995; Benjamin and Keshet 1997) and the production of death factors could mediate the latter process.

Developmentally programmed vessel regression has been studied in greater detail in the hyaloid capillary network in the eye during the first 3 weeks after birth (Lang et al. 1994; Lang and Bishop 1993). A macrophage-related cell, called the hyalocyte, induces apoptosis in some of the endothelial cells by secreting death factors such as tumour necrosis factor (TNF)-α, nitric oxide and reactive oxygen species (initiating apoptosis). Shedding of dead cells results in obstruction of the lumen (Meeson et al. 1996), inducing synchronous endothelial cell death (secondary apoptosis) due to deprivation of survival factors [VEGF, FGF2 or platelet-derived growth factor (PDGF)], or altered shear stress-dependent gene expression (PDGF, FGF2). Changes in periendothelial cell support also play a role in vessel regression, as PDGF-induced emigration of pericytes induces endothelial cell death in the retina (Benjamin and Keshet, personal communication). Interestingly, ANG2 has been associated with vessel regression, but presumably only at low VEGF levels (Hanahan 1997; Maisonpierre et al. 1997). Integrin $a_v\beta_3$ also provides essential survival function for angiogenic, but not for quiescent, endothelial cells (Brooks et al. 1994).

Another characteristic of this phase of blood-vessel development is that endothelial cells acquire particular heterogeneity due to specialised differentiation, which appears to be dependent on interactions with local parenchymal cells (Risau 1995; Wilting and Christ 1996) (Fig. 1). For example, astrocytes appear responsible for the induction of the blood–brain barrier, whereas choroidal epithelial cells induce the formation of fenestrated capillaries. Although the molecular mechanisms of such organ-specific differentiation remain largely unknown, the different VEGF-A splice forms have been suggested to be involved in the formation and the maintenance of a fenestrated endothelium (RISAU 1995; WILTING and CHRIST 1996).

## 2 The VEGF Family

### 2.1 VEGF-A: Structure and Gene Organization

VEGF-A, unlike other known angiogenic factors, has a unique combination of properties: (a) it is produced by cells in close vicinity of endothelial cells, suggesting paracrine regulation of blood vessel formation (BREIER et al. 1995; DUMONT et al. 1995); (b) it is secreted and exerts a direct and largely restricted effect on endothelial cells via interaction with cellular receptors VEGFR-1 and VEGFR-2 (TERMAN and DOUGHER VERMAZEN 1996); (c) it induces a pleiotropic response, allowing endothelial cells to proliferate, migrate, assemble into tubes and survive, and is one of the most potent permeability factors (DVORAK et al. 1995; FERRARA and DAVIS-SMYTH 1997; KLAGSBRUN and D'AMORE 1996); (d) its expression is highly regulated by hypoxia, providing a physiological feedback mechanism to accommodate insufficient tissue oxygenation by promoting blood-vessel formation (DOR and KESHET 1997); and (e) it is a potent growth factor since its over- or under-expression significantly affects blood-vessel formation in vivo (CARMELIET et al. 1996; DRAKE and LITTLE 1995; FERRARA et al. 1996; FLAMME et al. 1995).

VEGF-A is transcribed from a single gene that is alternatively processed in various isoforms (NEUFELD et al. 1996). The shortest form (VEGF-A$^{121}$) is freely diffusible in the surrounding extracellular milieu, whereas the longer isoforms (VEGF-A$^{165}$, VEGF-A$^{189}$ and VEGF-A$^{206}$) display an increasing binding to heparin-rich extracellular matrix (DVORAK et al. 1995; FERRARA and DAVIS-SMYTH 1997; KLAGSBRUN and D'AMORE 1996). The mouse has only three isoforms, which lack one amino acid residue (VEGF-A$^{120}$, VEGF-A$^{164}$ and VEGF-A$^{188}$). VEGF-A$^{165}$ can be liberated from the extracellular matrix by the serine proteinase plasmin which cleaves it into a mitogenic carboxy-terminal polypeptide (VEGF-A$^{111\ 165}$) and an amino-terminal VEGF-A$^{110}$ fragment with reduced binding to the VEGFR-1 and -2 (KEYT et al. 1996).

Notably, VEGF-A$^{110}$ and VEGF-A$^{121}$ have more than 100-fold reduced endothelial cell mitogenic potency than VEGF-A$^{165}$, indicating that the carboxy-terminal residues are critical for mitogenic potency. Since removal of the carboxy-terminal domain [whether due to alternative splicing of messenger ribonucleic acid (mRNA) or to proteolysis] increases the solubility but reduces the bioactivity of VEGF-A, it is possible that these different VEGF-A isoforms provide a spatial gradient of patterning information during blood vessel formation, e.g. the longer matrix-attached isoforms (close to the site of VEGF-A production) provide a stronger mitogenic signal than the shorter, more diffusible VEGF-A isoforms. Such a mechanism would attract endothelial cells to ischaemic or avascular sites. The various isoforms exhibit a tissue-specific and temporal pattern of expression but their differential (patho)physiologicial role in vivo remains largely undetermined (BASIC et al. 1995).

## 2.2 VEGF-A: Biological Role

The pleiotropic actions of VEGF-A render this molecule ideally suited to modulate blood-vessel formation. VEGF-A stimulates endothelial proliferation (DVORAK et al. 1995; FERRARA and DAVIS-SMYTH 1997) and reduces endothelial death (ALON et al. 1995). The survival function of VEGF-A may be important, as reduced levels have been associated with endothelial apoptosis, shedding into and causing obstruction of the lumen, resulting in vascular collapse with haemorrhaging and vessel regression (BENJAMIN and KESHET 1997). VEGF-A can stimulate endothelial migration via an effect on motility, proteolysis, cell-adhesion and -junction, and permeability. It increases the release of matrix metalloproteinases (MMPs) and plasminogen activators (PAs) (MIGNATTI and RIFKIN 1996). Proteolytically active MMP2 binds the $a_v\beta_3$ integrin, thereby linking matrix adhesion to regulation of proteinase activity (BROOKS et al. 1996). In addition, recent evidence suggests that urokinase-type plasminogen activator (u-PA), its cellular receptor (u-PAR) and the plasminogen activator inhibitor-1 (PAI-1) may control cell migration in a proteolysis-independent interaction with vitronectin and its $a_v\beta_3$ integrin receptor, although the precise mechanisms and the consequences on cellular migration are still unclear.

Indeed, proteolytically inactive u-PA may promote migration by displacing PAI-1 from vitronectin, allowing the latter to bind to the $a_v\beta_3$ integrin (STEFANSSON and LAWRENCE 1996). It is possible that u-PAR, which colocalises with $a_v\beta_5$ and possibly with other integrins on migrating cells (REINARTZ et al. 1995), is also involved in this interaction, as u-PA alters the conformation of u-PAR, favouring interaction of the latter with vitronectin (DENG et al. 1996; WEI et al. 1996). VEGF-A also induces endothelial expression of the $a_v\beta_3$ integrin (and presumably also of the $a_v\beta_5$ integrin), which are receptors for vitronectin and/or fibronectin, fibrinogen and osteopontin (FRIEDLANDER et al. 1995; VARNER et al. 1995). VEGF-stimulated microvascular endothelial cells produce increased amounts of osteopontin, which promotes endothelial migration (SENGER et al. 1996). In addition, VEGF-A may coordinate endothelial-cell migration by increasing intercellular communication via induced connexin 43 expression (PEPPER and MEDA 1992).

An important property of VEGF-A is its ability to increase vascular permeability, possibly via rapid formation of capillary fenestrations (ROBERTS and PALADE 1995), the assembly of vesiculo-vacuolar organelles, or the opening of intercellular junctions or trans-cellular gaps (DVORAK et al. 1996). As a result, plasma coagulation factors and other plasma proteins (fibronectin, vitronectin, fibrinogen) leak out into the perivascular compartment. Since VEGF-A induces expression of tissue factor (the cellular initiator of blood coagulation) (CLAUSS et al. 1990), a fibrin seal is formed which, together with the other plasma proteins, provides a scaffold for migrating cells (SENGER et al. 1996). In addition, tissue factor may promote smooth muscle migration directly (CARMELIET et al. 1996; SATO et al. 1996) and indirectly through generation of thrombin (COUGHLIN 1994). Through a feedback induction of VEGF-A release, tissue factor may further increase leakage (ZHANG et al. 1994, 1996). VEGF-A also promotes expression of VCAM-1 and ICAM-1 on endothelial cells, resulting in the adhesion of leucocytes (MELDER et al. 1996). As the latter cells may influence endothelial permeability (DEJANA 1996), and release a myriad of angiogenic growth factors (WEIHRAUCH et al. 1995), VEGF-A establishes a direct link between inflammation, permeability and angiogenesis.

The role of VEGF-A may not be entirely restricted to endothelial cells since VEGF receptors are also found on other cell types. Indeed, VEGF-A may control vessel formation via recruitment of macrophages (BARLEON et al. 1996; CLAUSS et al. 1990, 1996), which produce a variety of angiogenic factors. This process may be particularly relevant during inflammation or myocardial angiogenesis (SUNDERKOTTER et al. 1994). In addition, by stimulating pericyte recruitment or smooth muscle cell migration (GROSSKREUTZ et al. 1996; NOMURA et al. 1995; TAKAGI et al. 1996), VEGF-A may aid the maturation of the endothelial tubes, which, once surrounded by pericytes, may become less responsive to hypoxia regulation (Benjamin and Keshet, personal communication). VEGF-A increases the number of muscular vessels in the ischaemic heart, presumably via stimulation of endothelial release of PDGF (BANAI et al. 1994). It may also induce vasodilation of arterioles via release of endothelial nitric oxide, contributing to the hyperaemia in tissues undergoing neovascularization (HARIAWALA et al. 1996; VAN DER ZEE et al. 1997). In contrast, VEGF-2 (or VEGF-C; another VEGF homologue; see Sect. 2.9 below) appears to inhibit PDGF-stimulated proliferation of smooth muscle cells (JINGSHAN et al. 1997), suggesting distinct functions for each of the VEGF family members.

Other cell types are also responsive to VEGF-A. It affects differentiation of osteoblasts and of haematopoietic cells, regulates the production of insulin in pancreatic β-cells, and has an as yet unidentified function on melanoma cells (BROXMEYER et al. 1995; DVORAK et al. 1995; KATOH et al. 1995). It also inhibits maturation of host antigen-presenting cells (dendritic cells), suggesting that it may facilitate tumour growth by allowing the tumour to avoid the induction of an immune response (GABRILOVICH et al. 1996).

The biological role of VEGF-A appears to be contextual, that is, its effect is determined by the presence of other signals such as FGF2, TGF-β1, PDGF, ANGs etc. (HANAHAN 1997; MAISONPIERRE et al. 1997; WALTENBERGER 1997). Some of these molecules amplify the role of VEGF-A (FRIEDLANDER et al. 1995; PERTOVAARA et al. 1994; STAVRI et al. 1995), whereas others, such as angiostatin and probably other (yet to be defined) factors, inhibit its action. Moreover, VEGF-induced permeability of vessels is different in various tissues, suggesting an important role of the microenvironment in the overall angiogenic effect of VEGF-A (DELLIAN et al. 1996). In certain tissues, such as the mammary gland during gestation, VEGF-A is present, but does not seem to correlate with the vascularization response.

## 2.3 VEGF-A: Expression

The pattern of expression of the VEGF-ligands and their receptors during embryogenesis suggests a role for these molecules in vascular development. VEGF-A is produced by cells in close vicinity to the developing endothelial cells, such as for example in the visceral endoderm cells in the yolk sac or in the ventricular zone of the developing brain (BREIER et al. 1995). The VEGF receptors (see Sect. 2.11 below) have distinct, but overlapping, temporo-spatial expression patterns during embryogenesis (DUMONT et al. 1995; YAMAGUCHI et al. 1993). Both receptors are present in the mesenchyme of the yolk sac, the amnion and the chorion (7.5 days of gestation). By 9.5 days of gestation, they are expressed in the angioblasts in the blood islands of the yolk sac, and in the endothelial cells of the dorsal aorta and of the perineural plexus. Subsequently, by 11.5–17.5 days of gestation, they are detected in most capillaries in the developing organs, such as the intersomitic regions, the heart, the lung, the thymus, the intestines and the brain.

In all these tissues, VEGF-A is expressed by adjacent cells, suggesting paracrine regulation of VEGF-dependent blood-vessel formation. In the placenta, VEGFR-1 and -2 are differentially expressed. The syncytiotrophoblast cells, expressing VEGF-A and placental growth factor (PlGF; another VEGF homologue, see Sect. 2.7 below) (KHALIQ et al. 1996; VUORELA et al. 1997) are in close proximity to the VEGFR-2-expressing endothelium, but are further away from the VEGFR-1-expressing spongiotrophoblast layer (BREIER et al. 1995; DUMONT et al. 1995; YAMAGUCHI et al. 1993). This suggests that VEGFR-1 may interact with soluble VEGF-A and/or PlGF isoforms, or that other VEGF-related ligands may interact with VEGFR-1.

In contrast to the minimal levels of VEGF-A gene expression in most adult tissues, VEGF-A expression is detectable in the adult kidney, possibly implicating a role in the maintenance of endothelial cell homeostasis and/or in their fenestration (DVORAK et al. 1995; FERRARA and DAVIS-SMYTH 1997). In the adult, VEGF-A expression can be induced by macrophages, T-lymphocytes, astrocytes, osteoblasts, smooth muscle cells, fibroblasts, endothelial cells, cardiomyocytes, skeletal muscle cells and keratinocytes (DVORAK et al. 1995; FERRARA and DAVIS-SMYTH 1997). The expression of VEGF-A by these cells and of its receptors by endothelial cells is significantly induced during pathological conditions of myocardial ischaemia (DOR and KESHET 1997), atherosclerosis (KUZUYA et al. 1995), diabetic and ischaemic retinopathy (ALON et al. 1995; KLAGSBRUN and D'AMORE 1996), tumorigenesis (ALBINI et al. 1996; GABRILOVICH et al. 1996; MELDER et al. 1996; PLATE and RISAU 1995), arthritis and wound healing (FRANK et al. 1995). This suggests that the VEGF:VEGF receptor gene family is implicated in pathological neovascularization.

## 2.4 VEGF-A: Transgenesis

Targeted inactivation of a single VEGF-A allele resulted in haploinsufficiency with embryonic lethality due to abnormal blood vessel development around 9–10 days of gestation (CARMELIET and COLLEN 1997; CARMELIET et al. 1996; FERRARA et al. 1996). The dorsal aorta had a much smaller lumen. In addition, sprouting of the vessels in the intersomitic regions, in the head mesenchyme and in the neuroepithelium was reduced, and connections of the large blood vessels with the heart (in the outflow region) appeared abnormal. In the yolk sac, large vitello-embryonic blood vessels (that connect the yolk sac with the embryo) were absent, and only an irregular plexus of enlarged capillaries was present. Significant vascular defects were also present in the placenta, but haematopoiesis appeared normal (FERRARA et al. 1996). Blood vessel development was more affected in homozygous VEGF-A-deficient embryos, generated by aggregation of homozygous VEGF-A-deficient embryonic stem cells with tetraploid embryos, than in heterozygous VEGF-A-deficient embryos (Fig. 2). Expression of early endothelial-cell markers (VEGFR-1, VEGFR-2, TIE) was also lower in homozygous than in heterozygous VEGF-A-deficient embryos, suggesting that endothelial-cell development was delayed, but not aborted.

These data suggest a tight gene dose-dependent relationship and production of only minimally required VEGF-A levels during embryonic development. Indeed, at 8.5 days post-coitum, there were only some scattered endothelial cells within the homozygous VEGF-A-deficient embryo that failed to organize into a pair of dorsal aortas. A significant VEGF-A dose dependence is also suggested by studies involving the overexpression of VEGF-A in the developing avian embryos, resulting in the formation of a hyperfused network of vessels, and the development of vessels in areas that are usually avascular (DRAKE and LITTLE 1995). Overexpression of VEGF-A in the developing-chick limb bud resulted in local increases of vascular density with no other anomalies (FLAMME et al. 1995), whereas conditional overexpression of VEGF-A in implanted C6 gliomas resulted in haemangioblastoma-like vessels (BENJAMIN and KESHET 1997).

In considering these data, VEGF-A is not essential for initiating the differentiation of angioblasts to early endothelial cells, but affects further blood-vessel formation at different levels. Threshold levels of VEGF-A appear to be required for the continued differentiation or, possibly, the survival of endothelial cells, and for their fusion into a vessel around a large lumen. In addition, VEGF-A affects sprouting (and probably also other forms of) angiogenesis, and controls remodelling of emerging vessels into an interconnected network (as for example evidenced by the abnormal yolk sac vasculature in heterozygous VEGF-A-deficient embryos). Whether VEGF-A also affects maturation of vessels via periendothelial cell recruitment (such as occurs in the retina; Benjamin and Keshet, personal communication), remains to be determined.

More recently, we have used the Cre/LoxP system to generate other VEGF-A transgenic mice that only express the VEGF-A$^{120}$ isoform. Initial analysis indicates that homozygous VEGF-A$^{120}$ embryos die shortly after birth, possibly because of ischaemic cardiomyopathy, increased vascular fragility and haemorrhaging in several tissues. Thus, VEGF-A[120] is able to rescue the defective embryonic vascular development of heterogyous VEGF-A-deficient embryos, but appears to be insufficient to sustain a functional vasculature in a homozygous form (unpublished observations in collaboration with P. D'AMORE et al, Harvard Medical School, Boston, USA). In addition, a transgenic mouse model of intra- and subretinal neovascularization was generated by overexpression of VEGF-A in the photoreceptors (OKAMOTA et al. 1997). This may allow the study of the role of VEGF-A in the pathological neovascularization during diabetic and ischaemic retinopathy.

## 2.5 VEGF-A: Regulation by Hypoxia

Hypoxia is an important regulator of angiogenesis. In contrast to the other VEGF homologues PlGF, VEGF-B and VEGF-C (see Sect. 2.9 below), which are unresponsive or are even inhibited by hypoxia (CAO et al. 1996b; ENHOLM et al. 1997; GLEADLE et al. 1995; VIGLIETTO et al. 1995), expression of VEGF-A is markedly upregulated by hypoxia (DOR and KESHET 1997). Hypoxic regulation of VEGF-A gene expression is mediated by HIFs. To date, HIF-1α (SEMENZA 1996; WENGER and GASSMANN 1997), HIF-2α (also named EPAS-1, HRF or HLF) (EMA et al. 1997; FLAMME et al. 1997; TIAN et al. 1997; WENGER and GASSMANN 1997), and HIF-1β (or ARNT) (SEMENZA 1996; WENGER and GASSMANN 1997) have been identified, all of which bind an enhancer HIF-response element in the promoter region of VEGF-A. In addition, hypoxia stabilises the VEGF-A mRNA by interaction of RNA binding proteins with sequences in the 3' untranslated region (DOR and KESHET 1997). The VHL gene product has been implicated in the latter process (LEVY et al. 1997).

We and others have embarked on targeting and inactivating these hypoxia-inducible factors that regulate VEGF-A gene expression. For example, embryos deficient in the ARNT die at around 10.5 days of gestation due to defective vascular development, similar to VEGF-A-deficient embryos (MALTEPE et al. 1997). However, different from the latter, vascular defects were observed in the yolk sac and in some, but not all, regions within the embryo itself. ARNT-deficient embryonic cells produced significantly less VEGF-A under hypoxia. Somatic ARNT-deficient cells also had an impaired hypoxic induction of gene expression and failed to develop highly vascularized tumours in vivo (WOOD et al. 1996). In another ARNT inactivation study, ARNT-deficient embryos died at around 10 days of gestation, due to placental haemorrhaging, delayed and abnormal neurogenesis and visceral arch development (KOZAK et al. 1997). Despite seemingly normal chorioallantoic fusion, the chorionic capillary plexus was underdeveloped. The yolk sac vasculature, however, appeared normal.

Since ARNT is co-expressed with VEGF-A in the trophoblast giant cells of the developing placenta, it is conceivable that loss of ARNT gene function reduced VEGF-A levels to below the critical threshold required for proper vascular development in the placenta (BREIER et al. 1995; KOZAK et al. 1997). Alternatively, or additionally, VEGF-A levels may have been insufficient to provide an autocrine survival or differentiation function for the trophoblasts (KHALIQ et al. 1996). It remains to be determined whether the neural defects are secondary to the placental and circulatory defects, or related to the impaired heterodimerization with other ARNT partners.

Deficiency of the VHL gene resulted in embryonic lethality at around 10.5–12.5 days of gestation due to abnormal vascularization in the placenta (GNARRA et al. 1997). VHL is primarily expressed in the labyrinth trophoblasts, in the allantoic mesoderm and in some embryonic endothelial cells of the placenta. At a lower level, VHL is expressed in the yolk sac visceral endoderm and in other regions within the embryo proper at around 10.5 days of gestation. VEGF-A is also expressed at several of these sites (see Sect. 2.3 above). Both mutant embryo and placenta developed normally until 9.5 days of gestation.

At the time of chorioallantoic fusion and expansive growth of the placental labyrinth, VHL-deficient embryos lacked embryonic blood vessels in the placental labyrinth, but presumably, the actual embryo appeared normal. The trophoblasts failed to develop into syncytiotrophoblasts, and by 11.5–12.5 days, the placental labyrinth showed great disruption, loss of normal structure, necrosis and haemorrhage. In addition, the yolk sac in some embryos appeared atrophic and had reduced numbers of blood islands. Surprisingly, immunocytochemical VEGF-A levels in the placental-labyrinth trophoblasts, at 10.5 days, were reduced in the mutant embryos, contrary to the expectation that loss of an inhibitory regulator of VEGF-A expression would increase VEGF-A expression levels.

VHL appears to play an essential role in vasculogenesis in certain but not all embryonic regions, and the consequences of its gene loss appear to correlate with its level and pattern of expression. An intriguing question is whether hypoxic regulation of VEGF-A expression by VHL is opposite to that in renal carcinoma cell lines.

Analysis of embryonic stem cells lacking HIF-1$\alpha$ revealed that hypoxic or hypoglycaemic induction of several target genes (VEGF-A, phosphoglycerokinase-1, lactate dehydrogenase-A etc.) was significantly reduced or absent when cells were cultured as embryoid bodies or monolayers in vitro (Carmeliet et al. 1998b; Iyer et al. 1998; Ryan et al. 1998). Its role is specific, as other genes including the cellular receptor tissue factor (a haemostatic factor also involved in blood-vessel formation) are upregulated by hypoglycaemia, but independently of HIF-1$\alpha$. In addition, HIF-1$\alpha$-deficient teratocarcinomas were significantly more avascular when grown in nude mice in vivo. Interestingly, assembly of endothelial cells to endothelial cords and capillaries occurred normally, but formation of larger blood vessels was significantly impaired. VEGF was highly upregulated in the wild-type tumours, especially around sites of necrosis. As expression of VEGF was significantly reduced in the HIF-1$\alpha$-deficient tumours compared with the wild-type tumours, the reduced vascularization in the mutant tumours likely resulted from reduced stress-induced VEGF expression. These data imply important roles for hypoxia-inducible mechanisms in control of angiogenic factors and blood vessel formation in vivo.

## 2.6 VEGF-A (Gene) Therapy

VEGF-A (gene) therapy has been used to improve blood flow in ischaemic limbs and the myocardium (ISNER et al. 1996; PEARLMAN et al. 1995; WARE and SIMONS 1997). Interestingly, the neovascularization response appears to occur preferentially in the ischaemic tissues, possibly because of specific upregulation of VEGF receptors by hypoxia (LI et al. 1996; TUDER et al. 1995). Although VEGF-A treatment of ischaemic myocardium increases capillary density, it also increases the number of muscular distribution vessels, presumably via production of PDGF by the VEGF-activated endothelium (BANAI et al. 1994). Nitric oxide may mediate the VEGF-A angiogenic response, as VEGF-A induces release of nitric oxide (VAN DER ZEE et al. 1997), and nitric-oxide synthase blockers or gene deletion impairs neovascularization in the ischaemic limb despite similar VEGF-A expression (MUROHARA et al. 1997). VEGF-A may also increase blood flow and tissue reoxygenation by causing nitric oxide-dependent vasodilatation (HARIAWALA et al. 1996; YANG et al. 1996).

VEGF-A has been used with variable success to improve re-endothelialization of denuded vessels and to reduce arterial stenosis after injury (ASAHARA et al. 1995; LINDNER and REIDY 1996). The latter may be due to "silencing" of the activated intimal smooth muscle cells by re-growing endothelial cells. Alternatively, endothelial nitric oxide (induced by VEGF-A) may inhibit platelet aggregation and smooth muscle cell proliferation and, in addition, increase the overall vessel size by vasodilation. In a negative-feedback loop, nitric oxide may shut down VEGF-A expression after re-endothelialization has been completed and endothelial cells have regained their quiescent state (KOUREMBANAS et al. 1993). Anti-VEGF-A (gene) therapy, whether by immunoneutralization, anti-sense gene inhibition or transfer of dominant negative VEGF receptors or VEGF chimeric toxins (VEGF$^{165}$ cross-linked to diphtheria toxin), is an effective means to suppress neovascularization during cancer and diabetic retinopathy (FERRARA and DAVIS-SMYTH 1997; KIM et al. 1993; KLAGSBRUN and D'AMORE 1996; MILLAUER et al. 1994; YUAN et al. 1996).

## 2.7 Placental Growth Factor

More recently, other VEGF-related factors with a conserved pattern of eight cysteine residues have been identified. The latter may form intra- and interchain disulphide bonds, generating biologically active anti-parallel dimeric molecules similar to PDGF. Homo- or heterodimerization of these ligands may determine their biological specificity (BIRKENHAGER et al. 1996; DISALVO et al. 1995; PARK et al. 1994). PlGF is expressed in the placenta and, to a lesser extent, in the heart, lung and thyroid gland (KHALIQ et al. 1996; MAGLIONE et al. 1991; ZICHE et al. 1997). PlGF is variably expressed in tumours (HATVA et al. 1996; VIGLIETTO et al. 1995).

In the placenta, PlGF may act in an autocrine fashion on trophoblast growth and differentiation, as mRNA synthesis of both PlGF and its receptor VEGFR-1 occur in the villous trophoblasts and syncytiotrophoblasts (KHALIQ et al. 1996;

VUORELA et al. 1997). Since these cells also produce VEGF-A, PlGF and VEGF-A may form heterodimers. Moderate immunostaining of PlGF was evident in the endothelial cells lining the villi and in the media of larger blood vessels (KHALIQ et al. 1996). Two alternatively transcribed PlGF mRNA's have been identified in man, of which the longest form has affinity for extracellular matrix components (CAO et al. 1997; MAGLIONE et al. 1991; VUORELA et al. 1997; ZICHE et al. 1997). Although the role of PlGF homodimers on endothelial cell proliferation is controversial, it may, via interaction with VEGF-A, modulate the mitogenic, chemotactic and vascular permeability-inducing properties of the latter (BIRKENHAGER et al. 1996; CAO et al. 1996; DISALVO et al. 1995; HAUSER and WEICH 1993; PARK et al. 1994; SAWANO et al. 1996). $PlGF^{129}/VEGF-A^{165}$ heterodimers induced corneal neovascularization with a maximal vessel length similar to $VEGF-A^{165}$, but with a marked decrease of vessel density, consistent with the observations that these heterodimers were equally potent in stimulating endothelial migration, and 20- to 50-fold less mitogenic than the $VEGF-A^{165}$ homodimers (CAO et al. 1996a,b).

PlGF may also affect the angiogenic response indirectly by stimulating monocyte recruitment (CLAUSS et al. 1996). Notably, in contrast to the marked hypoxic upregulation of VEGF-A by hypoxia, PlGF expression is not affected or even slightly reduced by hypoxia (CAO et al. 1996; VIGLIETTO et al. 1995). Nevertheless, because hypoxia induces the expression of VEGF-A, the formation of PlGF/VEGF heterodimers is under hypoxic control (CAO et al. 1996a,b). Deficiency of PlGF in transgenic mice did not compromise development, fertility or placentation (unpublished observations in collaboration with G. Persico Naples, Italy). This was not anticipated, in view of the presumed role of PlGF in establishing vascular connections in the placenta. However, initial analysis suggests that healing of skin wounds, formation of granulation tissue and vascular permeability after topical VEGF-A application are abnormal.

## 2.8 VEGF-B

VEGF-B has similar endothelial mitogenic potency to VEGF-A and is primarily expressed in the heart, skeletal muscle, brain and kidney (OLOFFSON et al. 1996). VEGF-A and VEGF-B are co-expressed in many tissues and are able to heterodimerize with each other. VEGF-B remains largely cell-associated, possibly providing spatial cues to outgrowing endothelial cells, or acting as a releasable pool to induce endothelial cell regeneration after injury. Expression of VEGF-B is not responsive to hypoxia, PDGF, TGF-β1 or epidermal growth factor and, in fact, the half-life of VEGF-B mRNA is unusually stable ( > 8 h), suggesting chronic rather than acute regulation (ENHOLM et al. 1997). Recent data suggest that VEGF-B-deficient mice develop normally and are healthy (Eriksson et al. personal communication).

## 2.9 VEGF-C

VEGF-C [also called VEGF-related factor (VRP) or VEGF-2] stimulates migration and proliferation of endothelial cells, although with a lower potency than VEGF-A (JING-SHAN et al. 1997; JOUKOV et al. 1996; KUKK et al. 1996; LEE et al. 1996). At high doses, VEGF-C is able to interact with the VEGF receptor-2 (JOUKOV et al. 1996; KUKK et al. 1996). In the adult, it is most abundantly expressed in the heart, placenta, lung, kidney, muscle, ovary and small intestine. During embryonic development, VEGF-C is first detectable around 8.5 days of gestation in the cephalic mesenchyme, along the somites, in the tail region, and in the allantois (JOUKOV et al. 1996; KUKK et al. 1996).

The VEGF-C receptor (VEGF receptor-3 or FT4; see Sect. 2.11 below) is expressed at sites immediately adjacent to those of VEGF-C expression, e.g. in the angioblasts of the head mesenchyme, between the developing somites, in the allantois and in the endothelial cells of venous lacunae in the placenta, but not in the blood islands of the yolk sac (KAIPAINEN et al. 1995; KUKK et al. 1996). In 12.5-day-old embryos, VEGF-C mRNA is particularly abundant in the anterior paravertebral and intervertebral tissues, whereas FLT4 expression was high in the adjacent anterior veins and intervertebral vessels. Similar paracrine co-expression of VEGF-C and VEGFR-3 were observed around the developing metanephros and, at lower levels, in the lung and in the cephalic region. Taken together, VEGF-C and VEGFR-3 may, initially, be involved in the development of the venous system, whereas at later periods they co-localise in the perinephric, mesenterial and jugular regions where the first lymphatic vessels sprout from venous sac-like structures. Mice overexpressing VEGF-C in the skin specifically develop hyperplasia of lymphatic vessels (JELTSCH et al. 1997). Since VEGF-C binds to both VEGFR-2 and VEGFR-3 the precise mechanism of the specific effect of VEGF-C on lymphatic but not on endothelial cell function remains to be determined. Deficiency of VEGF-C (as well as of VEGFR-3) results in early embryonic lethality, the precise cause of which needs to be further determined (K. Alitalo, personal communication).

## 2.10 VEGF-D

c-fos-induced growth factor (FIGF; also called VEGF-D) is a VEGF homologue, cloned by differential display between normal and c-fos knockout mouse fibroblasts (ORLANDINI et al. 1996). It was independently retrieved from EST libraries as a VEGF homologue, with close structural relationship to VEGF-C (YAMADA et al. 1997). VEGF-D gene expression is induced by c-fos and is more abundant in serum-starved quiescent cells, similar to the expression of the growth-arrested *gas-6* gene (ORLANDINI et al. 1996). In vivo, VEGF-D is expressed abundantly in the lung, heart, small intestine and fetal lung. It was also expressed at lower levels in skeletal muscle, in the colon and in the pancreas (YAMADA et al. 1997). Expression in tumour lines was undetectable. Although the structural similarities between VEGF-C and VEGF-D suggest similar functions, their expression patterns differ (for example, VEGF-D is more abundantly expressed in the lung than VEGF-C) (YAMADA et al. 1997). Although VEGF-D is able to induce proliferation and morphological transformation of fibroblasts (ORLANDINI et al. 1996), its possible effects on endothelial cells, and its regulation by hypoxia remain, to date, unknown. The differential role of these VEGF homologues in governing vascular development during embryonic development or adult pathology is largely unknown.

## 2.11 VEGF Receptors

Three receptor tyrosine kinases with seven immunoglobulin domains that bind the VEGF family members with different specificity and affinity have, thus far, been identified (TERMAN and DOUGHER VERMAZEN 1996): the VEGF receptor-1 (VEGFR-1or FLT1) binds VEGF-A and PlGF (DE VRIES et al. 1992; PARK et al. 1994; SAWANO et al. 1996); the VEGF receptor-2 (VEGFR-2 or FLK1) binds VEGF-A and VEGF-C (possibly with a lower affinity, although controversial) (JOUKOV et al. 1996; QUINN et al. 1993); and the VEGF receptor-3 (VEGFR-3 or FLT4) binds VEGF-C (GALLAND et al. 1993; JOUKOV et al. 1996). In addition, neuropilin has been identified as a low-affinity surface-associated receptor that selectively binds VEGF-A$^{165}$ via the exon 7-encoded domain in certain tumour cells (SOKER et al. 1996).

A receptor for VEGF-D has not yet been characterised, but because of its structural similarities to VEGF-C, VEGFR-3 may be a candidate. It has been suggested that the stability of VEGF–heparan-sulphate receptor complexes contributes to effective signal transduction. Following stimulation with VEGF-A, VEGFR-2- overexpressing endothelial cells undergo changes in cell morphology, actin reorganization, membrane ruffling, chemotaxis and mitogenicity, whereas VEGFR-1–overexpressing cells lack such responses (WALTENBERGER 1997; WALTENBERGER et al. 1994). VEGFR-2 is autophosphorylated upon stimulation of intact endothelial cells with VEGF-A and VEGF-C (JOUKOV et al. 1996; SEETHARAM et al. 1995), whereas VEGFR-1 is only phosphorylated in vitro in response to VEGF-A (SEETHARAM et al. 1995). Recent evidence suggests, however, that VEGFR-1 is a signalling receptor in monocytes, involved in their chemoattraction (CLAUSS et al. 1996), and in trophoblasts, mediating nitric oxide release (KHALIQ et al. 1996). VEGFR-3 is autophosphorylated upon binding of VEGF-C, but not of VEGF-A or PlGF (JOUKOV et al. 1996; LEE et al. 1996). The second Ig-like domain of VEGFR-1 contains critical determinants required for the interaction with VEGF-A and PlGF (DAVIS SMYTH et al. 1996). These results suggest that VEGFR-2 and -3, but not VEGFR-1, confer the mitogenic potential of VEGF-A and VEGF-C to endothelial cells. Alanine-scanning mutagenesis of VEGF-A confirmed this hypothesis, since VEGF-A mutants lacking VEGFR-2 affinity are not mitogenic, whereas VEGF-A mutants with reduced VEGFR-1 affinity induce normal endothelial cell proliferation (KEYT et al. 1996).

Deficiency of the VEGF receptor VEGFR-2 resulted in embryonic lethality around 10 days of gestation due to abnormal vascular development (SHALABY et al.

1995). Histological examination revealed complete absence of organized blood vessels and necrosis in the actual mutant embryo. More recent analysis of chimeric wild-type, homozygous, VEGFR-2-null mutant mice indicated that endothelial cells were always wild type, demonstrating a cell-autonomous requirement for VEGFR-2 in endothelial cell differentiation (SHALABY et al. 1997). In addition, VEGFR-2 appears to ensure the recruitment of mesodermal precursors from the posterior primitive streak to the yolk sac, where they differentiate into endothelial progenitors. LacZ expression (by the targeted VEGFR-2 promoter) was only detected in putative endothelial cell precursors. Thus, the developing mesoderm cannot complete differentiation into blood islands in the absence of a functional receptor. Interestingly, the observation that VEGF-A deficiency did not result in a similar phenotype to the VEGFR-2 deficiency, may suggest the presence other VEGF-related ligands for VEGFR-2 (possibly VEGF-C; JOUKOV et al. 1996; KUKK et al. 1996), or rescue by maternal VEGF-A.

Targeting of the VEGFR-1 gene also resulted in embryonic lethality at around 10 days of gestation (FONG et al. 1995). Staining for LacZ (expressed by the targeted VEGFR-1 promoter) revealed the presence of numerous differentiated endothelial cells, which, however, failed to form an organized vascular network. These vascular structures were abnormally large and fused, and contained enclosed LacZ-positive endothelial cells. These findings suggest a possible role of VEGFR-1 in contact inhibition of endothelial cell growth, or in endothelial cell assembly via controlling adhesion between endothelial cell precursors.

## 3 Conclusion

Targeted gene manipulation has resulted in a better understanding of the molecular mechanisms governing endothelial cell function, the formation of endothelial cell-lined channels and the assembly of these endothelial cells into a connecting vasculature. Members of the VEGF gene family have proven to be essential regulators of vascular development during both embryogenesis and adulthood. Future studies are required to extend our current knowledge on the biology of VEGF, including the distinct roles of the different VEGF isoforms and homologues, and their regulation by hypoxia in supply of nutrients.

*Acknowledgements.* The authors are grateful to the members of the Center for Transgene Technology and Gene Therapy, to all external collaborators who contributed to these studies and to M. Deprez for artwork.

# References

Albini A, Soldi R, Giunciuglio D, Giraudo E, Benelli R, Primo L, Noonan D, Salio M, Camussi G, Rockl W, Bussolino F (1996) The angiogenesis induced by HIV-1 tat protein is mediated by the Flk-1/KDR receptor on vascular endothelial cells. Nat Med 2:1371–1375

Alon T, Hemo I, Itin A, Pe'er J, Stone J, Keshet E (1995) Vascular endothelial growth factor acts as a survival factor for newly formed retinal vessels and has implications for retinopathy of prematurity. Nat Med 1:1024–1028

Asahara T, Bauters C, Pastore C, Kearney M, Rossow S, Bunting S, Ferrara N, Symes JF, Isner JM (1995) Local delivery of vascular endothelial growth factor accelerates reendothelialization and attenuates intimal hyperplasia in balloon-injured rat carotid artery. Circulation 91:2793–2801

Banai S, Jaklitsch MT, Shou M, Lazarous DF, Scheinowitz M, Biro S, Epstein SE, Unger EF (1994) Angiogenic-induced enhancement of collateral blood flow to ischemic myocardium by vascular endothelial growth factor in dogs. Circulation 89:2183–2189

Barleon B, Sozzani S, Zhou D, Weich HA, Mantovani A, Marme D (1996) Migration of human monocytes in response to vascular endothelial growth factor (VEGF) is mediated via the VEGF receptor flt-1. Blood 87:3336–3343

Basic M, Edwards NA, Merrill MJ (1995) Differential expression of vascular endothelial growth factor (vascular permeability factor) forms in rat tissues. Growth Factors 12:11–15

Beck L, D'Amore P (1997) Vascular development: cellular and molecular recognition. FASEB J 11:365–373

Benjamin LA, Keshet E (1997) Conditional switching of vascular endothelial growth factor (VEGF) expression in tumors: induction of endothelial cell shedding and regression of hemangioblastoma-like vessels by VEGF withdrawal. Proc Natl Acad Sci USA 94:8761–8766

Birkenhager R, Schneppe B, Rockl W, Wilting J, Weich HA, McCarthy JE (1996) Synthesis and physiological activity of heterodimers comprising different splice forms of vascular endothelial growth factor and placenta growth factor. Biochem J 316:703–707

Breier G, Clauss M, Risau W (1995) Coordinate expression of vascular endothelial growth factor receptor-1 (flt-1) and its ligand suggests a paracrine regulation of murine vascular development. Dev Dyn 204:228–239

Brooks PC, Montgomery AM, Rosenfeld M, Reisfeld RA, Hu T, Klier G, Cheresh DA (1994) Integrin alpha v beta 3 antagonists promote tumor regression by inducing apoptosis of angiogenic blood vessels. Cell 79:1157–1164

Brooks PC, Stromblad S, Sanders LC, von Schalscha TL, Aimes RT, Stetler Stevenson WG, Quigley JP, Cheresh DA (1996) Localization of matrix metalloproteinase MMP-2 to the surface of invasive cells by interaction with integrin alpha v beta 3. Cell 85:683–693

Broxmeyer HE, Cooper S, Li ZH, Song HY, Kwon BS, Warren RS, Donner DB (1995) Myeloid progenitor cells regulatory effects of vascular endothelial growth factor. Int J Hematol 62:203–215

Cao Y, Chen H, Zhou L, Chiang MK, Anand-Apte B, Weatherbee JA, Wang Y, Fang F, Flanagan JG, Tsang MLS (1996a) Heterodimers of placenta growth factor/vascular endothelial growth factor: enodthelial activity, tumor cell expression, and high affinity binding to Flk-1/KDR. J Biol Chem 271:3154–3162

Cao Y, Linden P, Shima D, Browne F, Folkman J (1996b) In vivo angiogenic activity and hypoxia induction of heterodimers of placenta growth factor/vascular endothelial growth factor. J Clin Invest 98:2507–2511

Cao Y, Ji WR, Rosin A, Cao Y (1997) Placenta growth factor: identification and characterization of a novel isoform generated by RNA alternative splicing. Biochem Biophys Res Commun 235:493–498

Carmeliet P, Collen D (1996) Gene manipulation and transfer of the plasminogen system and coagulation system in mice. Semin Thromb Hemost 22:525–542

Carmeliet P, Collen D (1997) Genetic analysis of blood vessel formation. Role of endothelial versus smooth muscle cells. Trends Cardiovasc Med 7:271–281

Carmeliet P, Collen D (1997) Insights into vascular biology via targeted gene inactivation and adenovirus-mediated gene transfer of the plasminogen system. In: Feuerstein GZ (ed) Coronary restenosis. From genetics to therapeutics. Dekker, New York, pp 225–240

Carmeliet P, Collen D (1998a) Vascular development and disorders: molecular analysis and pathogenetic insights. Kidney Int (in press)

Carmeleit P, Dor Y, Herbert JM, Fukuwara D, Brusselinans K, Swderchun M, Nilman M, Bonof, Abramovitch R, Maxwell P, Koch CJ, Ratchiffi P, Moons L, Jain RK, Collen D, Kershet E. (1998b). Role of HiF-1α in hypoxia-mediated apoptosis, cell proliferetion and tumour anjiogenesis. Nature 394:487–490

Carmeliet P, Ferreira V, Breier G, Pollefeyt S, Kieckens L, Gertsenstein M, Fahrig M, Vandenhoeck A, Kendraprasad H, Eberhardt C, Declercq C, Pawling J, Moons L, Collen D, Risau W, Nagy A (1996a) Abnormal blood vessel development and lethality in embryos lacking a single vascular endothelial growth factor allele. Nature 380:435–439

Carmeliet P, Mackman N, Moons L, Luther T, Gressens P, Van Vlaenderen I, Demunck H, Kasper M, Breier G, Evrard F, Müller M, Risau W, Edgington T, Collen D (1996b) Role of tissue factor in embryonic blood vessel development. Nature 383:73–75

Clauss M, Gerlach M, Gerlach H et al (1990) A tumor-derived polypeptide that induces endothelial cell and monocyte procoagulant activity, and promotes monocyte migration. J Exp Med 172:1535–1545

Clauss M, Weich H, Breier G, Knies U, Rockl W, Waltenberger J, Risau W (1996) The vascular endothelial growth factor receptor Flt-1 mediates biological activities. Implications for a functional role of placenta growth factor in monocyte activation and chemotaxis. J Biol Chem 271:17629–17634

Coughlin SR (1994) Molecular mechanisms of thrombin signaling. Semin Hematol 31:270–277

Davis Smyth T, Chen H, Park J, Presta LG, Ferrara N (1996) The second immunoglobulin-like domain of the VEGF tyrosine kinase receptor Flt-1 determines ligand binding and may initiate a signal transduction cascade. EMBO J 15:4919–4927

Dejana E (1996) Endothelial adherens junctions: implications in the control of vascular permeability and angiogenesis. J Clin Invest 98:1949–1953

Dellian M, Witwer BP, Salehi HA, Yuan F, Jain RK (1996) Quantitation and physiological characterization of angiogenic vessels in mice. Effects of basic fibroblast growth factor, vascular endothelial growth factor/vascular permeability factor, and host microenvironment. Am J Pathol 149:59–71

Deng G, Curriden SA, Wang S, Rosenberg S, Loskutoff DJ (1996) Is plasminogen activator inhibitor-1 the molecular switch that governs urokinase receptor-mediated adhesion and release? J Cell Biol 134:1563–71

De Vries C, Escobedo JA, Ueno H, Houck K, Ferrara N, Williams LT (1992) The fms-like tyrosine kinase, a receptor for vascular endothelial growth factor. Science 255:989–991

Dickson MC, Martin JS, Cousins FM, Kulkarni AB, Karlsson S, Akhurst RJ (1995) Defective haematopoiesis and vasculogenesis in transforming growth factor-beta 1 knock out mice. Development 121:1845–54

DiSalvo J, Bayne ML, Conn G, Kwok PW, Trivedi PG, Soderman DD, Palisi TM, Sullivan KA, Thomas KA (1995) Purification and characterization of a naturally occurring vascular endothelial growth factor.placenta growth factor heterodimer. J Biol Chem 270:7717–7723

Dor Y, Keshet E (1997) Ischemia-driven angiogenesis. Trends Cardiovasc Med 7:289–294

Drake CJ, Cheresh DA, Little CD (1995) An antagonist of integrin avß3 prevents maturation of blood vessels during embryonic neovascularization. J Cell Sci 108:2655–2661

Drake CJ, Little CD (1995) Exogenous vascular endothelial growth factor induces malformed and hyperfused vessels during embryonic neovascularization. Proc Natl Acad Sci USA 92:7657–7661

Dumont DJ, Fong GH, Puri MC, Gradwohl G, Alitalo K, Breitman ML (1995) Vascularization of the mouse embryo: a study of flk-1, tek, tie, and vascular endothelial growth factor expression during development. Dev Dyn 203:80–92

Dvorak AM, Kohn S, Morgan ES, Fox P, Nagy JA, Dvorak HF (1996) The vesiculo-vacuolar organelle (VVO)- a distinct endothelial cell structure that provides a transcellular pathway for macromolecular extravasation. Leuk Biol 59:100–115

Dvorak HF, Brown LF, Detmar M, Dvorak AM (1995) Vascular permeability factor/vascular endothelial growth factor, microvascular hyperpermeability, and angiogenesis. Am J Pathol 146:1029–1039

Ema M, Taya S, Yokotani N, Sogawa K, Matsuda Y, Fujii-Kuriyama Y (1997) A novel bHLH-PAS factor with close sequence similarity to hypoxia-inducible factor 1 alpha regulates VEGF expression and is potentially involved in lung and vascular development. Proc Natl Acad Sci USA 94:4273–4278

Enholm B, Paavonen K, Ristimäki A, Kumar V, Gunji Y, Klefstrom J, Kivinen L, Laiho M, Olofsson B, Joukov V, Eriksson U, Alitalo K (1997) Comparison of VEGF, VEGF-B, VEGF-C, Ang-1 mRNA regulation by serum, growth factors, oncoproteins and hypoxia. Oncogene 14:2475–2483

Ferrara N, Davis-Smyth T (1997) The biology of vascular endothelial growth factor. Endocr Rev 18:4–25

Ferrara N, Carver Moore K, Chen H, Dowd M, Lu L, O'Shea KS, Powell Braxton L, Hillan KJ, Moore MW (1996) Heterozygous embryonic lethality induced by targeted inactivation of the VEGF gene. Nature 380:439–442

Flamme I, Risau W (1992) Induction of vasculogenesis and hematopoiesis in vitro. Development 116:435–439

Flamme I, von Reutern M, Drexler HC, Syed Ali S, Risau W (1995) Overexpression of vascular endothelial growth factor in the avian embryo induces hypervascularization and increased vascular permeability without alterations of embryonic pattern formation. Dev Biol 171:399–414

Flamme I, Fröhlich T, von Reutern M, Kappel A, Damert A, Risau W (1997) HRF, a putative basic helix-loop-helix-PAS-domain transcription factor is closely related to hypoxia-inducible factor-1alpha and developmentally expressed in blood vessels. Mech Dev 63:51–63

Folkman J, D'Amore PA (1996) Blood vessel formation: what is its molecular basis? (Comment). Cell 87:1153–1155

Fong GH, Rossant J, Gertsenstein M, Breitman ML (1995) Role of the Flt-1 receptor tyrosine kinase in regulating the assembly of vascular endothelium. Nature 376:66–70

Frank S, Hubner G, Breier G, Longaker MT, Greenhalgh DG, Werner S (1995) Regulation of vascular endothelial growth factor expression in cultured keratinocytes. Implications for normal and impaired wound healing. J Biol Chem 270:12607–12613

Friedlander M, Brooks PC, Shaffer RW, Kincaid CM, Varner JA, Cheresh DA (1995) Definition of two angiogenic pathways by distinct alpha v integrins. Science 270:1500–1502

Gabrilovich DI, Chen HL, Girgis KR, Cunningham HT, Meny GM, Nadaf S, Kavanaugh D, Carbone DP (1996) Production of vascular endothelial growth factor by human tumors inhibits the functional maturation of dendritic cells. Nat Med 2:1096–1103

Galland F, Karamysheva A, Pebusque MJ, Borg JP, Rottapel R, Dubreuil P, Rosnet O, Birnbaum D (1993) The FLT4 gene encodes a transmembrane tyrosine kinase related to the vascular endothelial growth factor receptor. Oncogene 8:1233–1240

George EL, Georges Labouesse EN, Patel King RS, Rayburn H, Hynes RO (1993) Defects in mesoderm, neural tube and vascular development in mouse embryos lacking fibronectin. Development 119:1079–1091

Gleadle JM, Ebert BL, Firth JD, Ratcliffe PJ (1995) Regulation of angiogenic growth factor expression by hypoxia, transition metals, and chelating agents. Am J Physiol 268:C1362–C1368

Gnarra JR, Ward JM, Porter FD, Wagner JR, Devor DE, Grinberg A, Emmert-Buck MR, Westphal H, Klaussner RD (1997) Defective placental vasculogenesis causes embryonic lethality in VHL-deficient mice. Proc Natl Acad Sci USA 94:9102–9107

Grosskreutz CL, Anand-Apte B, Terman B, Quinn TP, D'Amore PA (1996) Vascular endothelial growth factor (VEGF) is a chemoattractant for vascular smooth muscle cells (SMC). Invest Opthalmol Vis Sci 37:5470–5475

Gurtner GC, Davis V, Li H, McCoy MJ, Sharpe A, Cybulsky MI (1995) Targeted disruption of the murine VCAM1 gene: essential role of VCAM-1 in chorioallantoic fusion and placentation. Genes Dev 9:1–14

Hanahan D (1997) Signaling vascular morphogenesis and maintenance. Science 277:48–50

Hariawala MD, Horowitz JJ, Esakof D, Sheriff DD, Walter DH, Keyt B, Isner JM, Symes JF (1996) VEGF improves myocardial blood flow but produces EDRF-mediated hypotension in porcine hearts. J Surg Res 63:77–82

Hatva E, Bohling T, Jasskelainen J, Persico MG, Haltia M, Alitalo K (1996) Vascular growth factors and receptors in capillary hemangioblastomas and hemangiopericytomas. Am J Pathol 148:763–775

Hauser S, Weich H (1993) A heparin-binding form of placenta growth factor (PlGF-2) is expressed in human umbilical vein endothelial cells and in placenta. Growth Factors 9:259–268

Isner JM, Peiczeck A, Schainfield R, Blair R, Haley L, Asahara T, Rosenfield K, Razvi S, Walsch K, Symes JF (1996) Clinical evidence of angiogenesis after arterial gene transfer of phVEGF165 in patient with ischaemic limb. Lancet 348:370–374

Iyer NV, Kotch LE, Agani E, Leung SW, Laughner E, Weugs RH, Fessmanin M, Fearhartch D, Lawler AM, Yu AY, Femenza GL (1998) Genes Acv. 12:149–162

Jeltsch M, Kappainen AA, Joukov V, Meng X, Lakso M, Rauvala H, Swartz M, Fukumura D, Jain RK, Alitalo K (1997) Hyperplasia of lymphatic vessels in VEGF-C transgenic mice. Science 276:1423–1425

Jing-Shan H, Hastings GA, Cherry S, Gentz R, Ruben S, Coleman TA (1997) A novel regulatory function of proteolytically cleaved VEGF-2 for vascular endothelial and smooth muscle cells. FASEB J 11:498–504

Joukov V, Pajusola K, Kaipainen A, Chilov D, Lahtinen I, Kukk E, Saksela O, Kalkkinen N, Alitalo K (1996) A novel vascular endothelial growth factor, VEGF-C, is a ligand for the Flt4 (VEGFR-3) and KDR (VEGFR-2) receptor tyrosine kinases. EMBO J 15:290–298

Kaipainen A, Korhonen J, Mustonen T, van Hinsbergh VW, Fang GH, Dumont D, Breitman M, Alitalo K (1995) Expression of the fms-like tyrosine kinase 4 gene becomes restricted to lymphatic endothelium during development. Proc Natl Acad Sci USA 92:3566–3570

Katoh O, Tauchi H, Kawaishi K, Kimura A, Satow Y (1995) Expression of the vascular endothelial growth factor (VEGF) receptor gene, KDR, in hematopoietic cells and inhibitory effect of VEGF on apoptotic cell death caused by ionizing radiation. Cancer Res 55:5687–5692

Keyt BA, Berleau LT, Nguyen HV, Chen H, Heinsohn H, Vandlen R, Ferrara N (1996) The carboxyl-terminal domain (111–165) of vascular endothelial growth factor is critical for its mitogenic potency. J Biol Chem 271:7788–7795

Khaliq A, Li XF, Shams M, Sisi P, Acevedo CA, Whittle MJ, Weich H, Ahmed A (1996) Localisation of placenta growth factor (PlGF) in human term placenta. Growth Factors 13:243–250

Kim KJ, Li B, Winer J, Armanini M, Gillett N, Phillips HS, Ferrara N (1993) Inhibition of vascular endothelial growth factor-induced angiogenesis suppresses tumour growth in vivo. Nature 362:841–844

Klagsbrun M, D'Amore PA (1996) Vascular endothelial growth factor and its receptors. Cytokine Growth Factor Rev 7:259–270

Kourembanas S, Mcquillan LP, Leung GK, Faller DV (1993) Nitric oxide regulates the expression of vasoconstrictors and growth factors by vascular endothelium under both normoxia and hypoxia. J Clin Invest 92:99–104

Kozak KR, Abbott B, Hankinson O (1997) ARNT-deficient mice and placental differentiation. Dev Biol 191:247–306

Kukk E, Lymboussaki A, Taira S, Kaipainen A, Jeltsch M, Joukov V, Alitalo K (1996) VEGF-C receptor binding and pattern of expression with VEGFR-3 suggests a role in lymphatic vascular development. Development 122:3829–3837

Kuzuya M, Satake S, Esaki T, Yamada K, Hayashi T, Naito M, Asai K, Iguchi A (1995) Induction of angiogenesis by smooth muscle cell-derived factor: possible role in neovascularization in atherosclerotic plaque. J Cell Physiol 164:658–667

Kwee L, Baldwin S, Stewart C, Buck C, Labow M (1995) Defective development of the embryonic and extraembryonic circulatory system in vascular cell adhesion molecule (VCAM-1) deficient mice. Development 121:489–503

Lang RA, Bishop JM (1993) Macrophages are required for cell death and tissue remodeling in the developing mouse eye. Cell 74:453–462

Lang R, Lustig M, Francois F, Sellinger M, Plesken H (1994) Apoptosis during macrophage-dependent ocular tissue remodelling. Development 120:3395–3403

Lee J, Gray A, Yuan J, Luoh SM, Avraham H, Wood WI (1996) Vascular endothelial growth factor-related protein: a ligand and specific activator of the tyrosine kinase receptor Flt4. Proc Natl Acad Sci USA 93:1988–1992

Levéen P, Pekny M, Gebre Medhin S, Swolin B, Larsson E, Betsholtz C (1994) Mice deficient for PDGF B show renal, cardiovascular, and hematological abnormalities. Genes Dev 8:1875–1887

Levy AP, Levy NS, Iliopoulos O, Jiang C, Kaelin WG, Goldberg MA (1997) Regulation of vascular endothelial growth factor by hypoxia and its modulation by the von Hippel-Lindau tumor suppressor gene. Kidney Int 51:575–578

Li J, Brown LF, Grossman JD, Morgan JP, Simons M (1996) VEGF, flk-1, and flt-1 expression in a rat myocardial infarction model of angiogenesis. Am J Physiol 270:H1803–811

Lindahl P, Johansson BR, Levéen P, Betsholtz C (1997) Pericyte loss and microaneurysm formation in PDGF-BB deficient mice. Science 277:242–245

Lindner V, Reidy MA (1996) Expression of VEGF receptors in arteries after endothelial injury and lack of increased endothelial regrowth in response to VEGF. Arterioscler Thromb Vasc Biol 16:1399–405

Maglione D, Guerriero V, Viglietto G, Delli Bovi P, Persico MG (1991) Isolation of a human placenta cDNA coding for a protein related to the vascular permeability factor. Proc Natl Acad Sci USA 88:9267–9271

Maisonpierre PC, Suri C, Jones PF et al (1997) Angiopoeitin-2, a natural antagonist for Tie2 that disrupts in vivo angiogenesis. Science 277:55–60

Maltepe E, Schmidt JV, Baunoch D, Bradfield CA, Simon CM (1997) Abnormal angiogenesis and responses to glucose and oxygen deprivation in mice lacking the protein ARNT. Nature 386:403–407

Meeson A, Palmer M, Calfon M, Lang R (1996) A relationship between apoptosis and flow during programmed capillary regression is revealed by vital analysis. Development 122:3929–3938

Melder RJ, Koenig GC, Witwer BP, Safabakhsh N, Munn LL, Jain RK (1996) During angiogenesis, vascular endothelial growth factor and basic fibroblast growth factor regulate natural killer cell adhesion to tumor endothelium (see comments). Nat Med 2:992–997

Mignatti P, Rifkin DB (1996) Plasminogen activators and matrix metalloproteinases in angiogenesis. Enzyme Protein 49:117–137

Millauer B, Shawver LK, Plate KH, Risau W, Ullrich A (1994) Glioblastoma growth inhibited in vivo by a dominant-negative Flk-1 mutant. Nature 367:576–579

Murohara T, Asahara A, Silver M, Kearny M, Magner M, Yang J, Chen D, Chen D, Symes JF, Huang PM, Isner JM (1997) Essential role of endothelial nitric oxide synthase in angiogenesis in vivo. Circulation 96:I-550 (abstract)

Nehls V, Drenckhahn D (1993) The versatility of microvascular pericytes: from mesenchyme to smooth muscle? Histochemistry 99:1–12

Neufeld G, Cohen T, Gitay Goren H, Poltorak Z, Tessler S, Sharon R, Gengrinovitch S, Levi BZ (1996) Similarities and differences between the vascular endothelial growth factor (VEGF) splice variants. Cancer Metastasis Rev 15:153–158

Noden DM (1989) Embryonic origins and assembly of blood vessels. Ann Rev Respir Dis 140:1097–1103

Nomura M, Yamagishi S, Harada S, Hayashi Y, Tamashima T, Ymashita J, Yamamoto H (1995) Possible participation of autocrine and paracrine vascular endothelial gorwth factors in hypoxia-induced proliferation of endothelial cells and pericytes. J Biol Chem 270:28316–28324

Okamota N, Tobe T, Hackett SF et al. (1997) Transgenic mice with increased expression of vascular endothelial growth factor in the retina. Am J Pathol 151:281–291

Oloffson B, Pajusola K, Kaipainen A, von Eulen G, Joukow V, Saksela O, Orpona A, Pettersoson RF, Alitalo K, Eriksson U (1996) Vascular endothelial growth factor B, a novel growth factor for endothelial cells. Proc Natl Acad Sci USA 93:2576–2581

Orlandini M, Marconcini L, Ferruzzi R, Oliviero S (1996) Identification of a c-fos-induced gene that is related to the platelet-derived growth factor/vascular endothelial growth factor family (corrected; erratum to be published). Proc Natl Acad Sci USA 93:11675–11680

Park JE, Chen HH, Winer J, Houck KA, Ferrara N (1994) Placenta growth factor. Potentiation of vascular endothelial growth factor bioactivity, in vitro and in vivo, and high affinity binding to Flt-1 but not to Flk-1/KDR. J Biol Chem 269:25646–25654

Patan S (1998) TIE1 and TIE2 receptor tyrosine kinases inversely regulate embryonic angiogenesis by the mechanism of intussusceptive microvascular growth (submitted)

Patan S, Alvarez MJ, Schittny JC, Burri PH (1992) Intussusceptive microvascular growth: a common alternative to endothelial sprouting. Arch Histol Cytol 55:65–75

Patan S, Haenni B, Burri PH (1996) Implementation of intussusceptive microvascular growth in the chicken chorioallantoic membrane (CAM): 1. pillar formation by folding of the capillary wall. Microvasc Res 51:80–98

Patan S, Haenni B, Burri PH (1997) Implementation of intussusceptive microvascular growth in the chicken chorio-allantoic membrane (CAM) 2. Pillar formation by capillary fusion. Microvasc Res 53:33–52

Pearlman JD, Hibberd MG, Chuang ML, et al. (1995) Magnetic resonance mapping demonstrates benefits of VEGF-induced myocardial angiogenesis. Nature Med 1:1085–1089

Pepper MS, Meda P (1992) Basic fibroblast growth factor increases junctional communication and connexin 43 expression in microvascular endothelial cells. J Cell Physiol 153:196–205

Pertovaara L, Kaipanen A, Mustonen T, Orpana A, Ferrara N, Saksela O, Alitalo K (1994) Vascular endothelial growth factor is induced in response to transforming growth factor-ß in fibroblastic and epithelial cells. J Biol Chem 269:6271–6274

Plate KH, Risau W (1995) Angiogenesis in malignant gliomas. Glia 15:339–347

Puri MC, Rossant J, Alitalo K, Bernstein A, Partanen J (1995) The receptor tyrosine kinase TIE is required for integrity and survival of vascular endothelial cells. EMBO J 14:5884–5891

Quinn TP, Peters KG, De Vries C, Ferrara N, Williams LT (1993) Fetal liver kinase 1 is a receptor for vascular endothelial growth factor and is selectively expressed in vascular endothelium. Proc Natl Acad Sci USA 90:7533–7537

Reinartz J, Schafer B, Batria R, Klein CE, Kramer MD (1995) Plasmin abrogates avß5-mediated adhesion of a human keratinocyte line (HaCat). Exp Cell Res 214:486–498

Risau W (1995) Differentiation of endothelium. FASEB J 9:926–933

Risau W (1997) Mechanisms of angiogenesis. Nature 386:671–674

Robert B, St John PL, Hyink DP, Abrahams DR (1996) Evidence that embryonic kidney cells expressing flk-1 are intrinsic, vasculogenic angioblasts. Am J Physiol 271:F744–F753

Roberts WG, Palade GE (1995) Increased microvascular permeability and endothelial fenestration induced by vascular endothelial growth factor. J Cell Sci 108:2369–2379

Ryan HE, Lo J, Johnson RS. (1998) HiF-1α is required for solid tumour formetin and emlyonic vasenlarizetion. Erbo J. 17:3005–3015

Sato TN, Tozawa Y, Deutsch U, Wolburg Buchholz K, Fujiwara Y, Gendron Maguire M, Gridley T, Wolburg H, Risau W, Qin Y (1995) Distinct roles of the receptor tyrosine kinases Tie-1 and Tie-2 in blood vessel formation. Nature 376:70–74

Sato Y, Asada Y, Marutsuka K, Hatakeyama K, Sumiyoshi A (1996) Tissue factor induces migration of cultured aortic smooth muscle cells. Thromb Haemost 75:389–392

Sawano A, Takahashi T, Yamaguchi S, Aonumura M, Shibuya M (1996) Flt-1 but not KDR/Flk-1 tyrosine kinase is a receptor for placenta growth factor, which is related to vascular endothelial growth factor. Cell Growth Differ 7:213–221

Seetharam L, Gotoh N, Maru Y, Neufeld G, Yamaguchi S, Shibuya M (1995) A unique signal transduction from FLT tyrosine kinase, a receptor for vascular endothelial growth factor VEGF. Oncogene 10:135–147

Semenza GL (1996) Transcriptional regulation by hypoxia-inducible factor-1. Trends Carciovasc Med 6:151–157

Senger DR, Ledbetter SR, Claffey KP, Papadopoulos-Sergiou A, Perruzzi CA, Detmar M (1996) Stimulation of endothelial cell migration by vascular permeability factor/vascular endothelial growth factor through cooperative mechanisms involving the avß3 integrin, osteopontin, and thrombin. Am J Pathol 149:293–305

Shalaby F, Rossant J, Yamaguchi TP, Gertsenstein M, Wu XF, Breitman ML, Schuh AC (1995) Failure of blood-island formation and vasculogenesis in Flk-1-deficient mice. Nature 376:62–66

Shalaby F, Ho J, Stanford WL et al. (1997) A requirement for Flk-1 in primitive and definitive hematopoiesis and vasculogenesis. Cell 89:981–990

Soker S, Fidder H, Neufeld G, Klagsbrun M (1996) Characterization of novel vascular endothelial growth factor (VEGF) receptors on tumor cells that bind VEGF165 via its exon 7-encoded domain. J Biol Chem 271:5761–5767

Stavri GT, Zachary IC, Baskerville PA, Martin JF, Erusalimsky JD (1995) Basic fibroblast growth factor upregulates the expression of vascular endothelial growth factor in vascular smooth muscle cells. Synergistic interaction with hypoxia. Circulation 92:11–14

Stefansson S, Lawrence DA (1996) The serpin PAI-1 inhibits cell migration by blocking integrin alpha V beta3 binding to vitronectin. Nature 383:441–443

Sunderkotter C, Steinbrink K, Goebeler M, Bhardwaj R, Sorg C (1994) Macrophages and angiogenesis. J Leukoc Biol 55:410–422

Suri C, Jones PF, Patan S, Bartunkova S, Maisonpierre PC, Davis S, Sato TN, Yancopoulos GD (1996) Requisite role of angiopoietin-1, a ligand for the TIE2 receptor, during embryonic angiogenesis. Cell 87:1171–1180

Takagi H, King GL, Aiello LP (1996) Identification and characterization of vascular endothelial growth factor receptor (Flt) in bovine retinal pericytes. Diabetes 45:1016–1023

Terman BI, Dougher Vermazen M (1996) Biological properties of VEGF/VPF receptors. Cancer Metastasis Rev 15:159–163

Tian H, McKnight SL, Russell DW (1997) Endothelial PAS domain protein 1 (EPAS1), a transcription factor selectively expressed in endothelial cells. Genes Dev 11:72–82

Tuder RM, Flook BE, Voelkel NF (1995) Increased gene expression for VEGF and the VEGF receptors KDR/Flk and Flt in lungs exposed to acute or to chronic hypoxia. Modulation of gene expression by nitric oxide. J Clin Invest 95:1798–1807

van der Zee R, Murohara T, Luo Z, Zollmann F, Passeri J, Lekutat C, Isner JM (1997) Vascular endothelial growth factor/Vascular permeability factor augments nitric oxide release from quiescent rabbit and human vascular endothelium. Circulation 95:1030–1037

Varner JA, Brooks PC, Cheresh DA (1995) Review: the integrin αvß3: angiogenesis and apoptosis. Cell Adhesion Commun 3:367–374

Viglietto G, Maglione D, Rambaldi M, Cerutti J, Romano A, Trapasso F, Fedele M, Ippolito P, Chiappetta G, Botti G (1995) Upregulation of vascular endothelial growth factor (VEGF) and down-regulation of placenta growth factor (PlGF) associated with malignancy in human thyroid tumors and cell lines. Oncogene 11:1569–1579

Vittet D, Buchou T, Schweitzer A, Dejana E, Huber P (1997) Targeted null-mutation in the vascular endothelial-cadherin gene impairs the organization of vascular-like structures in embryoid bodies. Proc Natl Acad Sci USA 94:6273–6278

Vuorela P, Hatva E, Lymboussaki A, Kaipainen A, Joukov V, Persico MG, Alitalo K, Halmesmaki E (1997) Expression of vascular endothelial growth factor and placenta growth factor in human placenta. Biol Reprod 56:489–494

Waltenberger J (1997) Modulation of growth factor action. Implications for treatment of cardiovascular diseases. Circulation 96:4083–4094

Waltenberger J, Claesson Welsh L, Siegbahn A, Shibuya M, Heldin CH (1994) Different signal transduction properties of KDR and Flt1, two receptors for vascular endothelial growth factor. J Biol Chem 269:26988–26995

Ware JA, Simons M (1997) Angiogenesis in ischemic heart disease. Nat Med 3:158–163

Wei Y, Lukashev M, Simon DI, Bodary SC, Rosenberg S, Doyle MV, Chapman HA (1996) Regulation of integrin function by the urokinase receptor. Science 273:1551–1555

Weihrauch D, Arras M, Zimmermann R, Schaper J (1995) Importance of monocytes/macrophages and fibroblasts for healing of micronecroses in porcine myocardium. Mol Cell Biochem 147:13–19

Wenger RH, Gassmann M (1997) Oxygen(es) and the hypoxia-inducible factor-1. Biol Chem 378:609–616

Wilting J, Christ B (1996) Embryonic angiogenesis: a review. Naturwissenschaften 83:153–164

Wilting J, Birkenhager R, Eichmann A, Kurz H, Martiny Baron G, Marme D, McCarthy JE, Christ B, Weich HA (1996) VEGF121 induces proliferation of vascular endothelial cells and expression of flk-1 without affecting lymphatic vessels of chorioallantoic membrane. Dev Biol 176:76–85

Wood SM, Gleadle JM, Pugh CQ, Hankinson O, Ratcliffe P (1996) The role of the aryl hydrocarbon receptor nuclear transporter (ARNT) in hypoxic induction of gene expression. J Biol Chem 271:15117–15123

Yamada Y, Nezu J, Shimane M, Hirate Y (1997) Molecular cloning of a novel vascular endothelial growth factor, VEGF-D. Genomics 42:483–488

Yamaguchi TP, Dumont DJ, Conlon RA, Breitman ML, Rossant J (1993) flk-1, an flt-related receptor tyrosine kinase is an early marker for endothelial cell precursors. Development 118:489–98

Yang JT, Rayburn H, Hynes RO (1993) Embryonic mesodermal defects in alpha 5 integrin-deficient mice. Development 119:1093–105

Yang JT, Rayburn H, Hynes RO (1995) Cell adhesion events mediated by alpha 4 integrins are essential in placental and cardiac development. Development 121:549–560

Yang R, Thomas GR, Bunting S, Ko A, Ferrara N, Keyt B, Ross J, Jin H (1996) Effects of vascular endothelial growth factor on hemodynamics and cardiac performance. J Cardiovasc Pharmacol 27:838–844

Yuan F, Chen Y, Dellian M, Safabakhsh N, Ferrara N, Jain RK (1996) Time-dependent vascular regression and permeability changes in established human tumor xenografts induced by an anti-vascular endothelial growth factor/vascular permeability factor antibody. Proc Natl Acad Sci USA 93:14765–14770

Zhang Y, Deng Y, Luther T, Muller M, Ziegler R, Waldherr R, Stern DM, Nawroth PP (1994) Tissue factor controls the balance of angiogenic and antiangiogenic properties of tumor cells in mice. J Clin Invest 94:1320–1327

Zhang Y, Deng Y, Liliensiek B, Bierhaus A, Greten J, He W, Chen B, Hach-Wunderle V, Waldherr R, Ziegler R, Männet D, Stern DM, Nawroth PP (1996) Intravenous somatic gene transfer with antisense tissue factor restores blood flow by reducing tumor necrosis factor-induced tissue factor expression and fibrin deposition in mouse Meth-A sarcomas. J Clin Invest 97:2213–2224

Zheng X, Saunders TL, Camper SA, Samuelson LC, Ginsburg D (1995) Vitronectin is not essential for normal mammalian development and fertility. Proc Natl Acad Sci USA 92:12426–12430

Ziche M, Maglione D, Ribatti D, Morbidelli L, Lago CT, Battisti M, Poaletti I, Barra A, Tucci M, Parise G, Vincenti V, Granger HJ, Viglietto G, Persico MG (1997) Placenta growth factor-1 is chemotactic, mitogenic and angiogenic. Lab Invest 76:517–531

# Functions of Tie1 and Tie2 Receptor Tyrosine Kinases in Vascular Development

J. Partanen[1] and D. J. Dumont[2]

| | | |
|---|---|---|
| 1 | Introduction | 159 |
| 2 | Tie1 and Tie2 Receptor Tyrosine Kinases | 161 |
| 2.1 | Cloning and Structure | 161 |
| 2.2 | Identification of Downstream Substrates | 162 |
| 2.3 | Mutational Analysis of Putative Downstream Substrates | 163 |
| 3 | Tie1 and Tie2 Gene Expression | 164 |
| 3.1 | Expression in the Endothelial Cell Lineage | 164 |
| 3.2 | Characterization of Endothelial Cell-Specific Regulatory Elements | 165 |
| 3.3 | Expression in the Haematopoietic System | 165 |
| 4 | Analysis of Tie1 and Tie2 Functions In Vivo | 166 |
| 4.1 | Functions of Tie1 and Tie2 in Vascular Development | 166 |
| 4.2 | Tie1 and Tie2 in the Haematopoietic System | 169 |
| References | | 169 |

## 1 Introduction

Vascularization of the mouse embryo is accomplished via the collaboration of two major cellular processes. Differentiation of vascular endothelial cells de novo from their precursors, called the angioblasts, has been termed vasculogenesis (Risau 1997). Somewhat confusingly, the term vasculogenesis has also been used in a broader sense, covering all aspects of vascular development (Dumont et al. 1994; Noden 1991; Sherer 1991; Wilting and Christ 1996). Subsequent expansion of the endothelium by remodelling, migration and proliferation is called angiogenesis. Vasculogenesis is known to occur intraembryonically within the splanchnopleural and paraxial mesoderm, as well as extraembryonically in the yolk sac mesoderm, and is responsible for laying down the primitive vascular network (Pardanaud

---

[1] Samuel Lunenfeld Research Institute, Mount Sinai Hospital, 600 University Avenue, Toronto, Ontario, Canada M5G 1X5
[2] Amgen and Ontario Cancer Institutes, Department of Medical Biophysics, University of Toronto, Toronto, 620 University Avenue, Ontario, Canada M5G 1C5

et al. 1996; WILTING and CHRIST 1996). Vasculogenesis is characterized by angioblast differentiation and by either the immediate aggregation of angioblast cells to give rise to endothelium, or the directed migration of these cells through the embryo to segregate eventually into vascular cords. Interestingly, angioblasts appear to have different characteristics depending on their site of origin (PARDANAUD et al. 1996). Vasculogenesis both in the yolk sac and in the splanchnic mesoderm appears closely linked to haematopoietic differentiation, and the two lineages may share a common precursor, the haemangioblast (PARDANAUD et al. 1989; SHALABY et al. 1995, 1997).

Angiogenesis is the growth or derivation of vascular structures from pre-existing vessels. Three forms of angiogenesis have been distinguished (reviewed by RISAU 1997; WILTING and CHRIST 1996). The first type of angiogenesis involves remodelling of vessels into smaller ones, while maintaining vascular perfusion. This type of endothelial growth or remodelling of the endothelium has been called intussusceptive vascular growth (BURRI and TAREK 1990; CADUFF et al. 1986). During this angiogenic process, endothelial cells sprout into the lumen of the vessel and produce pillars that traverse the lumen. These pillars fuse with the opposite side of the vessel, thereby creating a vascular loop. The second mode of angiogenesis can be regarded more as a replenishment of the existing endothelium. This type of growth would involve the endothelialization of a vessel that is growing both in length and width or else the re-endothelialization of a vessel following injury or a surgical procedure, such as balloon angioplasty.

The third mode of angiogenesis is the most common form, which is sprouting angiogenesis. Vessel sprouting can be divided into five different steps: (i) local degradation of vascular basement membrane; (ii) migration towards angiogenic stimuli; (iii) lumen formation; (iv) cell division at the base of the sprout; and (v) anastomoses of adjacent loops. Angiogenic sprouting appears to be the primary mode of vascularization during organogenesis in several tissues, including brain and kidney. It is also thought to be responsible for physiological and pathological neovascularization processes in the adult. Interestingly, however, there is recent evidence for the presence of angioblastic cells in the adult circulatory system (ASAHARA et al. 1997). The origin of these cells is not known; however, they have been shown to express haematopoietic stem cell/endothelial cell markers on their surface and to contribute to endothelium during neovascularization.

Once the early vascular network is established, it undergoes several maturation steps. One of these steps is thought to involve attraction of smooth muscle cells and pericytes around the endothelium. Interaction of endothelial and perivascular cells is thought to lead to inhibition of endothelial cell proliferation and to maturation of the vessel (BECK and D'AMORE 1997).

Ensuring that the growth and remodelling of the vasculature is carried out correctly is clearly a very complex feat. Five receptor tyrosine kinases (RTKs), the expression of which is almost exclusively restricted to the endothelial cell lineage, have been identified (MUSTONEN and ALITALO 1995). These RTKs are Flt-1 (also denoted vascular endothelial growth factor receptor-1; VEGER-1), Flk-1 (VEGFR-2), and Flt-4 (VEGFR-3) receptors for the VEGF family, and Tie-1 and Tie-2, receptors for the recently-identified angiopoietin family (DAVIS et al. 1996).

The restricted expression profiles suggest that the signalling pathways mediated by these RTKs play key roles in vascular development. In fact, engineered mutations of mouse genes coding for all five RTKs and for some of their ligands result in distinct vascular defects, demonstrating that the signalling pathways mediated by these RTKs play specific roles in endothelial cell biology (see BECK and D'AMORE 1997). Many of these RTKs are also expressed in certain haematopoietic cells, including haematopoietic stem cells. With the exception of VEGER-2, which appears to be required for the development of both endothelial and haematopoietic lineages in vivo (SHALABY et al. 1995, 1997), it is not clear whether this expression reflects shared lineage or whether it has a functional basis.

In the following, we will discuss the biochemical and biological functions of Tie1 and Tie2. Unlike the VEGF receptor family, the functions of Tie1 and Tie2 receptors are restricted primarily to angiogenic expansion and the survival of the endothelium.

## 2 Tie1 and Tie2 Receptor Tyrosine Kinases

### 2.1 Cloning and Structure

Tie1(TIE) and Tie2(TEE) were identified by several groups searching for novel tyrosine kinases expressed in different cell lines and tissues (DUMONT et al. 1992, 1993; HORITA et al. 1992; IWAMA et al. 1993; MAISONPIERRE et al. 1993; PARTANEN et al. 1990, 1992; SATO et al. 1993; SCHNURCH and RISAU 1993; ZIEGLER et al. 1993). Sequence analysis of these two receptors reveals high sequence conservation in the cytoplasmic region, with considerable divergence in the extracellular ligand binding region (Fig. 1). The extracellular region is a composite of three distinct structural motifs that set these receptors apart from other RTKs. These motifs include two immunoglobulin (Ig)-like loops that are separated by three tandem epidermal growth factor (EGF)-like repeats. The second Ig loop is followed by three domains

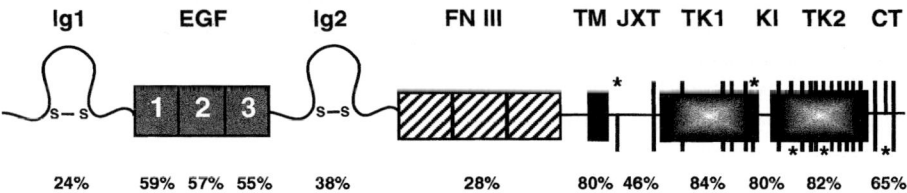

**Fig. 1.** Schematic structure of Tie2 and Tie1. Amino acid sequence identities between various domains of Tie2 and Tie1 are shown below the schematic receptor. *Bars* denote location of tyrosine residues in the intracellular region; *upper bars* denote location in Tie2 and *lower bars* location in Tie1. *Asterisks* denote differences at tyrosine residues. *Ig* immunoglobulin-like loop; *EGF* epidermal growth factor-like repeats; *FNIII* fibronectin type III-like repeat; *TM* transmembrane domain; *JXT* juxtamembrane domain; *TK* tyrosine kinase domain; *KI* kinase insert; *CT* carboxyl terminal tail

showing homology to fibronectin type-III repeats. Although the overall extracellular sequence identity between Tie1 and Tie2 is low, the EGF-like repeats show approximately 60% amino acid identity. This fairly high sequence conservation suggests that these domains probably serve similar functions, such as ligand binding, in both receptors. A ligand for Tie2, called angiopoietin-1, was recently reported (DAVIS et al. 1996). Despite sequence conservation, angiopoietin-1 does not bind Tie1, thereby demonstrating specificity of the ligand-binding domains.

## 2.2 Identification of Downstream Substrates

RTKs transduce their signals intracellularly through autophosphorylation that occurs with ligand-stimulated dimerization (PAWSON 1995). The phosphorylation of tyrosine residues on RTKs establishes high-affinity binding sites for src homology 2 (SH2), phosphotyrosine interaction or binding (PID/PTB), and for Met-binding domain-containing proteins (VAN DER GEER and PAWSON 1995; WEIDNER et al. 1996). The proteins that bind to a receptor are responsible for the transduction of a receptor-specific signal that, in turn, produces the biological function of the receptor.

The lack of ligands for Tie1 and Tie2 has been an obstacle in the study of signalling by these receptors. However, within the last year or so, researchers have started to unravel how Tie2 signals in endothelial cells. Much of what is known to date has been achieved by the application of molecular biological techniques that bypass the requirement for ligands. Several Tie2-binding partners have been identified by Dr. Kevin Peters' group, using an approach that was used to isolate different signalling molecules, such as GRB2, that bind to the EGF receptor (LOWENSTEIN et al. 1992). They screened an expression library with a recombinant form of Tie2 that was made using baculoviral expression system. The Tie2 from these insect cells is autophosphorylated, which allowed this group to isolate clones that have phosphotyrosine-dependent interactions. Using this approach, HUANG et al. (1995) identified the tyrosine phosphatase SHP-2 (Syp, SH-PTP2) and the adapter, GRB2, as Tie2-binding proteins. Subsequent binding studies with different Tie2-mutants expressed in, and purified from, insect cells and synthetic phosphopeptides, has allowed for the mapping of binding sites for GRB2 and SHP-2 on Tie2 to tyrosine residues 1101 and 1112, respectively (these sites correspond to 1100 and 1111 in mouse).

In an attempt both to bypass the requirement that a receptor be stimulated by ligands and to clone the complementary deoxyribonucleic acids (cDNAs) for the interacting proteins directly, we have used the yeast two-hybrid approach to identify Tie2-signalling partners (JONES AND DUMONT 1997). We constructed two yeast strains: one expressed a fusion between the DNA-binding domain of *lexA* and the cytoplasmic domain of Tie2; the other was identical, except that it carried a mutation within the lysine of the conserved adenosine triphosphate (ATP)-binding region, resulting in a kinase-inactive form of Tie2 (K853A) (DUMONT et al. 1994). Once introduced into yeast, the truncated receptor was catalytically active. This yeast strain was used to screen a cDNA library made from 12.5-day embryonic murine heart tissue. This tissue was expected to be a rich source of Tie2-signalling partners, given that Tie2 is highly expressed in embryonic heart (DUMONT et al. 1992).

Several clones were identified that exhibited a phosphotyrosine-dependent interaction with Tie2; many contained SH2 domains which validated our approach. Consistent with the results of HUANG et al. (1995) we also showed that SHP-2 and GRB2 interact with Tie2 in yeast. This result and other studies that have shown Gab1 binding to the Met receptor (WEIDNER et al. 1996), IRS1 and SHC binding to insulin and insulin-like growth factor receptors (TARTARE-DECKERT et al. 1995), and cloning of SHIP (LIOUBIN et al. 1996) validate the use of the yeast two-hybrid system to identify or characterize bonafide phosphotyrosine-dependent interactions among proteins.

We identified another clone in the yeast two-hybrid screen that contains a PTB domain and has sequence homology to the recently-cloned Abl- and rasGap-associated protein, Dok (CARPINO et al. 1997; YAMANASHI and BALTIMORE 1997). We have named this molecule Dok-R (Dok-related) (JONES AND DUMONT 1997). Dok-R binds to Tie2 both in vitro and in vivo in a phosphotyrosine-dependent manner, and immunoprecipitated Tie2 is able to use Dok-R as a substrate. Some of this work was performed by overexpressing Tie2 in 293T cells; this results in TIE2 activation (Jones and Dumont, 1998). Whether stimulation by angiopoietin-1 induces a similar phosphorylation pattern has yet to be determined.

## 2.3 Mutational Analysis of Putative Downstream Substrates

GRB2 is a small adapter protein which is complexed with the guanine nucleotide-exchange factor, Son of Sevenless (SOS) (PAWSON 1995). Once SOS is brought to the membrane by the binding of GRB2 to activated receptors, it is able to activate the ras pathway. The activation of this pathway results in cellular proliferation and/or differentiation. However, because mouse embryos homozygous for a Grb2-null allele die very early in utero (Alec Cheng, personal communication), the role of Grb2 in vascular development could not be determined.

It is thought that the binding of SHP-2 to the insulin, platelet-derived growth factor (PDGF) fibroblast growth factor (FGF), and EGF receptors plays an important role in mitogenic signalling by these receptors (BENNETT et al. 1994, 1996). The role of SHP-2 in Tie2 signalling may be inferred from the vascular defects seen in mice homozygous for a non-functional allele of SHP-2 (SAXTON et al. 1997). In SHP-2 mutant embryos allele, the vascular system is established; however, the elaboration and/or remodelling of the yolk sac vasculature is affected. The yolk sac defects resemble some aspects of the Tie2 phenotype (DUMONT et al. 1994; SATO et al. 1995). As might be expected for a molecule that is downstream of multiple RTK-mediated signalling pathways, pleiotropic defects are seen in SHP-2 mutant embryos. Thus, it is difficult to tell with certainty whether the impaired development of the yolk sac vasculature is a primary defect.

We have also shown that – like $p62^{DOK}$ – phosphorylated $p56^{Dok-R}$ will bind to rasGap (JONES AND DUMONT 1997). In endothelial cells, Tie2 signalling through $p56^{Dok-R}$ can perhaps be surmised from the vascular defects observed in Gap-mutant embryos (HENKEMEYER et al. 1995). The vascular defects in these embryos resemble the loss of vascular integrity seen in Tie2-mutant embryos (DUMONT et al. 1994; SATO et al. 1995), suggesting that some aspects of the Tie2-null phenotype are due, in part, to an imbalance in the active levels of ras (mediated by rasGap through $p56^{Dok-R}$).

To date, no reports have identified Tie1-signalling partners. Although the cytoplasmic region of Tie1 and Tie2 are quite homologous, not all tyrosine residues are conserved (Fig. 1). These differences may allow Tie1 and Tie2 receptors to bind to distinct cytoplasmic molecules on ligand stimulation. However, the number of overlapping tyrosine residues between Tie1 and Tie2 suggests that these receptors probably bind similar signalling partners and, as such, they may exert both similar and different biological effects. Interestingly, the Tie1-mutant embryos exhibit a similar loss of vascular integrity and endothelial cell number, as seen with the Tie2 mutants (see Sect. 4 below); this suggests that some aspects of the phenotypes of these embryos are due to defects in common signalling pathways in endothelial cells.

# 3 Tie1 and Tie2 Gene Expression

## 3.1 Expression in the Endothelial Cell Lineage

During embryonic development, Tie1 and Tie2 genes become transcriptionally activated in the endothelial cell precursors (called angioblasts) that are found in both the yolk sac and in the embryo itself (DUMONT et al. 1992; KORHONEN et al. 1994; MAISONPIERRE et al. 1993; SATO et al. 1993; SCHNURCH and RISAU 1993). Thereafter, Tie1 and Tie2 genes are expressed throughout the embryonic endothelium (except in the liver sinusoids). The expression of Tie1 and Tie2 follows that of the two receptors for VEGF, VEGFR-1 and -2 (DUMONT et al. 1995; YAMAGUCHI et al. 1993), which are required for the earliest stages of angioblast differentiation and blood vessel development (FONG et al. 1995; SHALABY et al. 1995). However, Tie1 and Tie2 expression in the developing endothelium still precedes the expression of more mature endothelial cell markers, such as von Willebrand factor (DUMONT et al. 1992).

The expression of Tie1 and Tie2 persists in quiescent adult endothelial cells, although at reduced levels in some organs, such as brain and heart (DUMONT et al. 1992; KORHONEN et al. 1994). Higher Tie1 expression seen in the adult lung might reflect differences in the turnover rates of various endothelial cell populations (HOBSON and DENEKAMP 1984). The expression of Tie1 is further induced during physiological and pathological neovascularization processes involved in ovarian-follicle maturation, wound healing, and tumor angiogenesis (HATVA et al. 1994; KAIPAINEN et al. 1994; KORHONEN et al. 1992). However, it is not clear whether the observed increase in the receptor expression is a cause or an effect of endothelial cell activation.

## 3.2 Characterization of Endothelial Cell-Specific Regulatory Elements

The transcriptional regulatory regions that drive angioblast and endothelial cell-specific reporter-gene expression in transgenic mice have been characterized for both Tie1 and Tie2 (KORHONEN et al. 1995; SCHLAEGER et al. 1995, 1997; M. Puri, unpublished). Only a short fragment of the proximal murine Tie1 promoter is required for endothelial cell-specific expression throughout development, whereas enhancer sequences from the first intron are needed for the Tie2 promoter to fully recapitulate the Tie2 expression pattern. It is therefore possible that endothelial cell-specific regulatory elements are differentially located in the two genes.

Similar binding sites for transcription factors, such as ETS-1 and PEA3, are found in both the Tie1 and Tie2 promoter regions, as well as in the intronic Tie2 enhancer. Furthermore, either mutation of the ETS-1 binding site or deletion of the PEA3 binding sites were both shown to inactivate the intronic Tie2 enhancer (SCHLAEGER et al. 1997). Consistent with a potential function in regulating Tie1 and Tie2, ETS-1 is also expressed in endothelial cell lineage during avian development (PARDANAUD and DIETERLEN-LIEVRE 1993) and, like Tie1 and Tie2, its expression is further induced in endothelial cells during angiogenic growth (WERNERT et al. 1992). Other transcription factors have also been implicated. An endothelial cell-specific transcription factor (EPAS-1) related to hypoxia-inducible factor was recently shown to activate transcription through the Tie2 promoter (TIAN et al. 1997). An intriguing possibility is that EPAS-1 or related factors (MALTEPE et al. 1997) mediate hypoxia-induced activation of Tie2 (and perhaps Tie1) during neovascularization and possibly also during embryonic development.

## 3.3 Expression in the Haematopoietic System

In addition to the angioblasts and endothelial cells, both Tie1 and Tie2 are expressed in certain cells of the haematopoietic lineage. Indeed, Tie1 was first isolated from a human chronic myeloid leukaemia cell line K562 (PARTANEN et al. 1990) and its expression is upregulated during in vitro megakaryoblastoid differentiation of certain leukaemia cell lines, including K562 (ARMSTRONG et al. 1993; HASHI-YAMA et al. 1996; PARTANEN et al. 1990). Tie1 expression is also found in vivo in megakaryocytes of fetal liver and adult bone marrow (BATARD et al. 1996; PARTANEN et al. 1996). In addition, Tie1-positive cells are found in haematopoietic stem cell populations isolated from both bone marrow and umbilical cord blood, as well as in some bone marrow B-cells (BATARD et al. 1996; HASHIYAMA et al. 1996; IWAMA et al. 1993). Tie2 also appears to be expressed in bone marrow haematopoietic stem cells (IWAMA et al. 1993). During embryonic development, both Tie1 and Tie2 are found in the yolk sac blood islands. However, the majority of primitive blood cells do not appear Tie1- or Tie2 positive (DUMONT et al. 1995; KORHONEN et al. 1994). This is consistent with the finding that Tie1 expression is downregulated on differentiation of isolated bone marrow haematopoietic stem cells (BATARD et al. 1996).

Whether the isolated Tie1 and Tie2 promoter elements discussed above drive expression in the haematopoietic system has not been reported. Several transcription factors are shared by endothelial and haematopoietic lineages. These include ETS-1, which has also been implicated in regulation of megakaryocyte-specific gene expression (DEVEAUX et al. 1996; NIMER et al. 1996). It was recently suggested that a subpopulation of circulating cells positive for haematopoietic stem cell markers are capable of differentiating into an endothelial-like phenotype (ASAHARA et al. 1997). Whether this population includes the Tie1- or Tie2-positive stem cells is not known. It also remains to be demonstrated whether these cells are truly bipotential cells, reminiscent of the hypothetical embryonic haemangioblasts. Outside the endothelial and haematopoietic cell lineages, Tie2 (but not Tie1) mRNA expression is detected in the mesodermal cell layer of the amnion (DUMONT et al. 1992; SATO et al. 1993).

# 4 Analysis of Tie1 and Tie2 Functions In Vivo

## 4.1 Functions of Tie1 and Tie2 in Vascular Development

Gene-targeting studies have provided insights into the functions of Tie1 and Tie2 during vascular development in the mouse. Unlike the VEGF-receptor system [in particular VEGFR-2, which is required for the very early differentiation of endothelial cells (SHALABY et al. 1995, 1997)], Tie2 and Tie1 support functions of the more mature endothelium (Fig. 2). In both Tie1- and Tie2-mutant mice, the sur-

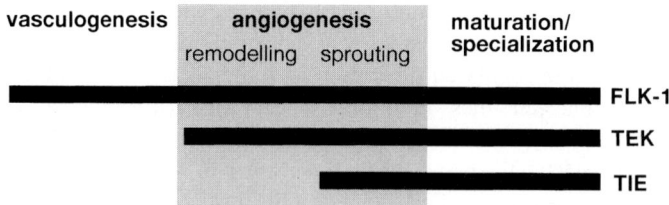

**Fig. 2.** Summary of FLK-1 (VEGFR-2), Tie2 and Tie1 functions during vascular development. *Bars* represent the requirement of these receptors in various phases of vascular development (see text for details)

vival of endothelium appears to be affected, but the phenotypes are manifested at different stages of embryonic development. Various endothelial cell populations are also affected differently by the two mutations.

Mouse embryos homozygous for a Tie2-null allele or expressing a dominant negative Tie2 receptor die around E9.5 and display defects throughout the vascular system. For examples, defects occur in the vascular endothelium of the yolk sac and in the embryonic endothelium of the heart, the dorsal aortae, and the microvasculature (DUMONT et al. 1994; SATO et al. 1995). Beginning from E8.5, the endothelial cells appear fewer in number, endothelial gene expression is downregulated, and the integrity of the endothelium is lost in the mutants (DUMONT et al. 1994). The heart endocardium appears to degenerate and loses its contacts with underlying myocardium, which, in turn, shows a reduced pattern of trabeculation. Vascular remodelling and sprouting into the neuroepithelium is also deficient in the mutant embryos (SATO et al. 1995), and the remaining capillaries appear dilated. These phenotypes suggest that Tie2 is dispensable for the early differentiation of the angioblasts, but is needed later for the survival and expansion of the endothelium. However, because of the intimate link between the cardiac and vascular defects and embryonic death, it is not possible to tell with certainty whether all the observed phenotypes are primary results of the loss of Tie2 function. Despite Tie2 expression in the amnionic mesoderm, no defects have been observed that suggest a role for Tie2 in the development of the amnion. The phenotypes observed and conclusions reached by both DUMONT et al. and SATO et al. are very similar. Unfortunately, the use of different terminology by these groups of researchers has been a source of confusion (SATO et al. 1995).

Mouse embryos homozygous for null alleles in the Tie1 locus die between E13.5 and P1 (PURI et al. 1995; SATO et al. 1995). There appear to be quantitative differences in the observed timing of the lethality. Whereas PURI et al. report that the majority of their homozygous mutants died prior to E15.5, SATO et al. report that the majority of their mutants survived to term. This discrepancy can likely be attributed to differences in the genetic backgrounds of the embryos analysed. Nevertheless, both groups of researchers have found that the Tie1-mutant embryos show defects in the integrity of the embryonic microvascular endothelium, resulting in oedema and haemorrhaging. In contrast to Tie2-mutants, in Tie1-mutants, the development of the extra-embryonic yolk sac vasculature and major embryonic vessels appears normal.

Analysis of chimeric embryos, which are mixtures of mutant and wild-type cells, provides a way to avoid secondary phenotypes caused by embryonic lethality. Chimeric analyses using embryonic stem cells lacking a functional Tie1 gene (marked with *LacZ* gene in the Tie1 locus) have shown that there is a primary, cell-autonomous requirement for Tie1 in several microvascular endothelial cell types (PARTANEN et al. 1996; PURI et al. 1995). In other words, the defect of Tie1-mutant cells cannot be rescued by a wild-type environment that includes the haematopoietic system and neighbouring endothelial cells. The chimeric analysis further demonstrates that Tie1 is not required for angioblast differentiation or vessel sprouting during early embryonic development.

In contrast, Tie1 is needed during organogenesis to promote angiogenic capillary growth; Tie1$^{-/-}$ endothelial cells contributed poorly to capillary beds of tissues and organs, such as brain, kidney and skin. The endothelium of the larger vessels, embryonic lung, and heart endocardium appeared relatively insensitive to a lack of Tie1 function, presumably reflecting the fact that these endothelia are generated primarily by vasculogenesis and non-sprouting angiogenesis. Analysis of adult chimeras showed that Tie1 also functions selectively in various capillary plexuses, for postnatal survival of endothelial cells. A different conclusion about the role of Tie1 in endocardium was reached by SATO et al., who showed ultrastructural alterations in the endocardium of Tie1-mutant embryos (SATO et al. 1995). These alterations might not affect the growth and survival of endocardium or might be secondary to other phenotypes of Tie1-mutant embryos.

The structural and phenotypic similarities between Tie1 and Tie2 suggest that these receptors may share functions in endothelial cells. The Tie1-Tie2- double-homozygous mutant embryos show a phenotype reminiscent of the Tie2 homozygotes (Mira Puri, JP, DJD, Alan Bernstein, unpublished results). The fact that even the lack of both Tie1 and Tie2 does not completely block angioblast differentiation further emphasises the role of these receptors in the later development of endothelial cells and the vascular system. In contrast, heterozygosity for the Tie2 mutation appears to enhance the phenotype of the homozygous Tie1-mutants. This finding suggests a functional redundancy for Tie1 and Tie2 in supporting later endothelial cell survival and angiogenic growth.

Although the studies described above indicate the developmental processes regulated by Tie1 and Tie2, the exact cellular mechanisms are still largely unknown. For example, the observed defects in angiogenesis could result from impaired endothelial cell migration, proliferation, morphogenesis, or target-finding. Recently, a ligand for Tie2 (angiopoietin-1) was isolated and shown to lack proliferative activity on endothelial cells in vitro (DAVIS et al. 1996). In contrast, based on the scarce perivascular cells that are seen in angiopoietin-1 mutant mice (SURI et al. 1996) and human patients with vascular dysmorphogenesis disorder resulting from an activating mutation in the Tie2 gene (SURI et al. 1996; VIKKULA et al. 1996), it was suggested that angiopoietin-1 and Tie2 would regulate the interactions of endothelial and perivascular smooth muscle cells as well as pericytes. However, it is somewhat puzzling how both loss-of-function and gain-of-function alleles can primarily inhibit the recruitment of the mesenchymal cells for the vascular wall. It is also possible that the reduced recruitment of perivascular cells in the Tie2 or angiopoietin-1 mutant embryos reflects the general downregulation of gene expression that is observed in the endothelium (DUMONT et al. 1994), and thus might be considered as a secondary defect to the impaired survival of the endothelium. The lack of endothelial cell–perivascular cell association can, in turn, contribute to enhanced endothelial cell death. Whatever the cellular mechanism, the cell-autonomous function of Tie1 in the endothelial cells (i.e. inability of the wild-type endothelial cell to rescue its mutant neighbour, see Sect. 4.1 above) strongly suggests that Tie1 (and likely also Tie2) is needed for more than simply inducing a secreted signal to attract perivascular cells around the endothelium. As suggested by SURI et al. (1996), it seems likely, that Tie1 and Tie2 collaborate with the VEGF receptors during cardiac and vascular development to promote the formation of mature endothelium, which is properly associated with its surroundings.

## 4.2 Tie1 and Tie2 in the Haematopoietic System

Because Tie1 and Tie2 are also expressed in certain haematopoietic cell types (see Sect. 3.3 above), it is of interest to know how their inactivation by gene-targeting affects haematopoietic development. This is an important issue given that a defect in haematopoiesis might contribute to the observed haemorrhaging phenotype and embryonic lethality (for example, see OKUDA et al. 1996). Both Tie1- and Tie2 mutants are able to produce blood cells. However, more subtle effects on haematopoiesis cannot be excluded based on the mutant phenotypes.

The chimeras resulting from fusion of Tie1$^{-/-}$ and wild-type morulae provide a way to study the haematopoietic function of Tie1, independently of vascular defects. In these chimeras, the haematopoietic cells derived from the blastomeres of Tie1$^{-/-}$ embryos were not selected against (PARTANEN et al. 1996). Because the Tie1$^{-/-}$ cells must compete with the wild-type host cells to populate the haematopoietic system, the role of Tie1 in haematopoiesis is tested very rigorously in these chimeras. Consistent with the lack of a role of Tie1 in hematopoiesis, in transplantation experiments, the livers of E15.5 Tie1-mutant embryos were found to contain haematopoietic progenitor and stem-cell activity (RODEWALD and SATO 1996).

Further experiments are needed to determine whether Tie2 also has a function in haematopoiesis. It is possible that Tie1 and Tie2 play redundant roles in the haematopoietic system. Haematopoietic cells are still seen in the Tie1-Tie2 double-mutant embryos (Mira Puri, JP, DJD, Alan Bernstein, unpublished results). However, a rescue from the vascular defects is needed in order to allow a thorough analysis of the haematopoietic function of Tie1 and Tie2 receptors. Further characterization of the angiopoietin family members, how they are expressed in the haematopoietic tissues, and how they affect the differentiation of haematopoietic cells in vitro should also shed light on the potential function of Tie1 and Tie2 in the haematopoietic system. In fact, TAKAKARU et al. (unpublished observations) have shown, using para-aortic splanopleural mesoderm (P-Sp) isolated from Tie2-null embryos, that they are defective for definitive haematopoiesis.

# References

Armstrong E, Korhonen J, Silvennoinen O, Cleveland JL, Lieberman MA, Alitalo R (1993) Expression of Tie receptor tyrosine kinase in leukemia cell lines. Leukemia 7:1585–1591

Asahara T, Murohara T, Sullivan A, Silver M, van der Zee R, Li T, Witzenbichler B, Schatteman G, Isner JM (1997) Isolation of putative progenitor endothelial cells for angiogenesis. Science 275:964–967

Batard P, Sansilvestri P, Scheinecker C, Knapp W, Debili N, Vainchenker W, Buhring HJ, Monier MN, Kukk E, Partanen J, Matikainen MT, Alitalo R, Hatzfeld J, Alitalo K (1996) The Tie receptor tyrosine kinase is expressed by human hematopoietic progenitor cells and by a subset of megakaryocytic cells. Blood 87:2212–2220

Beck L Jr, D'Amore PA (1997) Vascular development: cellular and molecular regulation (in process citation). FASEB J 11:365–373

Bennett AM, Tang TL, Sugimoto S, Walsh CT, Neel BG (1994) Protein-tyrosine-phosphatase SHPTP2 couples platelet-derived growth factor receptor beta to Ras. Proc Natl Acad Sci USA 91:7335–7339

Bennett AM, Hausdorff SF, O'Reilly AM, Freeman RM, Neel BG (1996) Multiple requirements for SHPTP2 in epidermal growth factor-mediated cell cycle progression. Mol Cell Biol 16:1189–1202

Burri PH, Tarek MR (1990) A novel mechanism of capillary growth in the rat pulmonary microcirculation. Anat Rec 228:35–45

Caduff JH, Fischer LC, Burri PH (1986) Scanning electron microscope study of the developing microvasculature in the postnatal rat lung. Anat Rec 216:154–164

Carpino N, Wisniewski D, Strife A, Marshak D, Kobayashi R, Stillman B, Clarkson B (1997) p62(dok): a constitutively tyrosine-phosphorylated, GAP-associated protein in chronic myelogenous leukemia progenitor cells. Cell 88: 197–204

Davis S, Aldrich TH, Jones PF, Acheson A, Compton DL, Jain V, Ryan TE, Bruno J, Radziejewski C, Maisonpierre PC, Yancopoulos GD (1996) Isolation of angiopoietin-1, a ligand for the Tie2 receptor, by secretion-trap expression cloning. Cell 87:1161–1169

Deveaux S, Filipe A, Lemarchandel V, Ghysdael J, Romeo PH, Mignotte V (1996) Analysis of the thrombopoietin receptor (MPL) promoter implicates GATA and Ets proteins in the coregulation of megakaryocyte-specific genes. Blood 87:4678–4685

Dumont DJ, Yamaguchi TP, Conlon RA, Rossant J, Breitman ML (1992) Tek, a novel tyrosine kinase gene located on mouse chromosome 4, is expressed in endothelial cells and their presumptive precursors. Oncogene 7:1471–1480

Dumont DJ, Gradwohl GJ, Fong GH, Auerbach R, Breitman ML (1993) The endothelial-specific receptor tyrosine kinase, Tek, is a member of a new subfamily of receptors. Oncogene 8:1293–1301

Dumont DJ, Gradwohl G, Fong GH, Puri MC, Gertsenstein M, Auerbach A, Breitman ML (1994) Dominant-negative and targeted null mutations in the endothelial receptor tyrosine kinase, Tek, reveal a critical role in vasculogenesis of the embryo. Genes Dev 8424:1897–1909

Dumont DJ, Fong GH, Puri MC, Gradwohl G, Alitalo K, Breitman ML (1995) Vascularization of the mouse embryo: a study of flk-1, Tek, Tie, and vascular endothelial growth factor expression during development. Dev Dyn 203:80–92

Fong G-H, Rossant J, Gertsenstein M, Breitman ML (1995) Role of the Flt-1 receptor tyrosine kinase in regulating the assembly of vascular endothelium. Nature 376:66–70

Hashiyama M, Iwama A, Ohshiro K, Kurozumi K, Yasunaga K, Shimizu Y, Masuho Y, Matsuda I, Yamaguchi N, Suda T (1996) Predominant expression of a receptor tyrosine kinase, Tie, in hematopoietic stem cells and B cells. Blood 87:93–101

Hatva E, Kaipainen A, Jaaskelainen J, Haltia M, Alitalo K (1994) Endothelial cell-specific receptor tyrosine kinases and growth factors in human gliomas and meningiomas. Am J Pathol 146:368–378

Henkemeyer M, Rossi DJ, Holmyard DP, Puri MC, Mbamalu G, Harpal K, Shih TS, Jacks T, Pawson T (1995) Vascular system defects and neuronal apoptosis in mice lacking ras GTPase-activating protein. Nature 377:695–701

Hobson B, Denekamp J (1984) Endothelial proliferation in tumors and normal tissues: continuous labelling studies. Br J Cancer 49:405–413

Horita K, Yagi T, Kohmura N, Tomooka Y, Ikawa Y, Aizawa S (1992) A novel tyrosine kinase, hyk, expressed in murine embryonic stem cells. Biochem Biophys Res Commun 189:1747–1753

Huang L, Turck CW, Rao P, Peters KG (1995) GRB2 and SH-PTP2: potentially important endothelial signalling molecules downstream of the Tek/Tie2 receptor tyrosine kinase. Oncogene 11:2097–2103

Iwama A, Hamaguchi I, Hashiyama M, Murayama Y, Yasunaga K, Suda T (1993) Molecular cloning and characterization of mouse Tie and Tek receptor tyrosine kinase genes and their expression in hematopoietic stem cells. Biochem Biophys Res Commun 195:301–309

Jones N, Dumont DJ (1998) The angiopoietin-1 receptor, Tek/Tie-2, signals through a novel Dok-related docking protein, p56dok-R. Oncogene (in press)

Kaipainen A, Vlaykova T, Hatva E, Bohling T, Jekunen A, Pyrhonen S, Alitalo K (1994) Enhanced expression of the Tie receptor tyrosine kinase mesenger RNA in the vascular endothelium of metastatic melanomas. Cancer Res 54:6571–6577

Koblizek TI, Runting AS, Stacker SA, Wilks AF, Risau W, Deutsch U (1997) Tie2 receptor expression and phosphorylation in cultured cells and mouse tissues. Eur J Biochem 244:774–779

Korhonen J, Partanen J, Armstrong E, Vaahtokari A, Elenius K, Jalkanen M, Alitalo K (1992) Enhanced expression of the Tie receptor tyrosine kinase in endothelial cells during neovascularization. Blood 80:2548–2555

Korhonen J, Polvi A, Partanen J, Alitalo K (1994) The mouse Tie receptor tyrosine kinase gene: expression during embryonic angiogenesis. Oncogene 9:395–403

Korhonen J, Lahtinen I, Halmekyto M, Alhonen L, Janne J, Dumont D, Alitalo K (1995) Endothelial-specific gene expression directed by the Tie gene promoter in vivo. Blood 86:1828–1835

Lioubin MN, Algate PA, Tsai S, Carlberg K, Aebersold A, Rohrschneider LR (1996) p150Ship, a signal transduction molecule with inositol polyphosphate-5-phosphatase activity. Genes Dev 10:1084–1095

Lowenstein EJ, Daly RJ, Batzer AG, Li W, Margolis B, Lammers R, Ullrich A, Skolnik EY, Bar-Sagi D, Schlessinger J (1992) The SH2 and SH3 domain-containing protein GRB2 links receptor tyrosine kinases to ras signalling. Cell 70:431–442

Maisonpierre PC, Goldfarb M, Yancopoulos GD, Gao G (1993) Distinct rat genes with related profiles of expression define a Tie receptor tyrosine kinase family. Oncogene 8:1631–1637

Maltepe E, Schmidt JV, Baunoch D, Bradfield CA, Simon MC (1997) Abnormal angiogenesis and responses to glucose and oxygen deprivation in mice lacking the protein ARNT. Nature 386:403–407

Mustonen T, Alitalo K (1995) Endothelial receptor tyrosine kinases involved in angiogenesis. J Cell Biol 129:895–898

Nimer S, Zhang J, Avraham H, Miyazaki Y (1996) Transcriptional regulation of interleukin-3 expression in megakaryocytes. Blood 88:66–74

Noden DM (1991) Development of craniofacial blood vessels. In: Feinberg RN, Sherer GK, Auerbach R (eds) The development of the vascular system. Karger, Basel, pp 1–24

Okuda T, van Deursen J, Hiebert SW, Grosveld G, Downing JR (1996) AML1, the target of multiple chromosomal translocations in human leukemia, is essential for normal fetal liver hematopoiesis. Cell 84:321–330

Pardanaud L, Yassine F, Dieterlen-Lievre F (1989) Relationship between vasculogenesis, angiogenesis and haemopoiesis during avian ontogeny. Development 105:473–485

Partanen J, Armstrong E, Makela TP, Korhonen J, Sandberg M, Renkonen R, Knuutila S, Huebner K, Alitalo K (1992) A novel endothelial cell surface receptor tyrosine kinase with extracellular epidermal growth factor homology domains. Mol Cell Biol 12:1698–1707

Pardanaud L, Dieterlen-Lievre F (1993) Expression of C-ETS1 in early chick embryo mesoderm: relationship to the hemangioblastic lineage. Cell Adhes Commun 1:151–160

Pardanaud L, Luton D, Prigent M, Bourcheix LM, Catala M, Dieterlen-Lievre F (1996) Two distinct endothelial lineages in ontogeny, one of them related to hemopoiesis. Development 122:1363–1371

Partanen J, Mäkelä TP, Alitalo R, Lehväslaiho H, Alitalo K (1990) Putative tyrosine kinases expressed in K-562 human leukemia cells. Proc Natl Acad Sci USA 87:8913–8917

Partanen J, Puri MC, Schwartz L, Fischer KD, Bernstein A, Rossant J (1996) Cell autonomous functions of the receptor tyrosine kinase Tie in a late phase of angiogenic capillary growth and endothelial cell survival during murine development. Development 122:3013–3021

Pawson T (1995) Protein modules and signalling networks. Nature 373:573–580

Puri MC, Rossant J, Alitalo K, Bernstein A, Partanen J (1995) The receptor tyrosine kinase Tie is required for integrity and survival of vascular endothelial cells. EMBO J 14:5884–5891

Risau W (1997) Mechanisms of angiogenesis. Nature 386:671–674

Rodewald HR, Sato TN (1996) Tie1, a receptor tyrosine kinase essential for vascular endothelial cell integrity, is not critical for the development of hematopoietic cells. Oncogene 12:397–404

Sato TN, Qin Y, Kozak CA, Audus KL (1993) Tie-1 and Tie-2 define another class of putative receptor tyrosine kinase genes expressed in early embryonic vascular system. Proc Natl Acad Sci USA 90:9355–9358

Sato TN, Tozawa Y, Deutsch U, Wolburg-Buchholz K, Fujiwara Y, Gendron-Maguire M, Gridley T, Wolburg H, Risau W, Qin Y (1995) Distinct roles of the receptor tyrosine kinases Tie-1 and Tie-2 in blood vessel formation. Nature 376:70–74

Saxton TM, Henkemeyer M, Gasca S, Shen R, Rossi DJ, Shalaby F, Feng G-S, Pawson T (1997) Abnormal mesoderm patterning in mouse embryos mutant for the SH2 tyrosine phosphatase Shp-2. EMBO J 16:2352–2364

Schlaeger TM, Qin Y, Fujiwara Y, Magram J, Sato TN (1995) Vascular endothelial cell lineage-specific promoter in transgenic mice. Development 121:1089–1098

Schlaeger TM, Bartunkova S, Lawitts JA, Teichmann G, Risau W, Deutsch U, Sato TN (1997) Uniform vascular-endothelial-cell-specific gene expression in both embryonic and adult transgenic mice. Proc Natl Acad Sci USA 94:3058–3063

Schnurch H, Risau W (1993) Expression of Tie-2, a member of a novel family of receptor tyrosine kinases, in the endothelial cell lineage. Development 119:957–968

Shalaby F, Rossant J, Yamaguchi TP, Gertsenstein M, Wu XF, Breitman ML, Schuh AC (1995) Failure of blood-island formation and vasculogenesis in Flk-1-deficient mice. Nature 376:62–66

Shalaby F, Ho J, Stanford WL, Fisher K-D, Schuh AC, Schwartz L, Bernstein A, Rossant J (1997) A requirement for FLK-1 in primitive and definitive hematopoiesis, and vasculogenesis. Cell 89:981–990

Sherer GK (1991) Vasculogenic mechanisms and epithelio-mesenchymal specificity in endodermal organs. In: Feinberg RN, Sherer GK, Auerbach R (eds) The development of the vascular system. Karger, Basel, pp 37–57

Suri C, Jones PF, Patan S, Bartunkova S, Maisonpierre PC, Davis S, Sato TN, Yancopoulos GD (1996) Requisite role of angiopoietin-1, a ligand for the Tie2 receptor, during embryonic angiogenesis. Cell 87:1171–1180

Tartare-Deckert S, Sawka-Verhelle D, Murdaca J, Van Obberghen E (1995) Evidence for a differential interaction of SHC and the insulin receptor substrate-1 (IRS-1) with the insulin-like growth factor-I (IGF-I) receptor in the yeast two-hybrid system. J Biol Chem 270:23456–23460

Tian H, McKnight SL, Russell DW (1997) Endothelial PAS domain protein 1 (EPAS1), a transcription factor selectively expressed in endothelial cells. Genes Dev 11:72–82

van der Geer P, Pawson T (1995) The PTB domain: a new protein module implicated in signal transduction. Trends Biochem Sci 20:277–280

Vikkula M, Boon LM, Carraway KLR, Calvert JT, Diamonti AJ, Goumnerov B, Pasyk KA, Marchuk DA, Warman ML, Cantley LC, Mulliken JB, Olsen BR (1996) Vascular dysmorphogenesis caused by an activating mutation in the receptor tyrosine kinase Tie2. Cell 87:1181–1190

Weidner KM, Di Cesare S, Sachs M, Brinkmann V, Behrens J, Birchmeier W (1996) Interaction between Gab1 and the c-Met receptor tyrosine kinase is responsible for epithelial morphogenesis. Nature 384:173–176

Wernert N, Raes MB, Lassalle P, Dehouck MP, Gosselin B, Vandenbunder B, Stehelin D (1992) c-ets1 proto-oncogene is a transcription factor expressed in endothelial cells during tumor vascularization and other forms of angiogenesis in humans. Am J Pathol 140:119–127

Wilting J, Christ B (1996) Embryonic angiogenesis: a review. Naturwissenschaften 83:153–164

Yamaguchi TP, Dumont DJ, Conlon RA, Breitman ML, Rossant J (1993) flk-1, an flt-related receptor tyrosine kinase is an early market for endothelial cell precursors. Development 118:489–498

Yamanashi Y, Baltimore D (1997) Identification of the Abl- and rasGAP-associated 62 kDa protein as a docking protein, Dok. Cell 88:205–211

Ziegler SF, Bird TA, Schneringer JA, Schooley KA, Baum PR (1993) Molecular cloning and characterization of a novel receptor protein tyrosine kinase from human placenta. Oncogene 8:663–670

# The Angiopoietins: Yin and Yang in Angiogenesis

S. Davis and G. D. Yancopoulos

| 1 | Introduction | 173 |
|---|---|---|
| 2 | Angiopoietin-1 | 174 |
| 2.1 | Expression Patterns of Angiopoietin-1 | 176 |
| 2.2 | Vascular Defects in Angiopoietin-1-Deficient Mice | 177 |
| 3 | Angiopoietin-2 | 179 |
| 3.1 | Expression Patterns of Angiopoietin-2 | 179 |
| 3.2 | Transgenic Overexpression of Angiopoietin-2 | 181 |
| 4 | Concluding Remarks | 181 |
| References | | 183 |

## 1 Introduction

Embryonic vascular development involves a complex series of events during which endothelial cells differentiate, proliferate, migrate, and undergo morphological organization in the context of their surrounding tissues (RISAU 1991, 1995). Vascular development is generally classified into two successive phases. The first, known as vasculogenesis, refers to the process whereby newly differentiated endothelial cells spontaneously coassemble into tubules that fuse to form a rather homogeneous primary vasculature in the embryo. Subsequent remodeling of this primary vascular network into large and small vessels brings into play a different process, termed angiogenesis. Angiogenesis in the embryo also leads to the sprouting of vessels into certain initially avascular organs, such as the brain. In the adult, angiogenesis accounts for neovascularization that accompanies the normal remodeling of the female reproductive organs during the menstrual cycle and pregnancy, in wound healing, and in various clinically significant pathological processes, such as tumor growth and diabetic retinopathy (FOLKMAN 1995; FERRARA 1995; HANAHAN and FOLKMAN 1996).

Intercellular signaling mechanisms that govern the formation of blood vessels have, only recently, begun to be studied at the molecular level. Two families of

---

Regeneron Pharmaceuticals, Inc., 777 Old Saw Mill River Road, Tarrytown, NY 10591, USA

receptor tyrosine kinases, the expression of which is largely restricted to endothelial cells, and that are essential for normal development of blood vessels, have been identified (Mustonen and Alitalo 1995). One family includes Flt-1 (also denoted vascular endothelial growth factor receptor-1; VEGFR-1), Flk-1/KDR (VEGFR-2) and Flt-4 (VEGFR-3) which all bind members of the VEGF family of growth factors. The roles of VEGFR-1 and VEGFR-2 during vascular development, as well as of VEGF, have been confirmed by analyzing genetically engineered mice that lack these proteins (Shalaby et al. 1995; Fong et al. 1995; Carmeliet et al. 1996; Ferrara et al. 1996). The more recently discovered TIE receptor family (Dumont et al. 1992; Partanen et al. 1992; Iwama et al. 1993; Maisonpierre et al. 1993; Sato et al. 1993; Schnurch and Risau 1993; Ziegler et al. 1993), consisting of TIE-1 and TIE-2 (also termed Tek), is likewise involved, critically, in the formation of the vasculature (Dumont et al. 1994; Sato et al. 1995; Puri et al. 1995).

Mice deficient in TIE-1 die between embryonic day 13.5 (E13.5) and birth, and display edema and hemorrhaging due to poor structural integrity of the endothelial cells (Sato et al. 1995; Puri et al. 1995). In contrast, mice deficient in TIE-2 show an earlier expression of their lethal phenotype, dying by E10.5 (Dumont et al. 1994; Sato et al. 1995). The most prominent defects observed in these mice include the failure of the endothelial lining of the heart to develop properly, the failure of remodeling of the primary capillary plexus into large and small vessels, and the lack of capillary sprouts into the neuroectoderm. In addition to its expression by endothelial cells, TIE-1 and TIE-2 are also expressed specifically in early hemopoietic stem cells (Iwama et al. 1993; Hashiyama et al. 1996; Batard et al. 1996), perhaps reflecting the origin of both lineages from a common hemangioblast precursor (Shalaby et al. 1995); unfortunately, the early death of mice lacking TIE-1 or TIE-2 has limited the use of these mice in elucidating the precise roles of the TIEs in hemopoiesis (Rodewald and Sato 1996).

## 2 Angiopoietin-1

The involvement of the TIE receptor family in critical aspects of angiogenesis, and possibly hemopoiesis as well, makes it important to identify and clone ligands that may activate these receptors. The initial identification of potential sources of ligands for the TIE-2 receptor (Davis et al. 1996) was accomplished by a fairly conventional route, parallel to that taken for the identification of ligands for many other "orphan" receptors (Stitt et al. 1995): a probe molecule containing the ectodomain of TIE-2 was coupled to the surface of a BIAcore sensor chip and used to screen conditioned media from a variety of cell lines for binding activity specific for TIE-2. Candidate cell lines were found, the conditioned media of which could bind specifically to TIE-2. In addition, conditioned media from these cell lines could induce tyrosine phosphorylation of the TIE-2 receptor in endothelial cells, indicating that the cells were releasing a ligand that could bind and activate the

TIE-2 receptor. Because the TIE-2 ectodomain did not bind to the surface of these cell lines, it was apparent that the ligand was a secreted molecule, and was not membrane bound.

Expression cloning methods have frequently been used to identify membrane-bound ligands (DAVIS et al. 1994). However, cloning secreted ligands for orphan receptors has traditionally proceeded not by expression cloning, but by affinity purification, using the soluble receptors as affinity reagents, followed by protein sequencing and cloning of the corresponding binding activity (STITT et al. 1995). Less often, expression cloning strategies have been employed for identifying secreted ligands that involve the construction and screening of "pooled expression libraries" (LOK et al. 1994). In these strategies, many small pools of clones are individually transfected into cells, and conditioned media from the individual transfections are then assayed separately for their ability to produce activities that stimulate receptor-bearing reporter cells; these strategies tend to be labor intensive and time consuming.

In the case of the TIE-2 receptor, a novel expression cloning strategy was developed to rapidly clone one of its ligands, based on a method utilized previously for the expression cloning of complementary deoxyribonucleic acids (cDNAs) encoding membrane-bound ligands. In this previously utilized method (DAVIS et al. 1994), an entire cDNA library is transfected into a single population of cells. Rare cells, expressing the target molecule on their surface and bearing plasmid DNA that encodes it, can be detected individually within a background of millions of other cells using soluble and epitope-tagged forms of the orphan receptor to decorate the cells for immunohistochemical detection; the cDNA encoding the ligand is subsequently rescued from the labeled cells (DAVIS et al. 1994). An extension of this method that allows for cloning of secreted ligands is based on the notion that a secreted ligand is, at least temporarily, trapped inside vesicular compartments of the cell until it is exocytosed. Thus, if the cells are appropriately fixed and permeabilized, it may be possible to detect the ligand en route to secretion. Such a strategy was adopted and led to the cloning of a ligand for the TIE-2 receptor, which was named angiopoietin-1 (Ang1) (DAVIS et al. 1996).

Ang1 contains 498 amino acids, including an amino-terminal secretory signal sequence; human and mouse Ang1 are 97.6% identical. Two regions within the coding sequence display homology to known proteins. The first region, consisting of residues 100–280, is weakly related to myosin and its relatives, in the regions of these proteins where they are known to possess coiled-coil quaternary structure. Algorithms that estimate the probability of coiled-coil structure in a given protein sequence give a high probability of coiled-coil structure in this region. The second region, consisting of residues 280–498, has strong similarity with the carboxy-terminal domain of fibrinogen (as well as related domains in tenascin, ficolin, HFREP, hpT49, and the Drosophila protein SCABROUS — see Fig. 1). It seems reasonable to conjecture that, like fibrinogen, Ang1 is a multimer, held together by coiled-coil structures and disulfide crosslinks. Recombinant Ang1 is a 70 kDa secreted glycoprotein that binds to the TIE-2 receptor with a $K_d$ of approximately

| | | COILED-COIL | FIBRINOGEN-LIKE |
|---|---|---|---|
| hANG-1 | | | |
| hANG-2 | 39% | 62% | 65% |
| mANG-3 | 28% | 40% | 58% |
| FIBRINOGEN-β | – | – | 35% |
| FIBRINOGEN-γ | – | – | 35% |
| FICOLIN-α | – | – | 38% |
| FICOLIN-β | – | – | 39% |
| TENASCIN | – | – | 31% |
| hFREP | – | – | 41% |
| hPT49 | – | – | 37% |
| SCABROUS | – | – | 25% |

**Fig. 1.** Similarity of the domains of the angiopoietins to one another and to related domains of other proteins. *Numbers* shown are the percentage identity between angiopoietin-1 and the indicated protein in the domain, shown in the graphical representation of angiopoietin-1. The *leftmost numbers* are the similarities in a small amino-terminal domain of about 50 amino acids that precedes the coiled-coil domain in the angiopoietins

3.7 nM, and induces tyrosine phosphorylation of TIE-2 in a variety of human and bovine endothelial cells (DAVIS et al. 1996).

Factors involved in angiogenesis, such as basic fibroblast growth factor (bFGF) and VEGF, typically have a variety of effects on cultured endothelial cells, such as proliferation and tubule formation. In the first in vitro experiments reported for Ang1, neither of these effects was observed, suggesting that Ang1 serves functions that are complementary to those served by other angiogenic factors.

## 2.1 Expression Patterns of Angiopoietin-1

The pattern of expression of Ang1 in the developing embryo points to an important role in developing vasculature (DAVIS et al. 1996; SURI et al. 1996). Early in development, at E9–11, Ang1 is found most prominently in the heart myocardium surrounding the endocardium. Later in development, Ang1 becomes distributed much more widely, most often in the mesenchyme surrounding developing vessels, and in close association with endothelial cells. Thus, the embryonic expression patterns of Ang1 suggest that it plays a particularly important role in the early development of the heart, and an increasingly widespread role as the rest of the vasculature matures.

## 2.2 Vascular Defects in Angiopoietin-1-Deficient Mice

The first experimental evidence demonstrating a direct physiological role for Ang1 came from mice in which the Ang1 gene was disrupted (SURI et al. 1996). These mice display distinctive deficits in vascular development that are unlike those seen in embryos lacking VEGF, but quite reminiscent of embryos lacking TIE-2 (SATO et al. 1995). Mouse embryos homozygous for the Ang1 gene disruption appear grossly abnormal by day 11 and die by E12.5. The most prominent defect in these embryos involves the heart; as in TIE-2-deficient embryos, these embryos have noticeably less complex ventricular infoldings and an almost collapsed endocardial lining. Whereas normal hearts exhibit complex myocardial trabeculations filling the ventricle, ventricles in mutant hearts are essentially devoid of trabeculae and, thus, consist of a simple open ventricular sac lined by a simple layer of endocardium. The less complex endocardial lining in the mutant hearts actually appears collapsed and retracted from the myocardial wall, as has also been noted in TIE-2-deficient embryos (SATO et al. 1995), and seems to reflect poor association between the endocardium and the underlying matrix. Because Ang1 is normally highly expressed in the myocardial wall surrounding the TIE-2-expressing endocardium, it appears as if the lack of Ang1 provided by the myocardium, in turn, prevents the endocardium from providing key signals required for normal trabecular formation, suggesting a crucial inter-dependency between these co-developing structures.

The blunted myocardial trabecular development that is observed in $Ang1^{-/-}$ mice is remarkably similar to heart defects described in mice lacking either neuregulin or its receptors (MEYER and BIRCHMEIER 1995; GASSMANN et al. 1995; LEE et al. 1995; KRAMER et al. 1996). In contrast to Ang1 and its receptor, which are expressed in the myocardium and endocardium, respectively, neuregulin and its receptor are expressed in a reciprocal fashion: neuregulin is made by the endothelium, while its receptors are expressed in the myocardium. Thus, it is intriguing to consider that the heart defect in $Ang1^{-/-}$ mice may simply represent a particular manifestation of the global disruption of proper interactions between endothelial cells and their surrounding mesenchyme and matrix. It seems likely that, in the heart, Ang1 and neuregulin may be two of the crucial reciprocal signals exchanged as part of a growth-factor loop that operates between the developing endocardium and myocardium; disruption of this loop at any point may lead to similar disturbances in the phenotype.

In addition to early heart defects, Ang1-deficient embryos also exhibit other defects that result in a generally less complex vascular network – even though the total number of endothelial cells seems unaffected. There is a reduced number of large vessels, these vessels are smaller in size than their normal counterparts and have fewer and straighter branches. The most striking example of a remodeling deficit is perhaps provided by the yolk sac vasculature, as was also noted for the $TIE-2^{-/-}$ embryos (SATO et al. 1995); in both cases, it appears as if remodeling of the initially homogeneous capillary network to form both large and small vessels does not occur.

The vascular defects observed in Ang1$^{-/-}$ are accompanied by defects at the ultrastructural level, as observed by electron microscopy (SURI et al. 1996). Differences between vessels in normal compared with Ang1$^{-/-}$ embryos are most apparent in "tissue folds" that split existing vessel lumens. These folds occur during vessel branching and remodeling, and are associated with the recruitment of periendothelial cells and the organization of matrix elements (such as collagen fibers) (PATAN et al. 1996a,b). The endothelial cells in folds from normal embryos appear thin and flattened, and are associated with the underlying periendothelial cells, as well as with dense clumps of collagen-like fibers. These are typical tissue folds thought to be involved in vessel branching (PATAN et al. 1996a,b). In contrast, the endothelial cells in folds from Ang1$^{-/-}$ embryos are less flattened, periendothelial cells are absent, the collagen-like fibers are individual and scattered, and the folds themselves appear shorter (SURI et al. 1996). In a survey of several tissue folds, it was noted that 0/12 tissue folds contained periendothelial cells in sections from the Ang1$^{-/-}$ embryos, while 12/12 tissue folds contained periendothelial cells in sections from normal embryos (SURI et al. 1996).

The ultrastructural analysis suggests that, in the absence of Ang1, endothelial cells are poorly associated with the underlying matrix and do not properly recruit and associate with periendothelial supporting cells. This failure to associate properly with underlying cells and matrix elements may result in the other observed ultrastructural defects – such as the lack of spreading and flattening of the endothelial cells, as well as poorer tissue folds. These ultrastructural defects may, in turn, explain some of the grosser defects observed, such as the failure in branching and vessel segmentation that leads to more homogenously dilated vessels.

Poor association with the underlying matrix may also adversely affect vessel integrity in mice that lack Ang1, and may prohibit appropriate stabilization and maturation of the vasculature. Such a problem may lead ultimately to vessel regression, as may be observed for the intersomitic vessels, and perhaps even to endothelial cell death. The possibility that Ang1 acts to regulate endothelial-matrix interactions is also consistent with the expression of Ang1 by abluminal cells tightly surrounding vessels (SURI et al. 1996; DAVIS et al. 1996). Interestingly, the ultrastructural defects noted in Ang1$^{-/-}$ vessels are somewhat reminiscent of those described recently for mice that lack tissue factor (TF) due to poor pericyte accumulation and function that causes problems in the developing vessel wall (CARMELIET et al. 1996); the defects in mice lacking TF are, like those of Ang1$^{-/-}$ mice, more prominent in the extra-embryonic vasculature, such as yolk-sac vessels. In addition, the defects in Ang1$^{-/-}$ and TF$^{-/-}$ mice may also be related to the progressive pericyte loss that seems to precede other vascular changes seen in diabetic retinopathy (FOLKMAN 1995). Thus, the interactions between endothelial cells and the surrounding matrix and mesenchyme appear to be of general importance for the formation of a stable vasculature and require the interaction of multiple systems (INGBER and FOLKMAN 1989). Integration may well involve intracellular signaling pathways; GTPase-activating protein (GAP) and the Ras pathway may serve as such critical downstream integrators, because the prominent remodeling defects seen in the yolk sac vasculature of both the Ang1$^{-/-}$ and TF$^{-/-}$ mice are also observed in mice lacking GAP (HENKEMEYER et al. 1995).

## 3 Angiopoietin-2

Like most biological phenomena, angiogenesis is influenced by both positive and negative regulation. Negative angiogenic regulators, such as proliferin-related protein (JACKSON et al. 1994), angiostatin (O'REILLY et al. 1994), and endostatin (O'REILLY et al. 1997), have been described (although their receptors, mechanisms of action, and physiological roles have yet to be defined). It is not surprising that the TIE system has negative regulation as well, but the discovery of angiopoietin-2 (Ang2) (MAISONPIERRE et al. 1997), a close relative of Ang1, provides an example of something unexpected: namely, an endogenous antagonist for a receptor tyrosine kinase.

cDNAs encoding mouse and human versions of Ang2 were isolated by low-stringency screening of human and mouse cDNA libraries, using the mouse Ang1 cDNA as a probe (MAISONPIERRE et al. 1997). The inferred Ang2 protein is comprised of 496 amino acids and has a secretion signal peptide. Human and mouse Ang2 are 85% identical in amino acid sequence, and ~60% identical to their Ang1 homologues (Fig. 1). Like Ang1, Ang2 has an amino-terminal coiled-coil domain and a carboxy-terminal fibrinogen-like domain. Eight of the nine cysteines in mature Ang1 are conserved in mature Ang2; one cysteine between the coiled-coil and fibrinogen-like domains in Ang1 is absent from Ang2.

In contrast to Ang1, which stimulates phosphorylation of TIE-2 in a variety of endothelial cell lines, Ang2 fails to elicit the same response. In fact, Ang2 blocks Ang1-stimulated activation of the TIE-2 receptor. This suggests that the natural role of Ang2 may be to antagonize Ang1 activation of TIE-2.

### 3.1 Expression Patterns of Angiopoietin-2

In contrast to Ang1, Ang2 is not readily detected in the developing heart, but is abundant in the dorsal aorta and major aortic branches, specifically in the smooth muscle layer immediately beneath the vessel endothelium (MAISONPIERRE et al. 1997). Both Ang2 and Ang1 are associated with the embryonic vasculature, although Ang2 has a more punctate pattern. For example, Ang2 transcripts in fetal liver are localized to cells at or close to the lumen of hepatic vessels, and are not present over all vessels in a given section. These Ang2-positive cells are likely to be endothelial cells or closely associated pericytes (HIRSCHI and D'AMORE 1996). The distinct but overlapping expression pattern of Ang1 and Ang2 is consistent with the possibility that Ang2 may regulate Ang1 function at particular sites and stages of vascular development.

In adult human tissue, Ang2 expression differs markedly from Ang1 expression. Whereas Ang1 is expressed widely, Ang2 is readily detectable only in ovary, placenta, and uterus, which are the three predominant sites of vascular remodeling in the normal adult. Further comparison of VEGF and angiopoietin expression patterns during adult vascular remodeling in the ovary was undertaken (MAISONPIERRE et al. 1997). Ovarian follicle development depends on sequential regulation of vascular outgrowth and vascular regression. At maturation, the follicle ruptures, expels the ovum, and then undergoes reorganization into a cell-dense secretory structure known as the corpus luteum. This process includes a wave of vascular sprouting and ingrowth that hypervascularizes the corpus luteum; these vessels eventually regress as the corpus luteum ages.

VEGF messenger ribonucleic acid (mRNA) is present in the pre-ovulatory follicle before vessel invasion, detectable first in cells of the stratum granulosum that surround the ovum, and later throughout the inner part of the granulosum layer. In the initial phase of vessel invasion, VEGF mRNA is abundant within the center of the developing corpus luteum, including regions where vessels have not yet formed. In contrast, Ang1 transcripts are closely associated with blood vessels, and appear to follow or coincide with, rather than precede, vessel ingrowth into the early corpus luteum (MAISONPIERRE et al. 1997). These observations are consistent with the hypothesis that Ang1 has a later role than VEGF in angiogenesis, perhaps involving vessel maturation and/or stabilization. The pattern of Ang2 expression, however, suggests that this factor plays an early role at the sites of vessel invasion (MAISONPIERRE et al. 1997). Initially, Ang2 transcripts are clustered in close association with blood vessels in the theca interna of the late pre-ovulatory follicle, and then are abundant at the front of vessels invading the developing corpus luteum. These expression patterns suggest that Ang2 may collaborate with VEGF at the front of invading vascular sprouts by blocking a constitutive stabilizing or maturing function of Ang1 and, thus, allowing vessels to revert to, and remain in, a more plastic state, where they may be more responsive to a sprouting signal provided by VEGF.

In follicular atresia, a condition in which large vesicular follicles fail to ovulate, there is no invasion of vessels into the stratum granulosum, and the surrounding vessels in the theca interna recede as the follicle regresses. VEGF mRNA is not detectable in atretic follicles, whereas Ang2 mRNA is present at uniformly high levels within the granulosum. Aged corpora lutei, in which vessels degenerate, also showed low levels of VEGF mRNA, but high levels of Ang2 mRNA. One interpretation of these expression patterns is that, in the presence of abundant VEGF, Ang2 can promote vessel sprouting by blocking a constitutive (stabilizing) Ang1 signal, whereas in the absence of VEGF, Ang2 inhibition of a constitutive Ang1 signal can contribute to vessel regression. Angiopoietin titration of the activation state of the TIE-2 receptor may regulate the plasticity of endothelial cells and thereby modify their requirement for, and responsiveness to, VEGF. Thus, while Ang1 may provide a maturation/stabilization signal via TIE-2, that can be blocked by Ang2, such inhibition may result in continued remodeling or the initiation of vascular sprouting in the context of simultaneous VEGF exposure, but frank vessel regression in the absence of VEGF (see Fig. 2).

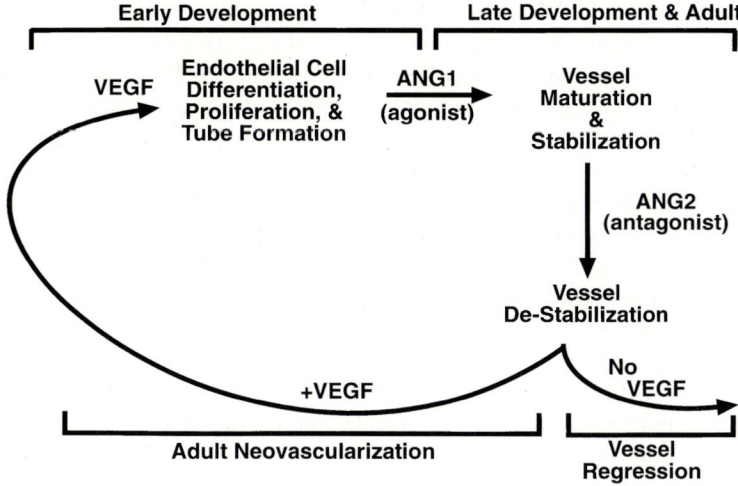

**Fig. 2.** Schematic depiction of proposed cooperative effects of vascular endothelial growth factor (VEGF), angiopoietin-1, and angiopoietin-2 on vascular structure

### 3.2 Transgenic Overexpression of Angiopoietin-2

Confirmation that Ang2 can act as a negative regulator of TIE-2 function in vivo comes from transgenic mice in which Ang2 is specifically overexpressed in their blood vessels (MAISONPIERRE et al. 1997). These mice were generated by use of an expression vector that utilizes TIE-2 transcriptional regulatory elements to direct expression of introduced genes to virtually all vascular structures in a developing mouse embryo. Since endogenous Ang2, unlike Ang1, is not expressed either in developing heart or around any other vascular beds at early embryonic stages (i.e., E9–10.5), vascular overexpression of a TIE-2 antagonist might be expected to mimic some of the defects observed at these ages in mouse embryos lacking Ang1 or its receptor. In fact, transgenic embryos overexpressing Ang2 were found to die at E9.5–10.5, and the defects in these embryos were similar to (and in some respects even more severe than) the vascular abnormalities seen in mice lacking either Ang1 or TIE-2 receptor.

## 4 Concluding Remarks

The study of angiopoietins has only just begun, and many challenges lie ahead in this emerging field. Additional members of the angiopoietin family have been discovered, and their properties have not yet been evaluated fully (Figs. 1 and 3). At the molecular level, much remains to be understood about the structure of the ligands and their mechanism of action. One of the most interesting issues is the antagonistic activity of Ang2 (Fig. 3). Intriguingly, while Ang2 is an antagonist on endothelial cells, it actually acts as an agonist for TIE-2 receptors expressed ectopically on non-endothelial cells (MAISONPIERRE et al. 1997; Fig. 3). This raises the possibility that endothelial cells contain a unique component(s) that interacts with TIE-2 and allows functional discrimination between the two angiopoietins. Equally mysterious is the role played by TIE-1; it is unclear whether TIE-1 has its own angiopoietin-like ligands, or whether it participates in heteromeric complexes with TIE-2. Further downstream, there is the question of signal transduction pathways activated by the angiopoietins, and how these signals are coordinated with or dependent on signals elicited by VEGF, integrins, or other components involved in angiogenic processes. It will also be of great interest to understand how these biochemical signaling events are subsequently translated into biological activities that can be measured in the in vitro assays commonly used to analyze angiogenic stimuli, such as proliferation and migration assays, tubule formation assays, etc., and in more sophisticated in vivo systems, such as animal tumor models.

Ultimately, it is hoped that this analysis will open new avenues for therapeutic manipulation of disease states. There is a variety of important clinical situations in which it would be desirable to promote angiogenic processes, such as for the in-

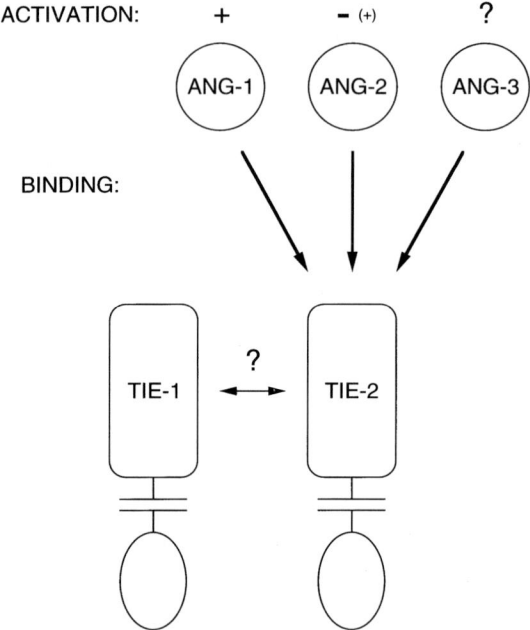

**Fig. 3.** Schematic outline of the known interactions between the angiopoietins and the TIE receptors. *Arrows* indicate that all the known angiopoietins bind to TIE-2, but not to TIE-1. *Symbols* above the angiopoietins indicate that Ang-1 is an agonist for activation of TIE-2, whereas Ang-2 can be an antagonist or an agonist, depending on the context. *Double arrow* between TIE-1 and TIE-2 indicates potential heterodimeric interactions between them

duction of collateral vascularization in an ischemic heart or limb (FERRARA et al. 1995). Conversely, there are pathologic situations in which preventing angiogenic processes could be used to starve a growing tumor or abate a raging inflammatory process (FOLKMAN 1995; FERRARA 1995; HANAHAN and FOLKMAN 1996). While many efforts in promoting or preventing angiogenesis have focused on the VEGF family, it now seems clear that any potential therapeutic treatments based on VEGF manipulation must be considered in the context of the angiopoietins, since these two growth-factor systems seem to act together normally for optimal blood-vessel formation. In support of this notion, administration of excess VEGF alone during embryogenesis seems to result in the formation of malformed and hyperfused vessels, quite analogous to those seen in embryos lacking Ang1 (DRAKE and LITTLE 1995). Ang1 and Ang2 provide for naturally occurring positive and negative regulators of angiogenesis that must be further tested for their ability to manipulate vessel formation (and possibly hemopoiesis as well), alone and in combination with other agents, in therapeutically beneficial ways.

# References

Batard P, Sansilvestri P, Scheinecker C, Knapp W, Deblie N, Vainchenker W, Buhring HJ, Monier MN, Kukk E, Partanen J, Matikainen MT, Alitalo R, Hatzfield J, Alitalo K (1996) The Tie receptor tyrosine kinase is expressed by human hematopoietic progenitor cells and by a subset of megakaryocytic cells. Blood 87:2212–2220

Carmeliet P, Ferreira V, Breier G, Pollefeyt S, Kieckens L, Gertsenstein M, Fahrig M, Vandenhoeck A, Harpal K, Eberhardt C, Declercq C, Pawling J, Moons L, Collen D, Risau W, Nagy A (1996) Abnormal blood vessel development and lethality in embryos lacking a single VEGF allele. Nature 380:435–439

Davis S, Gale NW, Aldrich TH, Maisonpierre PC, Lhotak V, Pawson T, Goldfarb M, Yancopoulos GD (1994) Ligands for the EPH-related receptor tyrosine kinases that require membrane attachment or clustering for activity. Science 266:816–819

Drake CJ, Little CD (1995) Exogenous vascular endothelial growth factor induces malformed and hyperfused vessels during embryonic neovascularization. Proc Natl Acad Sci USA 92:7657–7661

Dumont DJ, Yamaguchi TP, Conlon RA, Rossant J, Breitman ML (1992) Tek, novel tyrosine kinase gene located on mouse chromosome 4, is expressed in endothelial cells and their presumptive precursors. Oncogene 7:1471–1480

Dumont DJ, Gradwohl G, Fong G-H, Puri MC, Gerstenstein M, Auerbach A, Breitman ML (1994) Dominant-negative and targeted null mutations in the endothelial receptor tyrosine kinase, tek, reveal a critical role in vasculogenesis of the embryo. Genes Dev 8:1897–1909

Ferrara N (1995) The role of vascular endothelial growth factor in pathological angiogensis. Breast Cancer Res Treatment 36:127–137

Ferrara N, Heinsohn H, Walter CE, Buting S, Thomas GR (1995) The regulation of blood vessel growth by vascular endothelial growth factor. Ann NY Acad Sci 752:246–256

Ferrara N, Carver-Moore K, Chen H, Dowd M, Lu L, O'Shea KS, Powell-Braxton L, Hillan KJ, Moore MW (1996) Heterozygous embryonic lethality induced by targeted inactivation of the VEGF gene. Nature 380:439–442

Folkman J (1995) Angiogenesis in cancer, vascular, rheumatoid and other disease. Nat Med 1:27–31

Fong GH, Rossant J, Gertenstein M, Breitman ML (1995) Role of the Flt-1 receptor tyrosine kinase in regulating assembly of vascular endothelium. Nature 376:66–70

Gassmann M, Casagranda F, Orioli D, Simon H, Lai C, Klein R, Lemke G (1995) Aberrant neural and cardiac development in mice lacking the ErbB4 neuregulin receptor. Nature 378:390–394

Hanahan D, Folkman J (1996) Patterns and emerging mechanisms of the angiogenic switch during tumorigenesis. Cell 86:353-364

Hashiyama M, Iwama A, Ohshiro K, Kumorzumi K, Yasunaga K, Shimizu Y, Masuho Y, Matsuda I, Yamaguchi N, Suda T (1996) Predominant expression of a receptor tyrosine kinase, TIE, in hematopoietic stem cells and B cells. Blood 87:93-101

Henkemeyer M, Rossi DJ, Holmyard DP, Puri MC, Mbamalu G, Harpal K, Shih TS, Jacks T, Pawson T (1995) Vascular system defects and neuronal apoptosis in mice lacking Ras GTPase-activating protein. Nature 377:695-701

Hirschi KK, D'Amore PA (1996) Pericytes in the microvasculature. Cardiovasc Res 32:687-698

Ingber DE, Folkman J (1989) How does extracellular matrix control capillary morphogenesis. Cell 58:803-805

Iwama A, Hamaguchi I, Hashiyama M, Murayama Y, Yasunaga K, Suda T (1993) Molecular cloning and characterization of mouse Tie and Tek receptor tyrosine kinase genes and their expression in hematopoietic stem cells. Biochem Biophys Res Commun 195:301-309

Jackson D, Volpert OV, Bouck N, Linzer DI (1994) Stimulation and inhibition of angiogenesis by placental proliferin and proliferin-related protein. Science 266:1581-1584

Kramer R, Bucay N, Kane DJ, Martin LE, Tarpley JE, Theill LE (1996) Neuregulins with an Ig-like domain are essential for mouse myocardial and neuronal development. Proc Natl Acad Sci USA 93:4833-4838

Lee K-F, Simon H, Chen H, Bates B, Hung M-C, Hauser C (1995) Requirement for neuregulin receptor erbB2 in neural and cardiac development. Nature 378:394-398

Lok S, Kaushansky K, Holly RD, Kuljper JL, Lofton-Day CE, Oort PJ, Grant FJ, Helpel MD, Burkhead SK, Kramer JM, Bell LA, Sprecher CA, Blumberg H, Johnson R, Prunkard D, Ching AFT, Mathewes SL, Balley MC, Forstrom JW, Buddle MM, Osborn SG, Evans SJ, Sheppard PO, Presnell SR, O'Hara PJ, Hagen FS, Roth GJ, Foster DC (1994) Cloning and expression of murine thrombopoietin cDNA and stimulation of platelet production in vivo. Nature 369:565-568

Maisonpierre PC, Goldfarb M, Yancopoulos GD, Gao G (1993) Distinct rat genes with related profiles of expression define a TIE receptor tyrosine kinase family. Oncogene 8:1631-1637

Maisonpierre PC, Suri C, Jones PF, Bartunkova S, Wiegand SJ, Radziejewski C, Compton D, McClain J, Aldrich TH, Papadopoulos N, Daly TJ, Davis S, Sato TN, Yancopoulos GD (1997) Angiopoietin-2, a natural antagonist for Tie2 that disrupts in vivo angiogenesis. Science 277:55-60

Meyer D, Birchmeier C (1995) Multiple essential functions of neuregulin in development. Nature 378:386-390

Mustonen T, Alitalo K (1995) Endothelial receptor tyrosine kinases involved in angiogenesis. J Cell Biol 129:895-898

O'Reilly MS, Holmgren L, Shing Y, Chen C, Rosenthal RA, Moses M, Lane WS, Cao Y, Sage EH, Folkman J (1994) Angiostatin: a novel angiogenesis inhibitor that mediates the suppression of metastases by a Lewis lung carcinoma. Cell 79:315-328

O'Reilly MS, Boehm T, Shing Y, Fukai N, Vasios G, Lane WS, Flynn E, Birkhead JR, Olsen BR, Folkman J (1997) Endostatin: an endogenous inhibitor of angiogenesis and tumor growth. Cell 88:277-285

Partanen J, Armstrong E, Makela TP, Korhonen J, Sandberg M, Renkonen R, Knuutila S, Huebner K, Alitalo K (1992) A novel endothelial cell surface receptor tyrosine kinase with extracellular epidermal growth factor homology domains. Mol Cell Biol 12:1698-1707

Patan S, Haenni B, Burri PH (1996a) Implementation of intussusceptive microvascular growth in the chicken chorio-allantoic membrane (CAM): 1. Pillar formation by folding of the capillary wall. Microvasc Res 51:80-98

Patan S, Munn LL, Jain RK (1996b) Intussusceptive microvascular growth in a human colon adenocarcinoma xenograft: a novel mechanism of tumor angiogenesis. Microvasc Res 51:260-272

Puri MC, Rossant J, Alitalo K, Bernstein A, Partanen J (1995) The receptor tyrosine kinase TIE is required for integrity and survival of vascular endothelial cells. EMBO J 14:5884-5891

Risau W (1991) Vasculogenesis, angiogenesis and endothelial cell differentiation during embryonic development. In: Feinberg RN, Shere GK, Auerbach R (eds) The development of the vascular system. Karger, Basel, pp 58-68

Risau W (1995) Differentiation of endothelium. FASEB J 9:926-933

Rodewald H-R, Sato TN (1996) Tie1, a receptor tyrosine kinase essential for vascular endothelial cell integrity, is not critical for the development of hematopoietic cells. Oncogene 12:397-404

Sato TN, Qin Y, Kozak CA, Audus KL (1993) tie-1 and tie-2 define another class of putative receptor tyrosine kinase genes expressed in early embryonic vascular system. Proc Natl Acad Sci USA 90:9355–9358

Sato TN, Tozawa Y, Deutsch U, Wolburg-Buchholz K, Fujiwara Y, Gendron-Maguire M, Gridley T, Wolburg H, Risau W, Qin Y (1995) Distinct roles of the receptor tyrosine kinases Tie-1 and Tie-2 in blood vessel formation. Nature 376:70–74

Schnurch H, Risau W (1993) Expression of tie-2, a member of a novel family of receptor tyrosine kinases, in the endothelial cell lineage. Development 119:957–968

Shalaby F, Rossant J, Yamaguchi TP, Gertsenstein M, Wu XF, Breitman ML, Schuh AC (1995) Failure of blood-island formation and vasculogenesis in Flk-1-deficient mice. Nature 376:62–66

Stitt TN, Conn G, Gore M, Lai C, Bruno J, Radziejewski C, Mattson K, Fisher J, Gies DR, Jones PF, Masiakowski P, Ryan TE, Tobkes NJ, Chen DH, DiStefano PS, Long GL, Basilico C, Goldfarb MP, Lemke G, Glass DJ, Yancopoulos GD (1995) The anticoagulation factor protein S and its relative, Gas6, are ligands for the Tyro3/Axl family of receptor tyrosine kinases. Cell 80:661–670

Suri C, Jones PF, Patan S, Bartunkova S, Maisonpierre PC, Davis S, Sato TN, Yancopoulos GD (1996) Requisite role of TIE ligand-1 during embryonic angiogenesis. Cell 87:1171–1180

Ziegler SF, Bird TA, Schneringer KA, Schooley KA, Baum PR (1993) Molecular cloning and characterization of a novel receptor protein tyrosine kinase from human placenta. Oncogene 8:663–670

# Subject Index

**A**
albumin 113
Ang1 (*see* angiopoietin-1)
angioblasts 87, 160
angiogenesis 60, 70, 71, 89, 90, 93, 159, 173–183
– physiological 11
– therapeutic 18–20
angiopoietins 168, 173–183
ATF (activating transcription factor) 70
atherosclerosis 76
ATP analogues 77
atresia 180
autophosphorylation 46, 64

**B**
baculovirus system 64
blood islands 71
brown fat 45

**C**
c-Jun-NH2-terminal kinase 67
c-Kit 61, 69
CAM (chorioallontoic membrane) 90
capillary 168
caveolae 118
cell junctions 119
cell-association 43
chimeric analysis 167
chorioallontoic membrane (CAM) 90
choriocarcinoma 32
chromosome
– 4q34 46
– 11q13 45
– 12qD 38
– 14q24 38
colony-stimulating factor-1 (CSF-1) 61
cornea neovascularization 37
corpus luteum 108, 180
– angiogenesis 13
CRE (cyclic AMP response element) 70
CSF-1 (colony-stimulating factor-1) 61

**D**
delayed hypersensitivity 107
dextran 113
dimerization of receptor 64
dimerized ligands 61
DNA synthesis 90
Dok-R 163

**E**
EKR1 91
embryonic blood vessels 133–151
endocardium 167, 176, 177
endothelial cells 32, 98, 160
endothelium, lymphatic 49
enhancer 165
epidermal growth factor (EGF)-like repeats 161
Erk1 67, 87
Erk2 67, 87, 91
ETS motif 70
expression cloning 175

**F**
ferritin 113
fibrin 110
fibroblast growth factor (FGF) 68
– FGF-2 76
Flk-1 61
Flk-1/KDR 11–13
Flt-1 61
Flt-4 61, 69, 85
Fms family 61
fms-like tyrosine kinase 61

**G**
gaps 123
gene-targeting 166
germ cell tumours 33
glioblastoma 77
glycosylation 44
goitres 35
Grb2 67, 87, 91
GTPase-activating protein of Ras (RasGAP) 67

## H
haemangiomase 93
haematopoiesis 93
haematopoietic
- stem cells 166
- system 165
heart 45
heparin-binding site 64
hepatoma 32
heterodimers 44
high endothelial venules 87
HRP 113
hSOS 91
hydraulic conductivity 99
hyperpermeability in angiogenesis 97–125
hyperplasia 50
hypersensitivity, delayed 107
hypoglycaemia 76
hypoxia 106, 165
- inducible transcription factor 72
hypoxic conditions 72

## I
imaging 93
immunoglobulin (Ig)
- domain 61
- like folds 61
inflammatory skin disorders 108
inter-endothelial cell gaps 119

## J
JNK 67, 87, 91

## K
*Kaposi´s* sarcoma 91, 93
KDR/Fkl-1 61
kinase-insert 62

## L
lymph nodes 93
lymphangiogenesis 90, 93
lymphangiomatosis 93
lymphatic
- endothelium 49, 89
- system 89
- vessels 87, 90, 93

## M
MAP (mitogen-activated protein) 67
megakaryoblastoid 165
megakaryocytes 165
MEK 67
mesodermal yolk-sac blood-island propenitors 71
microvascular endothelium 167
migration 49, 75

Miles assay 100
mitogen-activated protein (MAP) 67
monocytes 72
myocardial
- infarcts 106
- wall 177
myocardium 176, 177

## N
Nck 67
net filtration rate 99
nitric oxid (NO) 110
notochord 88

## O
ovarian follicle 180
oxygen tension 6

## P
paxillin 91
PDGF (platelet-derived growth factor) 61
pericytes 89, 160, 179
periendothelial cells 178
phosphatidylinositol 3-kinase (PI3-kinase) 68
phospholipase C-$\gamma$ 67
phosphoryrosine 162
physiological angiogenesis 11
PKC (protein kinase C) 67
placenta growth factor (PlGF) 62, 75, 76, 99, 147, 148
plasminogen activators 75
platelet-derived growth factor (PDGF) 61
pores 123
promoter 165
protein kinase C (PKC) 67
proteolytic processing 48
psoriasis 108

## R
Raf kinase 67
RAFTK 87, 91
receptor tyrosine kinases 60, 160, 174
receptor-binding 44
retinopathies 106
rheumatoid arthritis 107

## S
Shc 87, 91
signal transduction 9, 10, 110
smooth muscle 89
Src Homology 2 (SH2)-domain 66
stem cell factor receptor 61
stress-induced kinase p38 67

## T
targeted inactivation 72
therapeutic angiogenesis 18–20

## Subject Index

thyroid  33
Tie-1  73, 159–169, 174–176
Tie-2  73, 174–176, 159–169
tissue
– factor  76
– folds  178
trabeculae  177
tracers  113
trans-endothelial
– openings  119
– pores  119
transcytosis  112
tube formation  76
tumour  33
– angiogenesis  14–16, 71
tyrosine-kinase domains  64
tyrosine phosphorylation  91

## V

vascular
– development  159–169
– dysmorphogenesis disorder  168
– endothelial growth factor (*see* VEGF)
– permeability  75, 89
– – factor  50
– – – /vascular endothelial growth factor (VPF/VEGF)  97–125
– plexus  88
– remodeling  180

vasculogenesis  70, 71, 159
VEGF  1–30, 41–55, 174
– A  140–147
– B  99, 148, 149
– C  62, 86, 88, 99, 149
– D  86, 88, 149, 150
– distribution  11
– family  140–155
– gene  4–6
– – expression, regulation  6–8
– isoforms  42
– pathological angiogenesis  14
– proteins  4–6
– receptors  8, 150, 151
– – Fms family  69
– – receptor-1 (VEGFR-1)  31, 37, 38
– – – soluble  64, 69
– – receptor-2 (VEGFR-2)  32, 37, 38, 86, 87
– – – dominant-negative  77, 78
– – receptor-2 (VEGFR-3)  86
vesiculo-vascular organelle (VVO)  113–118

## W

wound healing  105

## Y

yolk sac  177

# Current Topics in Microbiology and Immunology

Volumes published since 1989 (and still available)

Vol. 197: **Meyer, Peter (Ed.):** Gene Silencing in Higher Plants and Related Phenomena in Other Eukaryotes. 1995. 17 figs. IX, 232 pp. ISBN 3-540-58236-3

Vol. 198: **Griffiths, Gillian M.; Tschopp, Jürg (Eds.):** Pathways for Cytolysis. 1995. 45 figs. IX, 224 pp. ISBN 3-540-58725-X

Vol. 199/I: **Doerfler, Walter; Böhm, Petra (Eds.):** The Molecular Repertoire of Adenoviruses I. 1995. 51 figs. XIII, 280 pp. ISBN 3-540-58828-0

Vol. 199/II: **Doerfler, Walter; Böhm, Petra (Eds.):** The Molecular Repertoire of Adenoviruses II. 1995. 36 figs. XIII, 278 pp. ISBN 3-540-58829-9

Vol. 199/III: **Doerfler, Walter; Böhm, Petra (Eds.):** The Molecular Repertoire of Adenoviruses III. 1995. 51 figs. XIII, 310 pp. ISBN 3-540-58987-2

Vol. 200: **Kroemer, Guido; Martinez-A., Carlos (Eds.):** Apoptosis in Immunology. 1995. 14 figs. XI, 242 pp. ISBN 3-540-58756-X

Vol. 201: **Kosco-Vilbois, Marie H. (Ed.):** An Antigen Depository of the Immune System: Follicular Dendritic Cells. 1995. 39 figs. IX, 209 pp. ISBN 3-540-59013-7

Vol. 202: **Oldstone, Michael B. A.; Vitković, Ljubiša (Eds.):** HIV and Dementia. 1995. 40 figs. XIII, 279 pp. ISBN 3-540-59117-6

Vol. 203: **Sarnow, Peter (Ed.):** Cap-Independent Translation. 1995. 31 figs. XI, 183 pp. ISBN 3-540-59121-4

Vol. 204: **Saedler, Heinz; Gierl, Alfons (Eds.):** Transposable Elements. 1995. 42 figs. IX, 234 pp. ISBN 3-540-59342-X

Vol. 205: **Littman, Dan R. (Ed.):** The CD4 Molecule. 1995. 29 figs. XIII, 182 pp. ISBN 3-540-59344-6

Vol. 206: **Chisari, Francis V.; Oldstone, Michael B. A. (Eds.):** Transgenic Models of Human Viral and Immunological Disease. 1995. 53 figs. XI, 345 pp. ISBN 3-540-59341-1

Vol. 207: **Prusiner, Stanley B. (Ed.):** Prions Prions Prions. 1995. 42 figs. VII, 163 pp. ISBN 3-540-59343-8

Vol. 208: **Farnham, Peggy J. (Ed.):** Transcriptional Control of Cell Growth. 1995. 17 figs. IX, 141 pp. ISBN 3-540-60113-9

Vol. 209: **Miller, Virginia L. (Ed.):** Bacterial Invasiveness. 1996. 16 figs. IX, 115 pp. ISBN 3-540-60065-5

Vol. 210: **Potter, Michael; Rose, Noel R. (Eds.):** Immunology of Silicones. 1996. 136 figs. XX, 430 pp. ISBN 3-540-60272-0

Vol. 211: **Wolff, Linda; Perkins, Archibald S. (Eds.):** Molecular Aspects of Myeloid Stem Cell Development. 1996. 98 figs. XIV, 298 pp. ISBN 3-540-60414-6

Vol. 212: **Vainio, Olli; Imhof, Beat A. (Eds.):** Immunology and Developmental Biology of the Chicken. 1996. 43 figs. IX, 281 pp. ISBN 3-540-60585-1

Vol. 213/I: **Günthert, Ursula; Birchmeier, Walter (Eds.):** Attempts to Understand Metastasis Formation I. 1996. 35 figs. XV, 293 pp. ISBN 3-540-60680-7

Vol. 213/II: **Günthert, Ursula; Birchmeier, Walter (Eds.):** Attempts to Understand Metastasis Formation II. 1996. 33 figs. XV, 288 pp. ISBN 3-540-60681-5

Vol. 213/III: **Günthert, Ursula; Schlag, Peter M.; Birchmeier, Walter (Eds.):** Attempts to Understand Metastasis Formation III. 1996. 14 figs. XV, 262 pp. ISBN 3-540-60682-3

Vol. 214: **Kräusslich, Hans-Georg (Ed.)**: Morphogenesis and Maturation of Retroviruses. 1996. 34 figs. XI, 344 pp. ISBN 3-540-60928-8

Vol. 215: **Shinnick, Thomas M. (Ed.)**: Tuberculosis. 1996. 46 figs. XI, 307 pp. ISBN 3-540-60985-7

Vol. 216: **Rietschel, Ernst Th.; Wagner, Hermann (Eds.)**: Pathology of Septic Shock. 1996. 34 figs. X, 321 pp. ISBN 3-540-61026-X

Vol. 217: **Jessberger, Rolf; Lieber, Michael R. (Eds.)**: Molecular Analysis of DNA Rearrangements in the Immune System. 1996. 43 figs. IX, 224 pp. ISBN 3-540-61037-5

Vol. 218: **Berns, Kenneth I.; Giraud, Catherine (Eds.)**: Adeno-Associated Virus (AAV) Vectors in Gene Therapy. 1996. 38 figs. IX,173 pp. ISBN 3-540-61076-6

Vol. 219: **Gross, Uwe (Ed.)**: Toxoplasma gondii. 1996. 31 figs. XI, 274 pp. ISBN 3-540-61300-5

Vol. 220: **Rauscher, Frank J. III; Vogt, Peter K. (Eds.)**: Chromosomal Translocations and Oncogenic Transcription Factors. 1997. 28 figs. XI, 166 pp. ISBN 3-540-61402-8

Vol. 221: **Kastan, Michael B. (Ed.)**: Genetic Instability and Tumorigenesis. 1997. 12 figs.VII, 180 pp. ISBN 3-540-61518-0

Vol. 222: **Olding, Lars B. (Ed.)**: Reproductive Immunology. 1997. 17 figs. XII, 219 pp. ISBN 3-540-61888-0

Vol. 223: **Tracy, S.; Chapman, N. M.; Mahy, B. W. J. (Eds.)**: The Coxsackie B Viruses. 1997. 37 figs. VIII, 336 pp. ISBN 3-540-62390-6

Vol. 224: **Potter, Michael; Melchers, Fritz (Eds.)**: C-Myc in B-Cell Neoplasia. 1997. 94 figs. XII, 291 pp. ISBN 3-540-62892-4

Vol. 225: **Vogt, Peter K.; Mahan, Michael J. (Eds.)**: Bacterial Infection: Close Encounters at the Host Pathogen Interface. 1998. 15 figs. IX, 169 pp. ISBN 3-540-63260-3

Vol. 226: **Koprowski, Hilary; Weiner, David B. (Eds.)**: DNA Vaccination/Genetic Vaccination. 1998. 31 figs. XVIII, 198 pp. ISBN 3-540-63392-8

Vol. 227: **Vogt, Peter K.; Reed, Steven I. (Eds.)**: Cyclin Dependent Kinase (CDK) Inhibitors. 1998. 15 figs. XII, 169 pp. ISBN 3-540-63429-0

Vol. 228: **Pawson, Anthony I. (Ed.)**: Protein Modules in Signal Transduction. 1998. 42 figs. IX, 368 pp. ISBN 3-540-63396-0

Vol. 229: **Kelsoe, Garnett; Flajnik, Martin (Eds.)**: Somatic Diversification of Immune Responses. 1998. 38 figs. IX, 221 pp. ISBN 3-540-63608-0

Vol. 230: **Kärre, Klas; Colonna, Marco (Eds.)**: Specificity, Function, and Development of NK Cells. 1998. 22 figs. IX, 248 pp. ISBN 3-540-63941-1

Vol. 231: **Holzmann, Bernhard; Wagner, Hermann (Eds.)**: Leukocyte Integrins in the Immune System and Malignant Disease. 1998. 40 figs. XIII, 189 pp. ISBN 3-540-63609-9

Vol. 232: **Whitton, J. Lindsay (Ed.)**: Antigen Presentation. 1998. 11 figs. IX, 244 pp. ISBN 3-540-63813-X

Vol. 233/I: **Tyler, Kenneth L.; Oldstone, Michael B. A. (Eds.)**: Reoviruses I. 1998. 29 figs. XVIII, 223 pp. ISBN 3-540-63946-2

Vol. 233/II: **Tyler, Kenneth L.; Oldstone, Michael B. A. (Eds.)**: Reoviruses II. 1998. 45 figs. XVI, 187 pp. ISBN 3-540-63947-0

Vol. 234: **Frankel, Arthur E. (Ed.)**: Clinical Applications of Immunotoxins. 1999. 16 figs. IX, 122 pp. ISBN 3-540-64097-5

Vol. 235: **Klenk, Hans-Dieter (Ed.)**: Marburg and Ebola Viruses. 1999. 34 figs. XI, 225 pp. ISBN 3-540-64729-5

Vol. 236: **Kraehenbuhl, Jean-Pierre; Neutra, Marian R. (Eds.)**: Defense of Mucosal Surfaces: Pathogenesis, Immunity and Vaccines. 1999. 30 figs. IX, 296 pp. ISBN 3-540-64730-9

# Springer and the environment

At Springer we firmly believe that an international science publisher has a special obligation to the environment, and our corporate policies consistently reflect this conviction.

We also expect our business partners – paper mills, printers, packaging manufacturers, etc. – to commit themselves to using materials and production processes that do not harm the environment. The paper in this book is made from low- or no-chlorine pulp and is acid free, in conformance with international standards for paper permanency.

Printing: Saladruck, Berlin
Binding: Buchbinderei Lüderitz & Bauer, Berlin